MW00589525

Guide to Energy Management

Guide to Energy Management

by William J. Kennedy
Wayne C. Turner
Barney L. Capehart

Published by
THE FAIRMONT PRESS, INC.
700 Indian Trail
Lilburn, GA 30247

Library of Congress Cataloging-in-Publication Data

Kennedy, William J., 1938-
   Guide to energy management / by William J. Kennedy, Wayne C. Turner,
Barney L. Capehart.
   p. cm.
   Includes bibliographical references and index.
   ISBN 0-88173-185-4
   1. Energy conservation--Handbooks, manuals, etc. 2. Energy
consumption--Handbooks, manuals, etc. I. Turner, Wayne C., 1942- .
II. Capehart, B. L. (Barney L.) III. Title.
TJ163.3.K453   1994   621.042--dc20   93-49065
                                          CIP

*Guide To Energy Management by William J. Kennedy, Wayne C. Turner,
Barney L. Capehart.*

©1994 by The Fairmont Press, Inc. All rights reserved. No part of this
publication may be reproduced or transmitted in any form or by any means,
electronic or mechanical, including photocopy, recording, or any
information storage and retrieval system, without permission in writing from
the publisher.

Published by The Fairmont Press, Inc.
700 Indian Trail
Lilburn, GA 30247

Printed in the United States of America

10 9 8 7 6 5 4 3 2 1

ISBN 0-88173-185-4   FP

ISBN 0-13-147372-7   PH

While every effort is made to provide dependable information, the publisher, authors, and
editors cannot be held responsible for any errors or omissions.

Distributed by PTR Prentice Hall
Prentice-Hall, Inc.
A Paramount Communications Company
Englewood Cliffs, NJ 07632

Prentice-Hall International (UK) Limited, London
Prentice-Hall of Australia Pty. Limited, Sydney
Prentice-Hall Canada Inc., Toronto
Prentice-Hall Hispanoamericana, S.A., Mexico
Prentice-Hall of India Private Limited, New Delhi
Prentice-Hall of Japan, Inc., Tokyo
Simon & Schuster Asia Pte. Ltd., Singapore
Editora Prentice-Hall do Brasil, Ltda., Rio de Janeiro

Contents

Preface

This book is a culmination of many years of experience in energy management and is a result of our dedication to the concept of prudent energy utilization for maximum return per unit of energy consumed. We have seen organizations achieve remarkable results in energy management by applying sound engineering principles and utilizing creativity. Savings in energy and energy costs of 30% are usual. Savings can easily run 40-50% or more. Since energy efficiency technology continues to advance, these figures tend to remain fairly constant. Thus, the potential for future savings continues to be exciting and motivating.

The purpose of this book is to help you understand the objectives of energy management and to demonstrate and teach some of the basic techniques and tools. In the first Chapter, energy management is defined in depth. Then the proper tools for designing, initiating, and managing energy management programs are presented. In Chapter 2, we discuss energy auditing. Suggested forms, equipment and procedures are provided. Chapter 3 is a presentation of utility rate structures for various types of energy. Emphasis is on understanding and interpreting the various structures. Engineering economy and economic decision criteria are summarized in Chapter 4. Chapters 5-10 are treatments of energy management applications in lighting, HVAC systems, boilers, steam systems, controls, and maintenance. Chapter 11 is a discussion of insulation. In Chapter 12, we show some applications of energy management to industrial processes. Finally, in Chapter 13, we discuss solar energy use, other renewable energy sources, and water management.

As this book is being completed, it is the twentieth anniversary of the great oil embargo of 1973/74. That event radically changed our view that energy would always be a commodity in plentiful supply at a very low price. Energy management—through energy conservation and energy cost control—played a very important role in the reduction of oil imports, and in the increased energy efficiency of our buildings and our industries. Renewed commitment to energy efficiency and energy cost reduction has come about because of our national concerns for economic competitiveness, environmental quality, energy security and product quality. The Energy Policy Act of 1992 has provided a tremendous impetus to energy management in buildings and industry - both in the private sector and in government buildings and operations. The future for energy management looks very bright.

Guide to Energy Management is a substantially revised and expanded version of *Energy Management*, first published in 1984 by William J. "Biff" Kennedy and Wayne C. Turner. In late 1992, Barney L. Capehart was asked to help revise the book and update it to reflect some of the new approaches and technologies for energy management. He has enjoyed the opportunity to work with Biff Kennedy and Wayne Turner in expanding this pioneering book in Energy Management.

Particular thanks go to Ms. Lynne Capehart who painstakingly edited the entire book, and helped make many corrections and improvements to the manuscript. Thanks also go to Mr. Mark Spiller of Gainesville (Florida) Regional Utilities for substantially revising and updating Chapter 5 on Lighting. Mark also contributed many suggestions for improving the book.

Credit must also be given to the many students in the course EIN 4321, Energy Management, at the University of Florida who helped review, improve and update the various chapters. The students in the Spring 1993 class all helped in providing comments on both the original book, and on the draft manuscript. Two students who helped greatly were Ms. Holly Lloyd and Mr. David Enck.

Finally, we three believe strongly in America's ability to be economically competitive, to have an energy efficient economy, and to protect our environment. This book is dedicated to all of these in the hopes that each will be a little better off because of it.

W. J. K.
W. C. T.
B. L. C.

Chapter 1

Designing, Initiating, and Managing An Energy Management System

1.0 ENERGY MANAGEMENT

The phrase energy management means different things to different people, so our first objective is to provide our definition of the term. To us, energy management is:

The judicious and effective use of energy to maximize profits (minimize costs) and enhance competitive positions

Therefore, any management activity that affects the use of energy falls under this definition. This rather broad definition covers many operations from product and equipment design through product shipment. Waste disposal also presents many energy management opportunities. Without a whole systems viewpoint, many important activities will be overlooked or suboptimized. Presently, many businesses and industries are adopting a Total Quality Management (TQM) strategy for improving their operations. Any TQM approach should include an energy management component to reduce energy costs.

We've already stated that the primary objective of energy management is to maximize profits or minimize costs. Some desirable subobjectives of energy management programs include:

1. Conserving energy, thereby reducing costs

2. Cultivating good communications on energy matters

3. Developing and maintaining effective monitoring, reporting, and management strategies for wise energy usage

4. Finding new and better ways to increase returns from energy invest-
    ments through research and development

5. Developing interest in and dedication to the energy management
    program from all employees

6. Reducing the impacts of curtailments, brownouts, or any interruption
    in energy supplies

Although this list is not exhaustive, these six are sufficient for our
purposes. However, the sixth objective requires a little more explanation.

Curtailments occur when a major supplier of an energy source is
forced to reduce shipments or allocations (sometimes drastically) because
of severe weather conditions and/or distribution problems. For example,
natural gas is often sold to industry relatively inexpensively, but on an
interruptible basis. That is, residential customers and others on
noninterruptible schedules have priority, and those on interruptible
schedules receive what is left. This residual supply is normally sufficient
to meet industry needs, but periodically gas deliveries must be curtailed.

Even though curtailments do not occur frequently, the cost associated with them is so high—sometimes a complete shutdown is necessary—that management needs to be alert in order to minimize the negative effects. There are several ways of doing this, but the method most
often employed is the storage and use of a secondary or standby fuel.
Number 2 fuel oil is often stored on site and used in boilers capable of
burning either natural gas (primary fuel) or fuel oil (secondary fuel). Then
when curtailments are imposed, fuel oil can be used. Naturally, the cost of
equipping boilers with dual fire capability is high, as is the cost of storing
the fuel oil. However, these costs are minuscule compared to the cost of
forced shutdown. Other methods of planning for curtailments include
production scheduling to build up inventories, planned plant shutdowns,
or vacations during curtailment-likely periods, and contingency plans
whereby certain equipment, departments, etc., can be shut down so critical areas can keep operating. All these activities must be included in an
energy management program.

Although energy conservation is certainly an important part of energy management, it is not the only consideration. Curtailment-contin-
gency planning is certainly not conservation, and neither are load shedding or power factor improvement, both of which will be discussed later
on in this chapter. To concentrate solely on conservation would preclude
some of the most important activities—often those with the largest savings opportunity.

1.1 THE NEED FOR ENERGY MANAGEMENT

1.1.1 Economics

The American free enterprise system operates on the necessity of profits, or budget allocations in the case of nonprofit organizations. Thus, any new activity can be justified only if it is cost effective; that is, the net result must show a profit improvement or cost reduction greater than the cost of the activity. Energy management has proven time and time again that it is cost effective.

Energy cost savings of 5-15 percent are usually obtained quickly with little to no required capital expenditure when an aggressive energy management program is launched. Eventual savings of 30 percent are common and savings of 50, 60, and even 70 percent have been obtained. These savings all result from retrofit activities. New buildings designed to be energy efficient often operate on 20 percent of the energy (and a corresponding 80 percent savings) normally required by existing buildings. In fact, for most manufacturing and other commercial organizations *energy management is one of the most promising profit improvement-cost reduction programs available today.*

1.1.2 National Good

There are many reasons energy management programs are so vitally needed today. One of the most important is that energy management helps the nation face some of its biggest problems. Perhaps some statistics will help make this point.*

- It took 50 years (1900-1950) for total annual U.S. energy consumption to go from 4 million barrels of oil equivalent (MBOE) per day to 16 MBOE. It took only 20 years (1950-1970) to go from 16 to 32 MBOE. This rapid growth in energy use slowed in the early 1970's, but took a spurt in the late 1970's, reaching almost 40 MBOE in 1979. Energy use slowed again in the early 1980's and dropped to 35 MBOE in 1983. Economic growth in the mid 1980's returned the use to 40 MBOE in 1988. Since then, energy use has remained fairly steady at just over 40 MBOE.

- With only 5 percent of the world's population, the United States consumes about 25 percent of its energy and produces about 25 per-

---

*These statistics come from numerous sources, mostly government publications from the Energy Information Administration or from the U.S. Statistical Abstract.

cent of the world's gross national product (GNP). However, some nations such as Japan, West Germany, and Sweden produce the same or greater GNP per capita with significantly less energy than the United States.

- Domestic crude oil production peaked in 1970 at just over 10 million barrels per day (MBD), and has fallen slowly since then to just over 7 MBD in 1992. Domestic gas production peaked in 1973 at just over 24 trillion cubic feet (TCF) per year. Gas production fell about 3 TCF in the following few years, and now remains fairly steady (1988-1992) at about 21-22 TCF per year. Deregulation has improved our domestic production in the short run, but in the long run we continue to face decreasing domestic output.

- Annual average prices per barrel for imported crude oil rapidly escalated from $3.00 in the early 1970's to $12 in 1973-1974 and to $37 in 1981. Since 1981 prices have fallen from this peak, and dropped to $14 in 1986. Since 1986, prices have ranged from about $14 to $22 a barrel.

- The United States has been a net importer of oil since 1947. In 1970 the bill for this importation was only $3 billion; by 1978 it was $42 billion; by 1979, $60 billion; and by 1981, $80 billion, even though the volume imported was less than in 1979. This imported oil bill has severely damaged our trade balance and weakened the dollar in international markets. Since 1981 the bill for oil imports has fallen, reaching a low of $37 billion in 1985. It climbed to almost $64 billion in 1990, and it was around $55 billion in 1992.

    In addition to these discouraging statistics, there are a host of major environmental problems, as well as economic and industrial competitiveness problems, that came to the forefront of public concern in the late 1980's.

- Reducing acid rain. Lake acidification and deforestation have been the greatest effects of acid rain from the combustion of fossil fuels containing significant amounts of sulfur, such as coal and some oil. The Clean Air Act Amendments of 1990 will restrict the future emission of sulfur dioxide to the level emitted in 1980.

- Limiting global climate change. Carbon dioxide, the main contributor to potential global climate change, is produced by the combustion of fossil fuel, primarily to provide transportation and energy services. In 1992, many countries of the world adopted limitations on carbon dioxide emissions.

- Limiting ozone depletion. In the U.S., about half of the CFC's—which have been associated with ozone depletion—are used in providing energy services through refrigeration and air conditioning, and in manufacturing insulation. Recent international agreements will substantially phase out the use of CFC's in industrialized countries by the year 1996.

- Improving national security. Oil imports directly affect the energy security and balance of payments of our country. These oil imports must be reduced for a secure future, both politically and economically.

- Improving U.S. competitiveness. The U.S. spends about 9 percent of its gross national product for energy—a higher percentage than many of its foreign competitors. This higher energy cost amounts to a surtax on U.S. goods and services.

- The fall of the Berlin Wall in 1989 and the emergence of market economies in many Eastern European countries is leading to major changes in world energy supplies and demands. These changes significantly affect our nation, and provide us an economic impetus to help these countries greatly improve their own energy efficiencies and reduce their energy bills.

There are no easy answers, as the following points demonstrate:

- Many look to coal as the answer. Yet coal burning produces sulfur dioxide and carbon dioxide, which produce acid rain and potential global climate change.

- Synfuels require strip mining, incur large costs, and place large demands for water in arid areas. On-site coal gasification plants associated with gas-fired, combined-cycle power plants are presently being demonstrated by several electric utilities. However, it remains to be seen if these units can be built and operated in a cost-effective and environmentally acceptable manner.

- Solar-generated electricity, whether generated through photovoltaics or thermal processes, is still more expensive than conventional sources and has large land requirements. Technological improvements are occurring in both these areas, and costs are decreasing. Sometime in the near future, these approaches may become cost-effective.

- Biomass energy is also expensive, and any sort of monoculture would require large amounts of land. Some fear total devastation of forests. At best, biomass can provide only a few percentage points of our total needs without large problems.

- Wind energy has technological "noise" and aesthetic problems that probably can be overcome, but it too is very expensive. In addition, it is only feasible in limited geographic regions.

- Alcohol production from agricultural products raises perplexing questions about using food products for energy when large parts of the world are starving. Newer processes for producing ethanol from wood waste are just being tested, and may offer some significant improvements in this limitation.

- Fission has the well-known problems of waste disposal, safety, and a short time span with existing technology. Without breeder reactors we will soon run out of fuel, but breeder reactors dramatically increase the production of plutonium—a raw material for nuclear bombs.

- Fusion seems to be everyone's hope for the future, but many claim that we do not know the area well enough yet to predict its problems. When available commercially, fusion may very well have its own style of environmental-economical problems.

The preceding discussion paints a rather bleak picture. Our nation and our world are facing severe energy problems and there appears to be no simple answers.

Time and again energy management has shown that it can substantially reduce energy costs and energy consumption. This saved energy can be used elsewhere, so one energy source not mentioned in the preceding list is energy management. In fact, energy available from energy management activities has almost always proven to be the most economical source of "new" energy. Furthermore, energy management activities are more gentle to the environment than large-scale energy production, and they certainly lead to less consumption of scarce and valuable resources. Thus, although energy management cannot solve all the nation's problems, *perhaps it can ease the strain on our environment and buy enough time so that we can develop new energy sources.*

We feel the time for energy management has come. The world will have a greatly increased need for engineers who are adequately trained in the field of energy management. In fact, a large number of energy management jobs are available today. This text will help you prepare for a career which will be both exciting and challenging.

1.2 ENERGY BASICS FOR ENERGY MANAGERS

An energy manager must be familiar with energy terminology and units. Different energy types are measured in different units. Knowing how to convert from one measurement system to another is essential for making valid comparisons. The energy manager must also be informed about the national energy picture. The historical use patterns as well as the current trends are important to an understanding of options available to many facilities.

1.2.1 Energy Terminology, Units and Conversions

Knowing the terminology of energy use and the units of measure is essential to developing a strong energy management background. Energy represents the ability to do work, and the standard engineering measure for energy is the British thermal unit, or Btu. One Btu is the amount of energy needed to raise the temperature of one pound of water one degree Fahrenheit. In more concrete terms, one Btu is the energy released by burning one kitchen match head, according to the U.S. Energy Information Agency. The energy content of most common fuels is well known, and can be found in many reference handbooks. For example, a gallon of gasoline contains about 125,000 Btu and a barrel of oil contains about 5,100,000 Btu. A short listing of the average energy contained in a number of the most common fuels, as well as some energy unit conversions is shown below in Table 1-1.

Electrical energy is also measured by its ability to do work. The traditional unit of measure of electrical energy is the kilowatt-hour; in terms of Btu's, one kilowatt-hour (kWh) is equivalent to 3412 Btu. However, when electrical energy is generated from steam turbines with boilers

Table 1-1
Energy Units and Energy Content of Fuels

1 kWh	3412 Btu
1 ft^3 natural gas	1000 Btu
1 Ccf natural gas	100 ft^3 natural gas
1 Mcf natural gas	1000 ft^3 natural gas
1 therm natural gas	100,000 Btu
1 barrel crude oil	5,100,000 Btu
1 ton coal	25,000,000 Btu
1 gallon gasoline	125,000 Btu
1 gallon #2 fuel oil	140,000 Btu
1 gallon LP gas	95,000 Btu
1 cord of wood	30,000,000 Btu
1 MBtu	1000 Btu
1 MMBtu	10^6 Btu
1 Quad	10^{15} Btu
1 MW	10^6 watts

fired by fossil fuels such as coal, oil or gas, the large thermal losses in the process mean that it takes about 10,000 Btu of primary fuel to produce one kWh of electrical energy. Further losses occur when this electrical energy is then transmitted to its point of ultimate use. Thus, although the electrical energy at its point of end-use always contains 3412 Btu per kWh, it takes considerably more than 3412 Btus of fuel to produce a kWh of electrical energy.

1. 2. 2 Energy Supply and Use Statistics
Any energy manager should have a basic knowledge of the sources of energy and the uses of energy in the United States. Both our national energy policy and much of our economic policy are dictated by these supply and use statistics. Figure 1-1 shows the share of total U.S. energy supply provided by each major source. Figure 1-2 represents the percentage of total energy consumption by each major end-use sector.

1.2.3 Energy Use in Commercial Businesses
One question frequently asked by facility energy managers is "How

Figure 1-1
U.S. Energy Supply 1992 (100% = 82.35 Quads)
Source - U.S. Department of Energy EIA

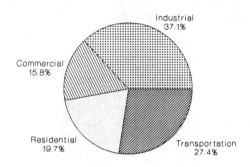

Figure 1-2
U.S. Energy Consumption 1992 (100% = 82.35 Quads)
⸞ ꭎurce U.S. Department of Energy

does energy use at my facility compare to other facilities in general, and to other facilities that are engaged in the same type of operation?" Figure 1-3 shows general energy usage in commercial facilities, and Figure 1-4 shows their electricity use. While individual facilities may differ significantly from these averages, it is still helpful to know what activities are likely to consume the most energy. This provides some basis for a comparison to other facilities—both energy wasting and energy efficient. In terms of priority of action for an energy management program, the largest areas of energy consumption should be examined first. The greatest savings will almost always occur from examining and improving the areas of greatest use.

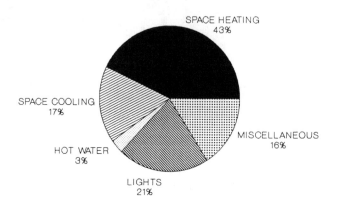

Figure 1-3
Commercial Energy Use 1990 (end-use basis)
Source - U.S. Department of Energy EIA

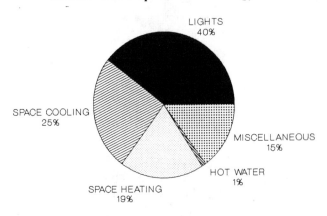

Figure 1-4
Commercial Electric Use 1990 (end-use basis)
Source - U.S. Department of Energy EIA

The commercial sector uses about 16 percent of all the primary energy consumed in the United States, at a cost of around 80 billion dollars each year [1]. On an end-use basis, natural gas and oil constitute about 50 percent of the commercial energy use, mainly for space heating. Over 40 percent of the energy use is in the form of electricity for lighting, air conditioning, ventilation, and some space heating. Although electricity provides less than half of the end-use energy used by a commercial facility, it represents well over half of the cost of the energy needed to operate the facility. Lighting is the predominant use of electricity in commercial buildings, and accounts for over one-third the cost of electricity.

Commercial activity is very diverse, and this leads to greatly varying energy intensities depending on the nature of the commercial facility. Recording energy use in a building or a facility of any kind and providing a history of this use is necessary for the successful implementation of an energy management program. A time record of energy use allows analysis and comparison so that results of energy productivity programs can be determined and evaluated.

1.2.4 Energy Use in Industry

The industrial sector—consisting of manufacturing, mining, agriculture and construction activities—consumes over one-third of the nation's primary energy use, at an annual cost of $100 billion [2]. Industrial energy use is shown in Figure 1-5 and industrial electricity use is shown in Figure 1-6.

Manufacturing companies, which use mechanical or chemical processes to transform materials or substances to new products, account for about 85 percent of the total industrial sector use. The "big three" in energy use are petroleum, chemicals and primary metals; these industries together consume over one-half of all industrial energy. The "big five," which adds the pulp and paper industry, as well as the stone, clay and glass group, together account for 70 percent of all industrial sector energy consumption.

According to the U.S. Energy Information Administration, energy efficiency in the manufacturing sector improved by 25 percent over the

Figure 1-5
Industrial Energy Use (end-use basis)
Source - U.S. Department of Energy EIA

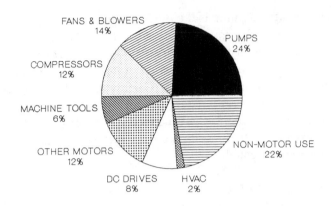

Figure 1-6
Industrial Electricity Use (end-use basis)
Source - Federal Energy Management Agency

period 1980 to 1985 [3]. During that time, manufacturing energy use declined 19 percent, and output increased 8 percent. These changes resulted in an overall improvement in energy efficiency of 25 percent. However, the "big five" did not match this overall improvement; although their energy use declined 21 percent, their output decreased by 5 percent—resulting in only a 17 percent improvement in energy efficiency during 1980-1985. This five year record of improvement in energy efficiency of the manufacturing sector came to an end, with total energy use in the sector growing by 10 percent from 1986 to 1988.

Restoring the record of energy efficiency improvements will require both re-establishing emphasis on energy management and making capital investments in new plant processes and facilities improvements. Reducing our energy costs per unit of manufactured product is one way that our country can become more competitive in the global industrial market. It is interesting to note that Japan—one of our major industrial competitors—has a law that every industrial plant must have a full-time energy manager [4].

1.3 DESIGNING AN ENERGY MANAGEMENT PROGRAM

1.3.1 Management Commitment

The most important single ingredient for successful implementation and operation of an energy management program is commitment to the program by top management. Without this commitment, the program will likely fail to reach its objectives. Thus, the role of the energy manager is crucial in ensuring that management is committed to the program.

Two situations are likely to occur with equal probability when designing an energy management program. In the first, management has decided that energy management is necessary and wants a program implemented. This puts you—the energy manager—in the *response* mode. In the second, you—an employee—have decided to convince management of the need for the program so you are in the aggressive mode. Obviously, the most desirable situation is the response mode as much of your sales effort is unnecessary; nonetheless, a large number of energy management programs have been started through the *aggressive* mode. Let's consider each of these modes.

In a typical scenario of the response mode, management has seen rapidly rising energy prices and/or curtailments, has heard of the results of other energy management programs, and has then initiated action to start the program. In this case, the management commitment already exists, and all that needs to be done is to cultivate that commitment periodically and to be sure the commitment is evident to all people affected by the program. We will discuss this aspect more when we talk about demonstrating the commitment.

In the aggressive mode, you, the employee, know that energy costs are rising dramatically and that sources are less secure. You may have taken a course in energy management, attended professional conferences, and/or read papers on the subject. At any rate, you are now convinced that the company needs an energy management program. All that remains is to convince management and obtain their commitment.

The best way to convince management is with facts and statistics. Sometimes the most startling way to show the facts is through graphs such as Figure 1-7. Note that different goals of energy cost reduction are shown. This graph can be done in total for all energy sources, or several graphs can be used—one for each source. The latter is probably better as savings goals can be identified by energy source. You must have accurate data. Past figures can use actual utility bills, but future figures call for forecasting. Local utilities and various state energy agencies can help you provide management with more accurate data.

Follow this data with quotes on programs from other companies showing these goals are realistic. Other company experiences are widely published in the literature; results can also be obtained through direct contacts with the energy manager in each company. Typical cost avoidance figures are shown in Table 1-2. However, as time progresses and the technology matures, these figures tend to change. For example, a short

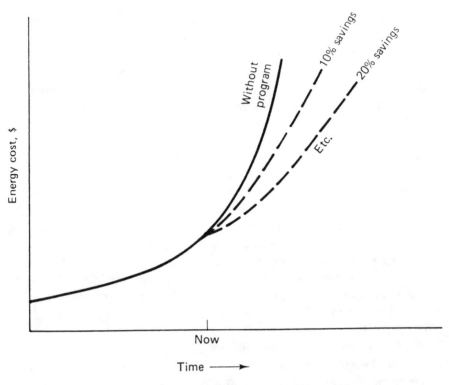

Figure 1-7 Energy costs—past and future.

time ago only a few people believed that an office building could reduce energy consumption by 70 percent or that manufacturing plants could operate on half the energy previously required, yet both are now occurring on a regular basis.

Table 1-2
Typical Energy Savings

Low cost, no cost changes	5-10%
Dedicated programs (3 years or so)	25-35%
Long-range goal	40-50%

As the proponent of an energy management program, you could then talk about the likelihood of energy curtailments or brownouts and what they would mean to the company. Follow this with a discussion of what the energy management program can do to minimize the impacts of curtailments and brownouts.

Finally, your presentation should discuss the competition and what they are doing. Accurate statistics on this can be obtained from trade and professional organizations as well as the U.S. Department of Energy. The savings obtained by competitors can also be used in developing the goals for your facility.

1.3.2 Energy Management Coordinator/Energy Manager

To develop and maintain vitality for the energy management program, a company must designate a single person who has responsibility for coordinating the program. If no one person has energy management as a specific part of his or her job assignment, management is likely to find that the energy management efforts are given a lower priority than other job responsibilities. Consequently, little or nothing may get done.

The energy management coordinator (EMC) should be strong, dynamic, goal oriented, and a good manager. Most important, management should support that person with resources including a staff. The energy management coordinator should report as high as possible in the organization without losing line orientation. A multiplant or multidivisional corporation may need several such coordinators—one for each plant and one for each level of organization. Typical scenarios are illustrated in Figure 1-8.

1.3.3 Backup Talent

Unfortunately, not all the talent necessary for a successful program resides in one person or discipline. For example, several engineering disciplines may be necessary to accomplish a full-scale study of the plant steam production, distribution, usage, and condensate return system. For this reason, most successful energy management programs have an energy management committee. Two subcommittees that are often desirable are the technical and steering subcommittees.

The technical committee is usually composed of several persons with strong technical background in their discipline. Chemical, industrial, electrical, civil, and mechanical engineers as well as others may all be represented on this committee. Their responsibility is to provide technical assistance for the coordinator and plant-level people. For example, the committee can keep up with developing technology and research into potential applications company-wide. The results can then be filtered down.

While the energy management coordinator may be a full-time position, the technical committee is likely to operate part-time, being called

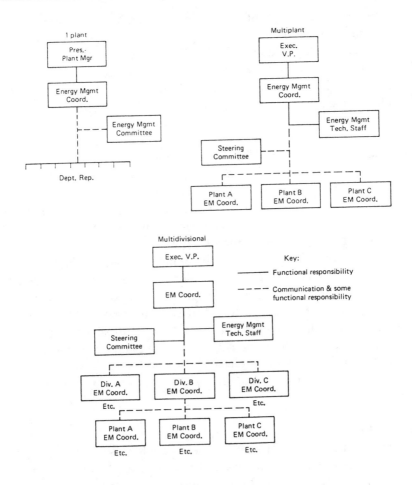

Figure 1-8
Typical organization designs for energy management programs

upon as necessary. In a multiplant or multidivisional organization, the technical committee may also be full time.

The steering committee has an entirely different purpose from the technical committee. It helps guide the activities of the energy management program and aids in communications through all organizational levels. The steering committee also helps ensure that all plant personnel are aware of the program. The steering committee members are usually chosen so that all major areas of the company are represented. A typical organization is presented in Figure 1-9.

Figure 1-9
Energy management steering committee.

Steering committee members should be selected because of their widespread interests and a sincere desire to aid in solving the energy problems. Departmental and hourly representatives can be chosen on a rotating basis. Such a committee should be able to develop a good composite picture of plant energy consumption which will help the energy management coordinator choose and manage his/her activities.

1.3.4 Cost Allocation

One of the most difficult problems for the energy manager is to try to reduce energy costs for a facility when the energy costs are accounted for as part of the general overhead. In that case, the individual managers and supervisors do not consider themselves responsible for controlling the energy costs. This is because they do not see any direct benefit from reducing costs that are part of the total company overhead. The best solution to this problem is for top management to allocate energy costs down to "cost centers" in the company or facility. Once energy costs are charged to production centers in the same way that materials and labor are charged, then the managers have a direct incentive to control those energy costs because this will improve the overall cost-effectiveness of the production center.

For a building, this allocation of energy costs means that each of the tenants are given information on their energy consumption, and that they individually pay for that energy consumption. Even if a large building is "master metered" to reduce utility fixed charges, there should be a division of the utility cost down to the individual customers.

1.3.5 Reporting and Monitoring

It is critical for the energy management coordinator and the steering committee to have their fingers on the "pulse of energy consumption" in the plant. This is best achieved through an effective and efficient system of energy reporting.

The objective of an energy reporting system is to measure energy consumption and compare it either to company goals or to some *standard of energy consumption.* Ideally, this should be done for each operation or production cost center in the plant, but most facilities simply do not have the required metering devices. Many plants only meter energy consumption at one place—where the various sources enter the plant. Most plants are attempting to remedy this, however, by installing additional metering devices when the opportunity arises (steam system shutdowns, vacation downtime, etc.). Systems that should be metered include steam, compressed air, and chilled and hot water.

As always, the reporting scheme needs to be reviewed periodically to ensure that only necessary material is being generated, that all needed data is available, and that the system is efficient and effective.

1.3.6 Training

Most energy management coordinators find that substantial training is necessary. This training can be broken down as shown in Figure 1-10.

Personnel involved	Type of necessary training	Source of required training
1. Technical committee	1. Sensitivity to EM	1. In house (with outside help ?)
	2. Technology developments	2. Professional societies universities, consulting groups, journals
2. Steering committee	1. Sensitivity to EM 2. Other Industries' experience	1. See 1 above 2. Trade journals, energy sharing groups, consultants
3. Plant-wide	1. Sensitivity to EM 2. What's expected, goals to be obtained, etc.	1. In house 2. In house

Figure 1-10
Energy Management Training.

Training cannot be accomplished overnight, nor is it ever "completed." Changes occur in energy management staff and employees at all levels, as well as new technology and production methods. All these precipitate training or retraining. The energy management coordinator must assume responsibility for this training.

1.4 STARTING AN ENERGY MANAGEMENT PROGRAM

Several items contribute to the successful start of an energy management program. They include:

1. Visibility of the program start-up
2. Demonstration of management commitment to the program
3. Selection of a good initial energy management project

1.4.1 Visibility of Start-up

To be successful, an energy management program must have the backing of the people involved. Obtaining this support is often not an easy task, so careful planning is necessary. The people must:

1. Understand why the program exists and what its goals are;
2. See how the program will affect their jobs and income;
3. Know that the program has full management support; and
4. Know what is expected of them.

Communicating this information to the employees is a joint task of management and the energy management coordinator. The company must take advantage of all existing communications channels while also taking into consideration the preceding four points. Some methods that have proven useful in most companies include:

•   Memos. Memos announcing the program can be sent to all employees. A comprehensive memo giving fairly complete details of the program can be sent to all management personnel from first-line supervision up. A more succinct one can be sent to all other employees that briefly states why the program is being formed and what is expected of them. These memos should be signed by local top management.

- News releases. Considerable publicity often accompanies the program start-up. Radio, TV, posters, newspapers, and billboards can all be used. The objective here is to obtain as much visibility for the program as possible and to reap any favorable public relations that might be available. News releases should contain information of interest to the general public as well as employees.

- Meetings. Corporate, plant, and department meetings are sometimes used, in conjunction with or in lieu of memos, to announce the program and provide details. Top management can demonstrate commitment by attending these meetings. The meeting agenda must provide time for discussion and interaction.

- Films, video tapes. Whether produced in-house or purchased, films and video tapes can add another dimension to the presentation. They can also be reused later for new employee training.

1.4.2 Demonstration of Management Commitment

As stressed earlier, management commitment to the program is essential, and this commitment must be obvious to all employees if the program is to reach its full potential. Management participation in the program start-up demonstrates this commitment, but it should also be emphasized in other ways. For example:

- Reward participating individuals. Recognition is highly motivating for most employees. An employee who has been a staunch supporter of the program should be recognized by a "pat on the back," a letter in the files, acknowledgment at performance appraisal time, etc. When the employee has made a suggestion that led to large energy savings, his/her activities should be recognized through monetary rewards, publicity, or both. Public recognition can be given in the company newsletter, on bulletin boards, or in plant-department meetings.

- Reinforce commitment. Management must realize that they are continually watched by employees. Lip service to the program is not enough—personal commitment must be demonstrated. Management should reinforce its commitment periodically, although the visibility scale can be lower than before. Existing newsletters, or a separate one for the energy management program, can include a short column or letter from management on the current results of the program and the plans for the future. This same newsletter can report on outstanding suggestions from employees.

- Fund cost-effective proposals. All companies have capital budgeting problems in varying degrees of severity, and unfortunately energy projects do not receive the same priority as front-line items such as equipment acquisition. However, management must realize that turning down the proposals of the energy management team while accepting others with less economic attractiveness is a sure way to kill enthusiasm. Energy management projects need to compete with others fairly. If an energy management project is cost effective, it should be funded. If money is not available for capital expenditures, then management should make this clear at the outset of the program and ask the team to develop a program which does not require capital expenditures.

1.4.3 Early Project Selection

The energy management program is on treacherous footing in the beginning. Most employees are afraid their heat is going to be set back, their air conditioning turned off, and their lighting reduced. If any of these actions do occur, it's little wonder employee support wanes. These things might occur eventually, but wouldn't it be smarter to have less controversial actions as the early projects.

An early failure can also be harmful, if not disastrous, to the program. Consequently, the astute energy management coordinator will "stack the deck" in his or her first set of projects. These projects should have a rapid payback, a high probability of success, and few negative consequences.

These ideal projects are not as difficult to find as you might expect. Every plant has a few good opportunities, and the energy management coordinator should be looking for them.

One good example involved a rather dimly lit refrigerator warehouse area. Mercury vapor lamps were used in this area. The local energy management coordinator did a relamping project. He switched from mercury vapor lamps to high pressure sodium lamps (a significantly more efficient source) and carefully designed the system to improve the lighting levels. Savings were quite large; less energy was needed for lighting; less "heat of light" had to be refrigerated; and, most important, the employees liked it. Their environment was improved since light levels were higher than before.

Other examples you should consider include:

1. Repairing steam leaks. Even a small leak can be very expensive over a year and quite uncomfortable for employees working in the area.

2. Insulating steam, hot water, and other heated fluid lines and tanks. Heat loss through an uninsulated steam line can be quite large, and the surrounding air may be heated unnecessarily.

3. Install high efficiency motors. This saves dramatically on the electrical utility cost in many cases, and has no negative employee consequences. However, the employees should be told about the savings since motor efficiency improvement has no physically discernible effect, unlike the lighting example above.

This list only begins to touch on the possibilities, and what may be glamorous for one facility might not be for another. All facilities, however, do have such opportunities. Remember, highly successful projects should be accompanied by publicity at all stages of the program—especially at the beginning.

1.5 MANAGEMENT OF THE PROGRAM

1.5.1 Establishing objectives in an Energy Management Program

Creativity is a vital element in the successful execution of an energy management program, and management should do all it can to encourage creativity rather than stifle it. Normally, this implies a laissez-faire approach by management with adequate monitoring. Management by objectives (MBO) is often utilized. If TQM is being implemented in a facility, then employee teams should foster this interest and creativity.

Goals need to be set, and these goals should be tough but achievable, measurable, and specific. They must also include a deadline for accomplishment. Once management and the energy management coordinator have agreed on the goals and established a good monitoring or reporting system, the coordinator should be left alone to do his/her job.

The following list provides some examples of such goals:

- Total energy per unit of production will drop by 10 percent the first year and an additional 5 percent the second.

- Within 2 years all energy consumers of 5 million British thermal units per hour (Btuh) or larger will be separately metered for monitoring purposes.

- Each plant in the division will have an active energy management program by the end of the first year.

- All plants will have contingency plans for gas curtailments of varying duration by the end of the first year.

- All boilers of 50,000 lb/hour or larger will be examined for waste heat recovery potential the first year.

The energy management coordinator must quickly establish the reporting systems to measure progress toward the goals and must develop the strategy plans to ensure progress. Gantt or CPM charting is often used to aid in the planning and assignment of responsibilities.

Some concepts or principles that aid the EMC in this execution are the following:

- Energy costs, not just Btus, are to be controlled. This means that any action that reduces energy costs is fair game. Demand shedding or leveling is an example activity that saves dollars but does not directly save Btus.

- Energy needs should be recognized and billed as a direct cost, not as overhead. Until the energy flow can be measured and charged to operating cost centers, the program will not reach its ultimate potential.

- Only the main energy functions need to be metered and monitored. The Pareto or ABC principle states that the majority of the energy costs are incurred by only a few machines. These high-use machines should be watched carefully.

1.5.2 A Model Energy Management Program

An excellent example of a successful energy management program in a large corporation is that of the 3M Company, headquartered in St. Paul, Minnesota [5]. 3M is a large, diversified manufacturing company with more than 50 major product lines; it makes some fifty thousand products at over fifty different factory locations around the country. The corporate energy management objective is to use energy as efficiently as possible in all operations; the management believes that all companies have an obligation to conserve energy and all other natural resources.

Energy productivity at 3M improved by 54 percent from 1973 to 1988. They saved $75 million in 1988 because of their energy management programs, and saved a total of $650 million in energy expenses from 1973 to 1988. Their program is staffed by six people who educate and motivate all levels of personnel on the benefits of energy management. The categories of programs implemented by 3M include: conservation, maintenance procedures, utility operation optimization, efficient new designs, retrofits through energy surveys, and process changes.

Energy efficiency goals at 3M are set and then the results are measured against a set standard in order to determine the success of the programs. The technologies that have resulted in the most dramatic improvement in energy efficiency include: heat recovery systems, high efficiency motors, variable speed drives, computerized facility management systems, steam trays maintenance, combustion improvements, variable air volume systems, thermal insulation, cogeneration, waste steam utilization, and process improvements. Integrated manufacturing techniques, better equipment utilization and shifting to non-hazardous solvents have also resulted in major process improvements.

The energy management program at 3M has worked very well, but management is not yet satisfied. They have set a goal of further improving energy efficiency at a rate of 3 percent per year for the next five years. They expect to substantially reduce their emissions of waste gases and liquids, to increase the energy recovered from wastes, and to constantly increase the profitability of their operations. 3M continues to stress the extreme importance that efficient use of energy can have on their industrial productivity.

1.6 ENERGY ACCOUNTING

Energy accounting is a system used to keep track of energy consumption and costs. "Successful corporate-level energy managers usually rank energy accounting systems right behind commitment from top corporate officials when they list the fundamentals of an ongoing energy conservation program. If commitment from the top is motherhood, careful accounting is apple pie."*

A basic energy accounting system has three parts: energy use monitoring, an energy use record, and a performance measure. The performance measure may range from a simple index of Btu/ft^2 or Btu/unit of production to a complex standard cost system complete with variance reports. In all cases, energy accounting requires metering. Monitoring the energy flow through a cost center, no matter how large or small, requires the ability to measure incoming and outgoing energy. The lack of necessary meters is probably the largest single deterrent to the widespread utilization of energy accounting systems.

---

*"Accounting of Energy Seen Corporate Must," *Energy User News*, Aug. 27, 1979, p. 1.

1.6.1 Levels of Energy Accounting

As in financial accounting, the level of sophistication or detail of energy accounting systems varies considerably from company to company. A very close correlation can be developed between the levels of sophistication of financial accounting systems and those of energy accounting systems. This is outlined in Figure 1-11.

Most companies with successful energy management programs have passed level 1 and are working toward the necessary submetering and reporting systems for level 2. In most cases, the subsequent data are compared to previous years or to a particular base year. However, few companies have developed systems that will calculate variations and find causes for those variations (level 3). Two notable exceptions are General Motors and Carborundum. To our knowledge, few companies have yet completely developed the data and procedures necessary for level 4, a standard Btu accounting system. Some examples of detailed energy accounting can be found in [6].

Financial	Energy
1. General accounting	1. Effective metering, development of reports, calculation of energy efficiency indices
2. Cost accounting	2. Calculation of energy flows and efficiency of utilization for various cost centers; requires substantial metering
3. Standard cost accounting historical standards	3. Effective cost center metering of energy and comparison to historical data; complete with variance reports and calculation of reasons for variation
4. Standard cost accounting engineered standards	4. Same as 3 except that standards for energy consumption are determined through accurate engineering models

Figure 1-11
Comparison between financial and energy accounting.

1. 6. 2 Performance Measures

1.6.2.1 Energy Utilization Index

A very basic measure of a facility's energy performance is called the Energy Utilization Index (EUI). This is a statement of the number of Btu's of energy used annually per square foot of conditioned space. To compute the EUI, all of the energy used in the facility must be identified, the total Btu content tabulated, and the total number of square feet of conditioned space determined. The EUI is then found as the ratio of the total Btu consumed to the total number of square feet of conditioned space.

Example 1.1—Consider a building with 100,000 square feet of floor space. It uses 1. 76 million kWh and 6.5 million cubic feet of natural gas in one year. Find the Energy Utilization Index (EUI) for this facility.

Solution: Each kWh contains 3412 Btu and each cubic foot of gas contains about 1000 Btu. Therefore the total annual energy use is:

$$\text{Total energy use} = (1.76 \times 10^6 \text{ kWh}) \times (3412 \text{ Btu/kWh})$$
$$+ (6.5 \times 10^6 \text{ ft}^3) \times (1000 \text{ Btu/ft}^3)$$

$$= 6.0 \times 10^9 + 6.5 \times 10^9$$
$$= 1.25 \times 10^{10} \text{ Btu/yr}$$

Dividing the total energy use by 10^5 ft^2 gives the EUI:

$$\text{EUI} = (1.25 \times 10^{10} \text{ Btu/yr})/(10^5 \text{ ft}^2)$$
$$= 125,000 \text{ Btu/ft}^2/\text{yr}$$

The average building EUI is 88,700 Btu/ft^2/yr; the average office building EUI is 106,000 Btu/ft^2/yr. Figure 1-12 below shows the range of energy intensiveness in 1000 Btu/ft^2/yr for the twelve different types of commercial facilities listed [7].

1.6.2.2 Energy Cost Index

Another useful performance index is the Energy Cost Index or ECI. This is a statement of the dollar cost of energy used annually per square foot of conditioned space. To compute the ECI, all of the energy used in the facility must be identified, the total cost of that energy tabulated, and

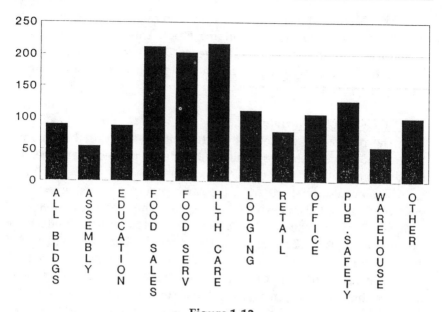

Figure 1-12
Building energy utilization index.
Source - U.S. Department of Energy EIA

the total number of square feet of conditioned space determined. The ECI is then found as the ratio of the total annual energy cost for a facility to the number of square feet of conditioned floor space of the facility.

Example 1.2 Consider the building in Example 1.1. The annual cost for electric energy is $115,000 and the annual cost for natural gas is $32,500. Find the Energy Cost Index (ECI) for this facility.

Solution: The ECI is the total annual energy cost divided by the number of conditioned square feet of floor space.

$$\text{Total energy cost} = \$115,000 + \$32,500 = \$147,500/\text{yr}$$

Dividing this total energy cost by 100,000 square feet of space gives:

$$\text{ECI} = (\$147,500/\text{yr})/(100,000 \text{ ft}^2) = \$1.48/\text{ft}^2/\text{yr}$$

The Energy Information Administration does not publish detailed statistics on ECI's for various building types. Thus, there is no similar figure to 1-12 to show for ECI's. However, the Energy Information Administration does provide a value of the ECI for the average building—it was reported as $1.12/ft^2/yr in 1992.

1.6.2.2 One-Shot Productivity Measures

The purpose of a one-shot productivity measure is illustrated in Figure 1-13. Here the energy utilization index is plotted over time, and trends can be noted.

Significant deviations from the same period during the previous year should be noted and explanations sought. This measure is often used to justify energy management activities or at least to show their effect. For example, in Figure 1-13 an energy management (EM) program was started at the beginning of year 2. Its effect can be noted by comparing peak summer consumption in year 2 to that of year 1. The decrease in peaks indicates that this has been a good program (or a mild summer, or both).

Table 1-3 shows some often-used indices. Some advantages and disadvantages of each index are listed, but specific applications will require careful study to determine the best index.

Table 1-4 proposes some newer concepts. Advantages and disadvantages are shown, but since most of these concepts have not been utilized in a large number of companies, there are probably other advantages and disadvantages not yet identified. Also, there are an infinite number of possible indices, and only three are shown here.

1.6.3 Energy Accounting Systems

Two major corporations with strong energy accounting systems are General Motors, which uses an energy responsibility accounting system, and Carborundum, which has a sophisticated analysis of variance system.

1.6.3.1 The GM system

According to General Motors, a good energy accounting system is implemented in three phases: (1) design and installation of accurate metering, (2) development of an energy budget, and (3) publication of regular performance reports including variances. Each phase is an important element of the complete system.

Phase 1—Metering. For execution of a successful energy accounting program, energy flow must be measured by cost center. The designing of

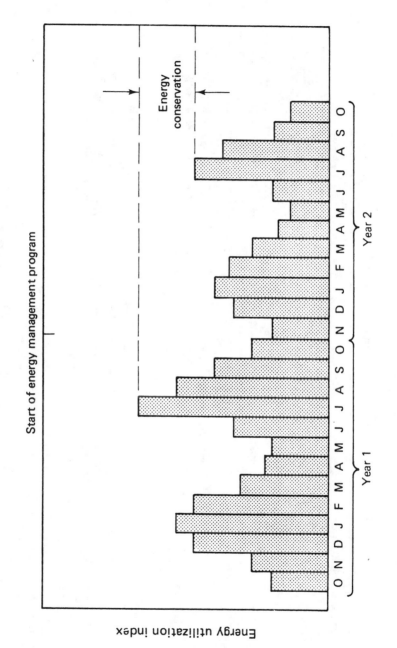

Figure 1-13
One-shot energy productivity measurement.

Table 1-3. Commonly Used Indices

Productivity indicator	Advantages	Disadvantages
1. Btu/unit of production	1. Concise, neat 2. Often accurate when process energy needs are high 3. Good for interplant and company comparison when appropriate	1. Difficult to define and measure "units" 2. Often not accurate (high HVAC* and lighting makes energy nonlinear to production)
2. Btu/degree day	1. Concise, neat, best used when HVAC* is a majority of energy bill 2. Often accurate when process needs are low or constant 3. Very consistent between plants, companies, etc. (all mfg can measure degree days)	1. Often not accurate (disregards process needs) 2. Thermally heavy buildings such as mfg plants usually do not respond to degree days
3. Btu/ft²	1. Concise, neat 2. Accurate when process needs are low or constant and weather is consistent 3. Very consistent (all mfg can measure square feet) 4. Expansions can be incorporated directly	1. No measure of production or weather 2. Energy not usually linearly proportional to floor space (piecewise linear?)
4. Combination, e.g., Btu/unit-degree day-ft² or Btu/unit-degree day	1. Measures several variables 2. Somewhat consistent, more accurate than above measures 3. More tailor-made for specific needs	1. Harder to comprehend

*Heating, ventilating, and air conditioning.

Table 1-4
Proposed Indices

Productivity indicator	Advantages	Disadvantages
1. Btu/sales dollar 2. $\dfrac{\$ \text{ energy}}{(\$ \text{ sales}) \text{ or } (\$ \text{ profit}) \text{ or } (\$ \text{ value added})}$	1. Easy to compute 1. Really what's desired 2. Inflation cancels or shows changing relative energy costs 3. Shows energy management results, not just conservation (e.g., fuel switching, demand leveling, contingency planning)	1. Impact of inflation 1. Very complex, e.g., lots of variables affect profit including accounting procedures 2. Not good for general employee distribution
3. Btu/DL hour (or machine hour or shift) where DL = direct labor	1. Almost a measure of production (same advantage as in Table 1-3) 2. Data easily obtained when already available 3. Comparable between plants or industries 4. Good for high process energy needs	1. More complex, e.g., can't touch a DL hour like a unit of production 2. Energy often not proportional to labor or machine input, e.g., high HVAC and lighting

cost center boundaries requires care; the cost centers must not be too large or too small. However, the primary design criterion is how much energy is involved. For example, a bank of large electric induction heat-treating furnaces might need separate metering even if the area involved is relatively small, but a large assembly area with only a few energy-consuming devices may require only one meter. Flexibility is important since a cost center that is too small today may not be too small tomorrow as energy costs change.

The choice of meters is also important. Meters should be accurate, rugged, and cost effective. They should have a good turndown ratio; a turndown ratio is defined as the ability to measure accurately over the entire range of energy flow involved.

Having the meters is not enough. A system must be designed to gather and record the data in a useful form. Meters can be read manually, they can record information on charts for permanent records, and/or they can be interfaced with microcomputers for real-time reporting and control. Many systems fail because the data collection system is not adequately utilized.

Phase 2—Energy Budget. The unique and perhaps vital aspect of General Motors' approach is the development of an energy budget. The GM energy responsibility accounting system is somewhere between levels 3 and 4 of Figure 1-11. If a budget is determined through engineering models, then it is a standard cost system and it is at level 4.

There are two ways to develop the energy budget: statistical manipulation of historical data or utilization of engineering models.

Statistical Model: By using historical data, the statistical model shows how much energy was utilized and how it compared to the standard year(s), but it does not show how efficiently the energy was used. For example, consider the data shown in Table 1-5.

The statistical model assumes that the base years are characteristic of all future years. Consequently, if 1981 produced 600 units with the same square footage and degree days as 1980, 1000 units of energy would be required. If 970 units of energy were used, the difference (30 units) would be due to conservation.

We could use multiple linear regression to develop the parameters for our model, given as follows:

Table 1-5
Energy Data for Statistical Model[a]

	1980	1981	1982
Total energy (units)	1,000	1,100	1,050
Production (units)	600	650	650
Square feet	150,000	150,000	170,000
Degree days (heating)	6,750	6,800	6,800

[a]Taken, in part, from R.P. Greene, (see the Bibliography).

energy forecast= a(production level) + b(ft^2 + c(degree days)   (1-1)

We can rewrite this in the following form:

$$X_4 = aX_1 + bX_2 + cX_3$$

where X_1= production (units)
X_2 = floor space (ft^2)
X_3 = weather data (degree days)
X_4 = energy forecast (Btu)

Degree days are explained in detail in Chapter Two, section 2.1.1.2. Their use provides a simple way to account for the severity of the weather, and thus the amount of energy needed for heating and cooling for a facility. Of course, the actual factors included in the model will vary between companies and need to be examined carefully.

Multiple linear regression estimates the parameters in the universal regression model in Equation 1-1 from a set of sample data. Using the base years, the procedure estimates values for parameters a, b, and c in Equation 1-1 in order to minimize the squared error where

squared error = $\displaystyle\sum_{\substack{\text{base} \\ \text{years}}} \left(X_4^i - X_4\right)^2$ (1-2) where X_4 = energy forecast by model

    X_4^i = actual energy usage

The development and execution of this statistical model is beyond the scope of this book. However, regardless of the analytical method used, the statistical model does not determine the amount of energy that ought to be used. It only forecasts consumption based on previous years' data.

The engineering model. The engineering model attempts to remedy the deficiency in the statistical model by developing complete energy balance calculations to determine the amount of energy theoretically required. By using the first law of thermodynamics, energy and mass balances can be completed for any process. The result is the energy required for production. Similarly, HVAC and lighting energy needs could be developed using heat loss equations and other simple calculations. Advantages of the engineering model include improved accuracy and flexibility in reacting to changes in building structures, production schedules, etc. Also, computer programs exist that will calculate the needs for HVAC and lighting.

Phase 3—Performance Reports. The next step is the publication of energy performance reports that compare actual energy consumption with that predicted by the models. The manager of each cost center should be evaluated on his or her performance as shown in these reports. The publication of these reports is the final step in the effort to transfer energy costs from an overhead category to a direct cost or at least to a direct overhead item. One example report is shown in Figure 1-14.

Sometimes more detail on variance is needed. For example, if consumption were shown in dollars, the variation could be shown in dollars and broken into price and consumption variation. Price variation is calculated as the difference between the budget and the actual unit price times the present actual consumption. The remaining variation would be due to a change in consumption and would be equal to the change in consumption times the budget price. This is illustrated in Example 1.3. Other categories of variation could include fuel switching, pollution control, and new equipment. (This will be examined more in the next section.)

	Actual	Budget	Variance	% variance
Department A				
Electricity	2000	1500	+500	+33.3%
Natural gas	3000	3300	−300	−9.1%
Steam	3500	3750	−250	−6.7%
Total	8500	8550	−50	−0.6%
Department B				
Electricity	1500	1600	−100	−6.2%
Natural gas	2000	2400	−400	−16.7%
Fuel oil	1100	1300	− 200	−15.4%
Coal	3500	3900	− 400	−10.2%
Total	8100	9200	− 1100	− 11.9%

Department C
-
-
-

Figure 1-14
Energy performance report (10^6 Btu)

Example 1.3

The table shown in Figure 1-15 portrays a common problem in energy management reporting. The energy management program in this heat treating department was quite successful. When you examine the totals, you see that the total consumption (at old prices) was reduced by $5631. The total energy cost, however, went up by $500, which was due to a substantial price variation of $6131. Consequently, total energy costs increased to $34,000.

Department (source)	[A] Actual $	[B] Budget $	[C] Unit price (budget)	[D] Unit price (actual)	[E] A - B^a variance	[F] (D - C)A^b price variance	[G] E - F or (B^b - A^b)C consumption variance
	10^6 Btu	10^6 Btu	$/10^6 Btu	$/10^6 Btu			
Heat treating (electricity)	$9,000 2,000	$8,500 2,125	$4.00 —	$4.50 —	+$500 —	+$1000 —	-$500 —
(natural gas)	15,000 4,808	16,000 6,400	2.50 —	3.12 —	-1000 —	+2980 —	-3980 —
(steam)	10,000 2,242	9,000 2,571	3.50 —	4.46 —	+1000 —	+2151 —	-1151 —
(total)	$34,000	$33,500	—	—	+$500	+$6131	-$5631

^{a}Measured in $
^{b}Measured in 10^6 Btu

Figure 1-15

Energy cost in dollars by department with variance analysis.

However, had energy consumption not been reduced, the total energy cost would have been:

$$2125(4.50) + 6400(3.12) + 2571(4.46) = \$40{,}997.$$

The total cost avoidance therefore was:

$$\$40{,}997 - \$34{,}000 = \$6997$$

which is the drop in consumption times the actual price or

$$(2125 - 2000)\,4.5 + (6400 - 4808)\,3.12 + (2571 - 2242)\,4.46 = \$6997$$

This problem of increased energy costs despite energy management savings can arise in a number of ways. Increased production, plant expansion, or increased energy costs can all cause this result.

1.6.3.2 The Carborundum system

The Carborundum energy accounting system carries the analysis of variance several steps further by breaking consumption variance down into volume-mix, weather, pollution, conservation, alternative fuels, and others. By stating the consumption variation in this manner, the Carborundum system provides more information for management action. The net impact would still be the algebraic sum of all the effects.

The calculation procedures for the different variances are quite simple.

Product Mix Effect. For each product line, the effect of changing production volume on energy consumption for a given source i s:

$$PM_{ij} = \left(Q_i - Q_i^1\right) E_{ij}^1$$

where PM_{ij} = product mix effect (10^6 Btu) of changing production levels on product i for energy source j

          Q_i = quantity of product i made to date

          Q_i^1 = quantity of product i made base year to date (or standard quantity)

          E_{ij}^1 = amount of energy of type j (10^6 Btu) required per unit of product i base year (or standard requirement)

To specify this in dollars, multiply PM_{ij} by last year's cost of energy of type j:

$$PM_{ij}(\$) = PM_{ij}C_j^1$$

where C_j^1 = cost of energy type j in base year (or standard price)

Heating Effect. The effect of varying heating demands can be determined from the following equation:

$$H_j = \left(DD_H - DD_H^1\right) E_{Hj}^1 \tag{1-5}$$

date (or standard quantity)

where H_j = heating effect on consumption of energy type j (10^6 Btu) this year

DD_H = heating degree days this year to day

DD_H^1 = heating degree days base year to date (or standard quantity)

E_{Hj}^1 = quantity type j energy required per degree day base year to

date (or standard quantity)

(Heating degree days are explained in Section 6.4.1.3.2.)

Specified in dollars, the heating effect is:

$$H_j\ (\$) = H_j C_j^1$$

(1-6)

Cooling Effect. The effect of varying cooling needs can be determined from

$$K_j = \left(DD_c - DD_c^1\right) E_{cj}^1$$

(1-7)

where K_j = cooling effect on consumption of energy type j (10^6 Btu) this year

DD_c = cooling degree days this year to day

DD_c^1 = cooling degree days base year to date (or standard quantity)

E_{cj}^1 = quantity type j energy required per cooling degree day base

year to date (or standard quantity)

(Cooling degree days are explained in Section 6.4.1.3.2.)

This cooling effect can be stated in dollars as:

$$K_j\ (\$) = K_j C_j^1$$

(1-8)

Other Effects. The effects of any other changes can be determined from the following equations repeated for as many effects as desired (e.g., pollution control, etc.) or one general account ("other"):

$$OE_j = O_j - O_j^1 \qquad\qquad (1\text{-}9)$$

$$OE_j\ (\$) = O_j\, C_j^1 \qquad\qquad (1\text{-}10)$$

where OE_j = effect of other changes on energy type j

O_j = energy type j effects from other changes

O_j^1 = energy type j effects from other changes base year to date

(or standard)

Example 1. 4

Ace Manufacturing Company has an active energy management program underway and has recently installed an energy accounting system of the type just developed. For the data given below comparing this year to last, calculate all variances and show the net impact:

	This year	Last year (base)
Product line 1 ($i = 1$)	1000 units	900 units
Product line 2 ($i = 2$)	2000 units	2200 units
Heating degree days	3570 °F days	3800 °F days
Cooling degree days	2200 °F days	2000 °F days
Price/10^6 Btu (nat. gas)	$3.50	$3.00
Price/10^6 Btu (electricity)	$5.00	$4.75
Nat. gas/heating degree day ($j = 1$)	25×10^6 Btu	30×10^6 Btu
Nat. gas/cooling degree day	0	0
Elec./heating degree day($j= 2$)	0	0
Elec./cooling degree day	60×10^6 Btu	70×10^6 Btu
Gas required/unit product 1 (E_{11})	80×10^6 Btu	100×10^6 Btu
Gas required/unit product 2 (E_{21})	45×10^6 Btu	50×10^6 Btu
Elec. required/unit product 1 (E_{12})	400×10^6 Btu	500×10^6 Btu
Elec. required/unit product 2 (E_{22})	200×10^6 Btu	300×10^6 Btu

Product mix effect:

$$PM_{11} = \left(Q_1 - Q_1^1\right) E_{11}^1$$
$$= (1000 - 900)\left(100 \times 10^6\right)$$
$$= +10,000 \times 10^6 \text{ Btu}$$

$$PM_{12} = \left(Q_1 - Q_1^1\right) E_{12}^1$$
$$= 100\left(500 \times 10^6\right)$$
$$= +50,000 \times 10^6 \text{ Btu}$$

$$PM_{21} = \left(Q_2 - Q_2^1\right) E_{21}^1$$
$$= (2000 - 2200)\left(50 \times 10^6\right)$$
$$= -10,000 \times 10^6 \text{ Btu}$$

$$PM_{22} = \left(Q_2 - Q_2^1\right) E_{22}^1$$
$$= -200\left(300 \times 10^6\right)$$
$$= -60,000 \times 10^6 \text{ Btu}$$

$$PM_{11} (\$) = PM_{11} C_i^1$$
$$= \left(10,000 \times 10^6 \text{ Btu}\right)\left(\$3.00/10^6 \text{ Btu}\right)$$
$$= +\$30,000$$

$$PM_{12} (\$) = PM_{12} C_2^1$$
$$= \left(50,000 \times 10^6 \text{ Btu}\right)\left(\$4.75/10^6 \text{ Btu}\right)$$
$$= +\$237,500$$

$$PM_{21} (\$) = PM_{21} C_1^1$$
$$= \left(-10,000 \times 10^6 \text{ Btu}\right)\left(\$3.00/10^6 \text{ Btu}\right)$$
$$= -\$30,000$$

$$PM_{22}\ (\$) = PM_{22}\ C_2^1$$
$$= \left(-60{,}000 \times 10^6\ \text{Btu}\right)\left(\$4.75/10^6\ \text{Btu}\right)$$
$$= -\$285{,}000$$

net product mix effect =
$$-\$285{,}000 - \$30{,}000 + \$237{,}500 + \$30{,}000 = -\$47{,}500$$

Heating effect:

$$H_1 = \left(DD_H - DD_H^1\right) E_{H1}^1$$
$$= (3570 - 3800)\left(30 \times 10^6\ \text{Btu}\right)$$
$$= -6900 \times 10^6\ \text{Btu}$$

$$H_2 = 0$$

$$H_1(\$) = H_1 C_1^1$$
$$= \left(-6900 \times 10^6\ \text{Btu}\right)\left(\$3.00/10^6\ \text{Btu}\right)$$
$$= -\$20{,}700$$

Cooling effect:

$$K_1 = 0$$

$$K_1 = \left(DD_c - DD_c^1\right) E_c^1$$
$$= (2200 - 2000)\left(70 \times 10^6\ \text{Btu}\right)$$
$$= 14{,}000 \times 10^6\ \text{Btu}$$

$$K_2\ (\$) = K_2\ C_2^1$$
$$= \left(14{,}000 \times 10^6\ \text{Btu}\right)\left(\$4.75/10^6\ \text{Btu}\right)$$
$$= \$66{,}500$$

Other consumption effects: None shown.

Price effects-natural gas: The price effect for natural gas is the difference in prices between this year and the base year times the quantity used this year (see the Section 1.5.3). The quantity of natural gas used this year is

(3570 degree days) (25 × 10^6 Btu/degree day) + (1000 units product 1) × (80 × 10^6 Btu/unit) + (2000 units product 2) (45 × 10^6 Btu/unit)
$$= 259{,}000 \times 10^6 \text{ Btu natural gas}$$

natural gas price effect   = (259,250 × 10^6 Btu) ($3.50 – $3.00)/$10^6$ Btu
                                      = + \$129,625

Electricity: Similarly the price effect for electricity is the consumption this year times the price differential. The consumption this year is

(2200 degree days)(60 × 10^6 Btu/degree day) + (1000 units product 1) (400 × 10^6 Btu/unit) + (2000 units product 2)(200 × 10^6 Btu/unit)
$$= 932{,}000 \times 10^6 \text{ Btu}$$

electricity price effect   = (932,000 × 10^6Btu)(\$5.00— 4.75)/10^6Btu
                                      = \$233,000

total price effect   = \$233,000 + \$129,625
                                 = + \$362,625

The total energy cost for this year is

(259,250 × 10^6 Btu nat. gas)(\$3.50/10^6 Btu)
+ (932,000 × 10^6 Btu elec.) (\$5.00/10^6 Btu)
$$= \underline{\$5{,}567{,}375}$$

To determine the total energy cost for the base year, we need the total consumption:

total consumption   =   (3800 degree days) (30 × 10^6 Btu/degree day)
natural gas   + (900 units product 1) (100 × 10^6 Btu/unit) +
                                    (2200 units product 2)(50 × 10^6Btu/unit)
                          =   314,000 × 10^6 Btu

total consumption $=$ (2000 degree days) (70 × 10^6 Btu/degree day)
electricity $+$ (900 units product 1) (500 × 10^6 Btu/unit)
                       $+$ (2200 units product 2)(300 × 10^{6}Btu/unit)
                    $=$ 1,250,000 × 10^6 Btu

total energy cost $=$ (314,000 × 10^6 Btu gas)($3.00/10^{6}Btu gas)
(base year) $+$ (1,250,000 × 10^6 Btu elec.)($4.75/10^6 Btu)
                    $=$ $6,879,500

A summary of effects is given in Table 1-6.

Table 1-6
Energy Effects Breakdown

	Base year	This year	Change
Energy consumption ($)	6,879,500	5,567,375	–$1,312,125
Price effect	—	—	+ $362,625
Cooling effect	—	—	+ $66,500
Heating effect	—	—	– $20,700
Product mix effect	—	—	– $47,500
Conservation (balancing amount)	—	—	– $951,200

As can be seen from Table 1-6, the price and cooling effect changes were largely negative and offset the savings from heating and the product mix. Yet the total energy cost dropped by $1,312,125. Therefore, the balancing amount is due to conservation ($951,200). This is a large savings and signifies a successful program.

1.7 SUMMARY

This chapter has examined the design, initiation, and management of energy management programs. The chapter emphasizes energy accounting, especially cost center accounting and necessary submetering.

We defined an energy management activity as any decision that involves energy and affects the profit level. Anything that improves profits and/or enhances competitive positions is considered effective energy management, and anything else is poor energy management. The motivation for starting energy management programs is multi-faceted and varies among companies. The following outline lists the major reasons:

- Economic—Energy management will improve profits and enhance competitive positions.

- National good—Energy management is good for the U.S. economy as the balance of payments becomes more favorable and the dollar stronger.

  Energy management makes us less vulnerable to energy cutoffs or curtailments due to political unrest or natural disasters elsewhere.

  Energy management is kind to our environment as it eases some of the strain on our natural resources and may leave a better world for future generations.

In designing an energy management program, several ingredients are vital:

- Top Management commitment. Commitment from the top must be strong and highly visible.

- One-person responsibility. The responsibility for the energy management program must lie in one person who reports as high in the organization structure as possible.

- Committee backup. The energy management coordinator must have the support of two committees. The first is a steering committee, which provides direction for the program. The second is a technical committee, which provides technical backup in the necessary engineering disciplines.

- Reporting and monitoring. An effective monitoring and reporting system for energy consumption must be provided.

- Training. Energy management is a unique undertaking. Hence, training and retraining at all levels is required.

To successfully start an energy management program, some publicity must accompany the early stages. This can be achieved with news releases, films, plant meetings, or a combination of them. Early project selection is a critical step. Early projects should be visible, and should have good monetary returns, with few negative consequences.

Management and creative personnel are always critical components of an energy management program. Tough, specific, and measurable goals need to be developed. Once the goals are established, management should carefully monitor the results, but the energy management staff should be allowed to perform its functions. Staff and management need to realize that (1) energy costs, not consumption, are to be controlled (2) energy should be a direct cost—not an overhead item, and (3) only the main energy consumers need be metered and monitored closely.

Energy accounting is the art and science of tracing Btu and energy dollar flow through an organization. Cost center orientation is important, as are comparison to some standard or base and calculations of variances. Causes for variances must then be sought. General Motor's energy responsibility accounting system was discussed as was Carborundum's rather sophisticated approach to determining variances. However, no accounting system is a panacea, and any system is only as accurate as the metering and reporting systems allow it to be.

REFERENCES

1. The National Energy Strategy, Interim Report, *Chapter on Commercial Energy Use*, U.S. Department of Energy, Washington, D.C., April 2, 1990, p. 26.

2. The National Energy Strategy, Interim Report, *Chapter on Industrial Energy Use*, U.S. Department of Energy, Washington, D.C., April 2, 1990, p. 31.

3. Energy Information Administration, *Manufacturing Energy Consumption Survey: Changes in Energy Efficiency 1980-1985*, U.S. Department of Energy, Washington, D.C., January 17, 1990, p. vii.

4. The National Energy Strategy, *Chapter on Industrial Energy Use*, U.S. Department of Energy, Washington, D.C., April 2, 1990, p. xx.

5. Aspenson, R.L., *Testimony to the U.S. Department of Energy on the National Energy Strategy, Hearing on Energy and Productivity*, Providence, RI, December 1, 1989.

6. Applebaum, Bruce and Mark Oven, "Energy Management Skills for Energy Engineers," *Energy Engineering*, Vol. 89, No. 5, 1992, pp. 52-80.

7. Energy Information Administration, *Commercial Buildings Consumption and Expenditures 1989*, Energy Information Administration (EIA), U.S. Department of Energy, Washington, D.C., April, 1992.

BIBLIOGRAPHY

Barrett, John A., "Energy Accounting Systems for Existing Buildings," ASHRAE Journal, January 1979, p. 56.

"Btu Accounting," client newsletter, Seidman and Seidman, Certified Public Accountants, Grand Rapids, MI, 1980.

Capehart, Barney L., "Energy Management," Chapter 70 in the *Handbook of Industrial Engineering*, Second Edition, John Wiley and Sons, Inc., New York, 1992.

Capehart, Barney L. and Mark Spiller, "Energy Auditing," Chapter 3 in the *Energy Management Handbook*, Second Edition, Fairmont Press, 1993.

Grant, Eugene L. and Richard S. Leavenworth, Statistical Quality Control, Sixth Edition, McGraw-Hill Company, New York, NY, 1988.

Energy Conservation and Program Guide for Industry and Commerce, GPO, Washington, DC, September 1974.

Energy Conservation Trends—Understanding the Factors That Affect Conservation in the U.S. Economy, U.S. Department of Energy, DOE/PE-0092, September, 1989.

Fearing, Watson B., 3M Corporation, St. Paul, MN, personal correspondence, 1980.

Greene, Richard P., "Energy Responsibility Accounting, an Energy Conservation Tool for Industrial Facilities," unpublished working paper, General Motors Corporation, date unknown.

Howell, Edward D., "Energy Accounting and Analysis: The Carborundum System," unpublished working paper, the Carborundum Company, date unknown.

Millhone, John P., John A. Beldock, Barney L. Capehart, and Jerome Lamontagne, "United States Energy Efficiency: Strategy for the Future," Strategic Planning for Energy and the Environment, Vol. 11, No. 3, Winter, 1992.

Shirley, James M., and W.C. Turner, "Industrial Energy Management— First Steps Are Easy," Industrial Engineering, May 1978, pp. 3441.

Stobaugh, Robert, and Daniel Yergin, *Energy Future*, Second Edition, Random House, New York, 1983.

Total Energy Management, A Handbook prepared by the National Electrical Contractors Association and the National Electrical Manufacturers Association, Washington, DC, Third Edition, 1986.

Turner, W. C., Editor, *Energy Management Handbook*, Second Edition, Fairmont Press, Atlanta, GA, 1993.

Turner, W.C., and B.W. Tompkins, "Productivity Measures for Energy Management," in IIE National Meeting, Detroit, MI, Proceedings, May 1981.

Witte, Larry C., Phillip S. Schmidt, and David R. Brown, *Industrial Energy Management and Utilization*, Hemisphere Publishing Corporation, Washington, DC, 1988.

Chapter 2

The Energy Audit Process: An Overview

2.0 INTRODUCTION

Once a commercial or industrial facility has designated its energy manager and given that person the support and authority necessary to develop an adequate energy management program, the first step the energy manager should take is to conduct an energy audit. Also called an energy survey, energy analysis, or energy evaluation, the energy audit examines the ways energy is currently used in that facility and identifies some alternatives for reducing energy costs. The goals of the audit are:

- to clearly identify the types and costs of energy use, to understand how that energy is being used—and possibly wasted,
- to identify and analyze alternatives such as improved operational techniques and/or new equipment that could substantially reduce energy costs*, and
- to perform an economic analysis on those alternatives and determine which ones are cost-effective for the business or industry involved.

This chapter addresses the three phases of an energy audit: preparing for the audit visit; performing the audit survey and implementing the audit recommendations. In the first phase, data from the energy bills is analyzed in detail to determine where energy is being used and how the use varies with time. Information on the facility is compiled, the necessary

---

*In most cases the energy cost savings will result from reduced consumption, but occasionally a cost savings will be associated with increased energy use. For example, a thermal storage system for heating and/or cooling may save on electric bills, but may actually increase the use of electric energy due to the losses in the storage system. While the primary goal of energy management programs is to reduce energy costs, some proposed alternatives may not always produce greater energy efficiency. However, an overall improvement in a facility's energy efficiency should be the overriding goal for any company's energy management team.

auditing tools are gathered, and an audit team is assembled. After a safety briefing, that team performs walk-through inspections, looking carefully at each of the physical systems within the facility and recording the information for later use. As a result of the audit, the energy manager writes a report outlining the problem areas, analyzing alternatives, and recommending changes in energy use.

The implementation phase begins when the energy manager and the facility management agree on specific energy savings goals and initiate some or all of the actions recommended to achieve those goals. A monitoring system must be set up to allow management to assess the degree to which the chosen goals have been accomplished and to show which measures have been successful and which have failed. The results of the monitoring should feed back to the beginning of the audit cycle and thus potentially initiate more analysis, implementation, and monitoring.

2.1 PREPARING FOR AN ENERGY AUDIT

The energy manager begins the energy audit process by examining the historical and descriptive energy data for the facility. Specific data that should be gathered in the phase preceding the actual audit includes energy bills for the past twelve months, descriptive information about the facility such as a plant layout, and a list of each piece of equipment that affects the energy consumption. Before the audit begins, the auditor must know what special measurement tools will be needed. A briefing on safety procedures is also a wise precaution.

2.1.1 Gathering Preliminary Data on the Facility

Before performing the facility audit, the auditors must gather information on the historical energy use at the facility and on the factors likely to affect the energy use in the facility. Past energy bills, geographic location, weather data, facility layout and construction, operating hours, and equipment lists are all part of the data needed.

2.1.1.1 Analysis of Bills

The audit must begin with a detailed analysis of the energy bills for the previous twelve months. This is important for several reasons: the bills show the proportionate use of each different energy source when compared to the total energy bill; an examination of where energy is used can point out previously unknown energy wastes; and, the total amount spent on energy puts an obvious upper limit on the amount that can be saved.

The data from the energy bills can be conveniently entered onto a form such as shown in Figure 2-1. Note that the most significant billing factors are shown, including peak demand for electricity.

Location_____

From _____ to _____

	Electrical use			Gas use			Fuel oil	
Month	Peak kW	Usage: kWh	Cost	MMCF[a]	Dth[b]	Cost	Gallons	Cost
_____	_____	_____	___	_____	_____	____	_____	____
_____	_____	_____	___	_____	_____	____	_____	____
_____	_____	_____	___	_____	_____	____	_____	____
_____	_____	_____	___	_____	_____	____	_____	____
_____	_____	_____	___	_____	_____	____	_____	____
_____	_____	_____	___	_____	_____	____	_____	____

[a] 1 MMCF = 10^6 ft^3
[b] 1Dth = 10 therms = 10^6 Btu

Figure 2-1
Summary form for energy use.

The energy bill data must be analyzed by energy source and billed location. The data can be tabulated as illustrated in Figure 2-2 of Example 2-1.

Example 2-1. This example demonstrates the importance of analyzing energy bills. As Figure 2-2 shows, most of the gas used at this facility is used by the main heating plant. Therefore most of the energy management effort and money should be concentrated on the main heating plant.

Building	Energy Costs	Percentage of total
Heating plant	$38,742.34	83.2%
East dormitory	4,035.92	8.7%
Married student apartments	1,370.79	2.9%
Undergraduate dormitory	768.42	1.7%
Greenhouse	560.21	1.2%
Child development center	551.05	1.2%
President's home	398.53	0.9%
Art barn	104.77	0.2%

Figure 2-2. Natural gas bills for a small college.

The data in this example should raise some additional questions. The greenhouse appears to use a lot of energy; it uses more than the child development center and almost as much as the undergraduate dormitory. The greenhouse is not particularly large, so this data raised a red flag for the energy auditor. Because one should never make assumptions about what is actually using the energy, the auditor checked the gas consumption meter at the greenhouse to make sure it was not measuring gas consumption from somewhere else as well. Subsequent investigation revealed that the heating and cooling in the greenhouse were controlled by different thermostats. One thermostat turned the cooling on when the temperature got too high—but before the second thermostat turned the heat off! If it had turned out that the gas use from several other buildings had been metered by the same meter as the greenhouse, it would have been necessary to find a way to allocate gas consumption to each building.

In the example in Figure 2-2 the amount of energy used in the president's house could also be questioned; it uses nearly as much gas as the child development center. Perhaps the president's home is used for activities that would warrant this much gas use, but some equipment problem might also be causing this difference, so an energy audit of this facility might be worthwhile.

Another way to present the data is in graph form. A sample of the type of graph that should be made for each type of energy is shown in Figure 2-3. Each area of the country and each different industry type has a unique pattern of energy consumption, and presenting the data as shown in Figure 2-3 helps in defining and analyzing these patterns. In the facility from which this example came, natural gas is used in the winter for space heating, so the January peak is not surprising. For electrical consumption, if a peak demand charge is based on the *annual* peak, the energy auditor must know the time and size of this peak in order to address measures to reduce it.

A complete analysis of the energy bills for a facility requires a detailed knowledge of the rate structures in effect for the facility. To accurately determine the costs of operating individual pieces of equipment, the energy bills must be broken down into their components, such as demand charge and energy charges for the electric bill. This breakdown is also necessary to be able to calculate the savings from Energy Management Opportunities (EMOs) such as high-efficiency lights and high-efficiency motors, and off-peak electrical use by rescheduling some operations. This examination of energy rate structures is explained in detail in Chapter 3.

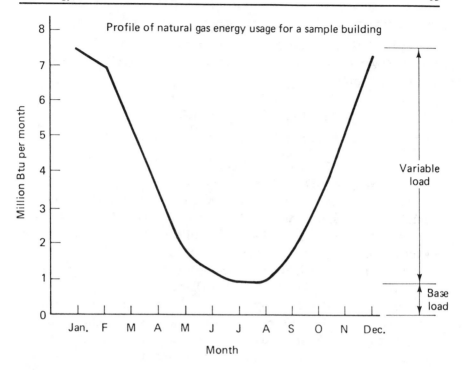

Figure 2-3
Graph of energy consumption over time. (Courtesy of U.S. Department of Energy Conservation, *Instructions for Energy Auditors*, DOE/CS-0041/ 12 and 0041/13, Sept. 1978.)

2.1.1.2 Geographic Location/Degree Days/Weather Data

The geographic location of the facility should be noted, together with the weather data for that location. The local weather station, the local utility or the state energy office can provide the average degree days for heating and cooling for that location for the past twelve months. This degree-day data will be very useful in analyzing the energy needed for heating or cooling the facility.

Heating degree days (HDD) and cooling degree days (CDD) are given separately, and are specific to a particular geographic location. The degree day concept assumes that the average building has a desired indoor temperature of 70°F, and that 5°F of this is supplied by internal heat sources such as lights, appliances, equipment, and people. Thus, the base for computing HDD is 65°F.

Example 2-2 If there were a period of three days when the outside temperature averaged 50°F each day, then the number of HDD for this three day period would be

$$HDD = (65° - 50°) \times 3 \text{ days} = 45 \text{ degree days.}$$

The actual calculation of HDD for an entire year is more involved than this example. HDD for a year are found by taking the outside temperature each hour of the heating season, subtracting that temperature from 65°F, and summing up all of these hourly increments to find the total number of degree hours. This total is then divided by 24 to get the number of HDD. Cooling degree days are similar, using 65°F as the base, and finding the number of hours that the outside temperature is above 65°F, and dividing this by 24 to get the total CDD [1].

Bin weather data is also useful if a thermal envelope simulation of the facility is going to be performed as part of the audit. Weather data for a specific geographic location has been statistically analyzed and the results grouped in cells or bins. These bins contain the number of hours that the outside temperature was within a certain range (e.g. 500 hours at 90-95°). This data is called bin weather data and is available for a large number of locations throughout the United States [2].

2.1.1.3 Facility Layout

Next the facility layout or plan should be obtained, and reviewed to determine the facility size, floor plan, and construction features such as wall and roof material and insulation levels, as well as door and window sizes and construction. A set of building plans could supply this information in sufficient detail. It is important to make sure the plans reflect the "as-built" features of the facility, since the original building plans seldom are completed without alterations.

2.1.1.4 Operating Hours

Operating hours for the facility should also be obtained. How many shifts does the facility run? Is there only a single shift? Are there two? Three? Knowing the operating hours in advance allows some determination as to whether any loads could be shifted to off-peak times. Adding a second shift can often reduce energy bills because the energy costs during second and third shifts are usually substantially cheaper. See Chapter Three for an explanation of on-peak and off-peak electric rates.

2.1.1.5 Equipment List

Finally, the auditor should get an equipment list for the facility and review it before conducting the audit. All large pieces of energy-consuming equipment such as heaters, air conditioners, water heaters, and specific process-related equipment should be identified. This list, together with data on operational uses of the equipment allows the auditor to gain a good understanding of the major energy-consuming tasks or equipment at the facility.

The equipment found at an audit location will depend greatly on the type of facility involved. Residential audits for single-family dwellings generally involve small lighting, heating, air conditioning and refrigeration systems. Commercial operations such as grocery stores, office buildings and shopping centers usually have equipment similar to residences, but the equipment is much larger in size and in energy use. Large residential structures such as apartment buildings have heating, air conditioning and lighting systems that are closer in size to commercial systems. Some commercial audits will require an examination of specialized business equipment that is substantially different from the equipment found in residences.

Industrial auditors encounter the most complex equipment. Commercial-scale lighting, heating, air conditioning and refrigeration, as well as office business equipment, is generally used at most industrial facilities. The major difference is in the highly specialized equipment used for the industrial production processes. For example, this can include equipment that is used for chemical mixing and blending, metal plating and treatment, welding, plastic injection molding, paper making and printing, metal refining, electronic assembly, or making glass.

2.1.2 Tools for the Audit

To obtain the best information for a successful energy cost control program, the auditor must make some measurements during the audit visit. The amount of equipment needed depends on the type of energy-consuming equipment used at the facility, and on the range of potential EMOs that might be considered. For example, if waste heat recovery is being considered, then the auditor must take substantial temperature measurement data from potential heat sources. Tools commonly needed for energy audits include the following:

- *Tape measures*—The most basic measuring device is the tape measure. A 25-foot tape measure and a 100-foot tape measure are used to check the dimensions of the walls, ceilings, doors, and windows, and the distances between pieces of equipment for purposes such as determining the length of a pipe for transferring waste heat from one piece of equipment to another.

- *Lightmeter*—A portable lightmeter that can fit into a pocket is extremely useful. This instrument is used to measure illumination levels in facilities. A lightmeter that reads in footcandles allows direct analysis of lighting systems and comparison with recommended light levels specified by the Illuminating Engineering Society [3]. Many areas in buildings and plants are still significantly over-lighted, and measuring this excess illumination allows the auditor to recommend a reduction in lighting levels through lamp removal programs or by replacing inefficient lamps with high efficiency lamps that may supply slightly less illumination than the old inefficient lamps.

- *Thermometers*—Several thermometers are generally needed to measure temperatures in offices and other worker areas, and to measure the temperature of operating equipment. Knowing process temperatures allows the auditor to determine process equipment efficiencies, and also to identify waste heat sources for potential heat recovery programs. Inexpensive electronic thermometers with interchangeable probes are available to measure temperatures in both these areas. Some common types include an immersion probe, a surface temperature probe, and a radiation-shielded probe for measuring true air temperature. Other types of infra-red thermometers and thermographic equipment are also available. An infra-red "gun" can measure temperatures of steam lines that are inaccessible without a ladder.

- *Voltmeter*—A voltmeter is useful for determining operating voltages on electrical equipment, and especially useful when the nameplate has worn off of a piece of equipment or is otherwise unreadable or missing. The most versatile instrument is a combined volt-ohm-ammeter with a clamp-on feature for measuring currents in conductors that are easily accessible. This type of multi-meter is convenient and relatively inexpensive.

- *Wattmeter/Power Factor Meter*—A portable hand-held wattmeter and power factor meter is very handy for determining the power consumption and power factor of individual motors and other inductive devices. This meter typically has a clamp-on feature which allows an easy connection to the current-carrying conductor, and has probes for voltage connections.

- *Combustion Analyzer*—Combustion analyzers are portable devices which estimate the combustion efficiency of furnaces, boilers, or other fossil fuel burning machines. Two types are available: digital analyzers and manual combustion analysis kits. Digital combustion analysis equipment performs the measurements and reads out combustion efficiency in percent. These instruments are fairly complex and expensive.

The manual combustion analysis kits typically require multiple measurements including the temperature, oxygen content, and carbon dioxide content of the exhaust stack. The efficiency of the combustion process can be calculated after determining these parameters. The manual process is lengthy and frequently subject to human error.

- *Airflow Measurement Devices*—Measuring air flow from heating, air conditioning or ventilating ducts, or from other sources of air flow is one of the energy auditor's tasks. Airflow measurement devices can be used to identify problems with air flows, such as whether the combustion air flow into a gas heater is correct. Typical airflow measuring devices include a velometer, an anemometer, or an airflow hood.

- *Blower Door Attachment*—Building or structure tightness can be measured with a blower door attachment. This device is frequently used in residences and in office buildings to determine the air leakage rate or the number of air changes per hour in the facility. This often helps determine whether the facility has substantial structural or duct leaks that need to be found and sealed.

- *Smoke Generator*—A simple smoke generator can be used in residences, offices and other buildings to find air infiltration and leakage around doors, windows, ducts and other structural features. Care must be taken in using this device, since the chemical "smoke" produced may be hazardous, and breathing protection masks may be needed.

- *Safety Equipment*—The use of safety equipment is a vital precaution for any energy auditor. A good pair of safety glasses is an absolute necessity for almost any audit visit. Hearing protectors may also be required on audit visits to noisy plants or in areas where high horsepower motors are used to drive fans and pumps. Electrically insulated gloves should be used if electrical measurements will be taken, and asbestos gloves should be used for working around boilers and heaters. Breathing protection masks may also be needed when hazardous fumes are present from processes or materials used. Steel-toe and steel-shank safety shoes may be needed on audits of plants where heavy, hot, sharp or hazardous materials are being used.

2.1.3 Safety Considerations

Safety is a critical part of any energy audit. The auditor and the audit team should have a basic knowledge of safety equipment and procedures. Before starting the facility tour, the auditor or audit team should be thoroughly briefed on any specialized safety equipment and procedures for the facility. They should never place themselves in a position where they could injure themselves or other people at the facility.

Adequate safety equipment should be worn at all appropriate times. Auditors should be extremely careful making any measurements on electrical systems, or on high temperature devices such as boilers, heaters, cookers, etc. Electrical gloves or asbestos gloves should be worn as appropriate.

The auditor should be cautious when examining any operating piece of equipment, especially those with open drive shafts, belts, gears, or any form of rotating machinery. The equipment operator or supervisor should be notified that the auditor is going to look at that piece of equipment and might need to get information from some part of the device. If necessary, the auditor may need to return when the machine or device is idle in order to get the data safely. The auditor should never approach a piece of equipment and inspect it without notifying the operator or supervisor first.

Safety Checklist

Electrical:
- Avoid working on live circuits, if possible.
- Securely lock circuits and switches in the off position before working on a piece of equipment.
- Always keep one hand in your pocket while making measurements on live circuits to help prevent accidental electrical shocks.

Respiratory:
- When necessary, wear a full face respirator mask with adequate filtration particle size.

- Use activated carbon cartridges in the mask when working around low concentrations of noxious gases. Change the cartridges on a regular basis.
- Use a self-contained breathing apparatus for work in toxic environments.

Hearing: • Use foam insert plugs while working around loud machinery to reduce sound levels by nearly 30 decibels.

2.2 THE FACILITY INSPECTION

Once all of the basic data has been collected and analyzed, the audit team should tour the entire facility to examine the operational patterns and equipment usage, and should collect detailed data on the facility itself as well as on all energy using equipment. This facility inspection should systematically examine the nine major systems within a facility, using portable instrumentation and common sense guided by an anticipation of what can go wrong. These systems are: the building envelope; the boiler and steam distribution system; the heating, ventilating, and air conditioning system; the electrical supply system; the lighting system, including all lights, windows, and adjacent surfaces; the hot water distribution system; the compressed air distribution system; the motors; and the manufacturing system. Together, these systems account for all the energy used in any facility; examining all of them is a necessary step toward understanding and managing energy utilization within the facility. We briefly describe these systems later in this chapter; we also cover most of them in detail in separate chapters.

The facility inspection can often provide valuable information on ways to reduce energy use at no cost or at a low cost. Actually, several inspections should be made at different times and on different days to discover if lights or other equipment are left on unnecessarily, or to target process waste streams that should be eliminated or minimized.* These

---

*Two preliminary energy inspections are sometimes performed to make sure that the full-blown audit will be worthwhile. The first is done under usual working conditions and is aimed at uncovering practices that are consistently expensive or wasteful. The second (the 2 a.m. survey) is done at midnight or later when energy consumption should be at a minimum. The objective of this second inspection is to find lights that are left on for no reason, motors that are running but not being used, rooms that are warm but not warming people, air being cooled unnecessarily, and air and steam leaks that might not be detected under the noise conditions of daily operations.

inspections can also help identify maintenance tasks that could reduce energy use. Broken windows should be fixed, holes and cracks should be filled, lights should be cleaned, and HVAC filters should be cleaned or replaced.

The facility inspection is an important part of the overall audit process. Data gathered on this tour, together with an extensive analysis of this data will result in an audit report that includes a complete description of the time-varying energy consumption patterns of the facility, a list of each piece of equipment that affects the energy consumption together with an assessment of its condition, a chronology of normal operating and maintenance practices, and a list of recommended energy management ideas for possible implementation.

2.2.1 Introductory Meeting

The audit person—or team—should start the audit by meeting with the facility manager and the maintenance supervisor. They should briefly explain the purpose of the audit and indicate the kind of information they need to obtain during the facility tour. If possible, a facility employee who is in a position to authorize expenditures or make operating policy decisions should be at this initial meeting.

2.2.2 Audit Interviews

Getting the correct information on facility equipment and operation is important if the audit is going to be most successful in identifying ways to save money on energy bills. The company philosophy towards investments, the impetus behind requesting the audit, and the expectations from the audit can be determined by interviewing the general manager, chief operating officer, or other executives. The facility manager or plant manager should have access to much of the operational data on the facility, and a file of data on facility equipment. The finance officer can provide any necessary financial records, such as utility bills for electric, gas, oil, other fuels, water and wastewater, expenditures for maintenance and repair, etc.

The auditor must also interview the floor supervisors and equipment operators to understand the building and process problems. Line or area supervisors usually have the best information on the times their equipment is used. The maintenance supervisor is often the primary person to talk to about types of lighting and lamps, sizes of motors, sizes of air conditioners and space heaters, and electrical loads of specialized process equipment. Finally, the maintenance staff must be interviewed to find the equipment and performance problems.

The auditor should write down these people's names, job functions and telephone numbers, since additional information is often needed after the initial audit visit.

2.2.3 Initial Walk-through Tour

An initial facility/plant tour should be conducted by the facility/ plant manager, and should allow the auditor or audit team to see the major operational and equipment features of the facility. The main purpose of the initial tour is to obtain general information, and to obtain a general understanding of the facility's operation. More specific information should be obtained from the maintenance and operational people during a second, and more detailed data collection tour.

2.2.4 Gathering Detailed Data

Following the initial facility or plant tour, the auditor or audit team should acquire the detailed data on facility equipment and operation that will lead to identifying the significant Energy Management Opportunities (EMOs) that may be appropriate for this facility. This data is gathered by examining the nine major energy-using systems in the facility.

As each of these systems are examined, the following questions should be asked:

1. What function(s) does this system serve?
2. How does this system serve its function(s)?
3. What is the energy consumption of this system?
4. What are the indications that this system is probably working?
5. If this system is not working, how can it be restored to good working condition?
6. How can the energy cost of this system be reduced?
7. How should this system be maintained?
8. Who has direct responsibility for maintaining and improving the operation and energy efficiency of this system?

As each system is inspected, this data should be recorded on individualized data sheets that have been prepared in advance. Manual entry data forms for handling this energy data are available from several sources, including the energy management handbook from the National Electrical Manufacturers Association [4]. Some energy analysis procedures in current use are computer-based, and data is entered directly into the computer.

2.2.5.1 The building envelope.

The building envelope includes all building components that are directly exposed to the outside environment. Its main function is to protect employees and materials from outside weather conditions and temperature variations; in addition, it provides privacy for the business and can serve other psychological functions. The components of the building envelope are outside doors, windows, and walls; the roof; and, in some cases, the floor. The heating and cooling loads for the building envelope are discussed in Chapter Six on the HVAC system.

As you examine the building envelope, you should record information on the insulation levels in the various parts of the facility, the condition of the roof and walls, the location and size of any leaks or holes, and the location and size of any door or windows that open from conditioned to unconditioned space. Insulation is discussed in Chapter Eleven. Figure 2-4 shows a sample data form for the building envelope.

System: Envelope

Component	Location	Maintenance condition	Est. air gap (total)
Door	North side	Poor	0.2 ft^2
	South	OK	0.05 ft^2
	Gymnasium	Good	None
Windows	North	Some broken	2.2 ft^2
	East	OK	None
Roof	Main building	No insulation	

Figure 2-4. Completed inspection form for building envelope.

2.2.5.2 The boiler and steam distribution system.

A *boiler* burns fuel to produce heat that converts water into steam, and the *steam distribution system* takes the steam from the boiler to the point of use. Boilers consume much of the fuel used in many production facilities. The boiler is thus the first place to look when attempting to reduce natural gas or oil consumption. The steam distribution system is also a very important place to look for energy savings, since every pound of steam lost is another pound of steam that the boiler must produce. A detailed description of boilers and the steam distribution system, including operating and maintenance recommendations, is covered in Chapters 7 and 8.

2.2.5.3 The heating, ventilating, and air conditioning system.

All *heating, air conditioning and ventilation* (HVAC) equipment should be inventoried. Prepared data sheets can be used to record type, size, model numbers, age, electrical specifications or fuel use specifications, and estimated hours of operation. The equipment should be inspected to determine the condition of the evaporator and condenser coils, the air filters, and the insulation on the refrigerant lines. Air velocity measurements may also be made and recorded to assess operating efficiencies or to discover conditioned air leaks. This data will allow later analysis to examine alternative equipment and operations that would reduce energy costs for heating, ventilating, and air conditioning. The HVAC system is discussed in detail in Chapter Six.

2.2.5.4 The electrical supply system.

This system consists of transformers, wiring, switches, and fuses—all the components needed to enable electricity to move from the utility-owned wires at the facility boundary to its point of use within the company. By our definition, this supply system does not include lights, motors, or electrical controls. Most energy problems associated with the distribution of electricity are also safety problems, and solving the energy problems helps to solve the safety-related problems.

Electricity from a utility enters a facility at a service transformer. The area around the transformer should be dry, the transformer fins should be free from leaves and debris so that they can perform their cooling function, and the transformer should not be leaking oil. If a transformer fails to meet any one of these conditions there is a serious problem which should justify a call to the local electrical utility, or, if the transformer is company-owned, to the person or department in charge of maintaining the electrical system.

A more detailed audit of transformers should also include drawing a small (1-pint) sample of transformer dielectric fluid and examining it both visually and for dielectric strength. If the fluid is brown, the dielectric has been contaminated by acid; if it is cloudy, it is contaminated with water. The dielectric strength should be 20,000 v/cm or equivalent. Both the color and the dielectric strength should be recorded for comparison against future readings.

In examining transformers, also check to see whether any company-owned transformer is serving an area that is not currently used. A transformer that is connected to the utility lines but not supplying power to the facility is wasting two to four percent of its rated capacity in core losses. These losses can be avoided by disconnecting the transformer or by installing switching between the transformer and the electrical lines from the utility.*

A person performing an energy audit should examine the electrical supply panels and switch boxes. Danger signs and symptoms of wasted energy include signs of arcing such as burned spots on contacts, burned insulation, arcing sounds, and frayed wire. Other concerns are warm spots around fuse boxes and switches and the smell of warm insulation. Any of these symptoms can indicate a fire hazard and should be checked in more detail immediately. Safety considerations are paramount when inspecting live electrical systems.

2.2.5.5 Lights, windows, and reflective surfaces.

The functions of this system are to provide sufficient light for necessary work, to enable people to see where they are going, to assist in building and area security at night, to illuminate advertising, and to provide decoration. Making a detailed inventory of all lighting systems is important. Data should be recorded on numbers of each type of light fixture and lamp, the wattages of the lamps, and the hours of operation of each group of lights. A lighting inventory data sheet should be used to record this data. See Figure 2-5 for a sample lighting data sheet. Lighting is discussed in detail in Chapter 5. Windows and reflective surfaces are discussed in Chapter Six on the HVAC System and in Chapter Thirteen on Renewable Energy Sources.

2.2.5.6 The hot water distribution system

The hot water system distributes hot water for washing, for use in industrial cleaning, and for use in kitchens. Its main components are hot water heaters, storage tanks, piping, and faucets. Electric boilers and radiators are also found in some facilities. Boilers and the hot water distribution system are discussed in Chapter Eight.

All water heaters should be examined, and data recorded on their type, size, age, model number, electrical characteristics or fuel use. What the hot water is used for, how much is used, and what time it is used should all be noted. The temperature of the hot water should be measured and recorded.

2.2.5.7 Air compressors and the air distribution system.

Air compressors and the *air distribution system* provide motive power

---

*Personal communication from Mr. Bryan Drennan, Customer Services, Utah Power and Light Company, Salt Lake City, Utah.

Figure 2-5 Data collection form for lighting system.

Area	Type of lighting (e.g., HPS)	Watts per fixture	Number of fixtures	Total kW	Operating hours	Operating days	kWh/month
Interior							
	___	___	___	___	___	___	___
	___	___	___	___	___	___	___
	___	___	___	___	___	___	___
	___	___	___	___	___	___	___
	___	___	___	___	___	___	___
Exterior							
	___	___	___	___	___	___	___
	___	___	___	___	___	___	___
	___	___	___	___	___	___	___
	___	___	___	___	___	___	___
	___	___	___	___	___	___	___

for tools and some machinery, and often provide air to operate the heating, ventilating, and air conditioning system. If you use compressed air to run equipment, look for leaks and for places where compressed air is purposely allowed to vent into the air. Such leaks can be expensive.

2.2.5.8 Motors

Electric motors account for about three-fourths of all the electric energy used by industry and about two-fifths of all electric energy use by commercial facilities. Replacement of existing motors with more efficient models is usually cost effective for applications where the motor is heavily used. Motors and drives are discussed in Chapter Twelve on Process Energy Improvements.

All electric motors over 1 horsepower should be inventoried. Prepared data sheets can be used to record motor size, use, age, model number, estimated hours of operation, other electrical characteristics, and possibly the operating power factor. Measurement of voltages, currents, and power factors may be appropriate for some motors. Notes should be taken on the use of motors, particularly recording those that are infrequently used and might be candidates for peak load control or shifting use to off-peak times. All motors over 1 hp and with times of use of 2000 hours per year or greater, are likely candidates for replacement by high-efficiency motors—at least when they fail and must be replaced.

2.2.5.9 Manufacturing processes

Each manufacturing process has opportunities for energy management, and each offers ways for the unwary to create operating problems in the name of energy management. The best way to avoid such operating problems is to include operating personnel in the energy audit process (see Section 2.2.2) and to avoid rigid insistence on energy conservation as the most important goal.

The generic industrial processes that use the most energy are combustion for process steam and self-generated electricity, electrolytic processes, chemical reactors, combustion for direct heat in furnaces and kilns, and direct motor drive. Process energy efficiency improvement is discussed in Chapter Twelve.

Any other equipment that consumes a substantial amount of energy should be inventoried and examined. Commercial facilities may have extensive computer and copying equipment, refrigeration and cooling equipment, cooking devices, printing equipment, water heaters, etc. Industrial facilities will have many highly specialized process and production operations and machines. Data on types, sizes, capacities, fuel use, electrical characteristics, age, and operating hours should be recorded for all of this equipment.

2.2.6 Preliminary Identification of
Energy Management Opportunities

As the audit is being conducted, the auditor should take notes on potential EMOs that are evident. As a general rule, the greatest effort should be devoted to analyzing and implementing the EMOs which show the greatest savings, and the least effort to those with the smallest savings potential. Therefore, the largest energy and cost activities should be examined carefully to see where savings could be achieved.

Identifying EMOs requires a good knowledge of the available energy efficiency technologies that can accomplish the same job with less energy and less cost. For example, over-lighting indicate a potential lamp removal or lamp change EMO, and inefficient lamps indicate a potential lamp technology change. Motors with high use times are potential EMOs for high efficiency replacements. Notes on waste heat sources should indicate what other heating sources they might replace, and how far away they are from the end use point. Identifying any potential EMOs during the walk-through will make it easier later on to analyze the data and to determine the final EMO recommendations.

2.2.7 The Energy Audit Report

The next step in the energy audit process is to prepare a report which details the final results of the energy analyses and provides energy cost saving recommendations. The length and detail of this report will vary depending on the type of facility audited. A residential audit may result in a computer printout from the utility. An industrial audit is more likely to have a detailed explanation of the EMOs and benefit-cost analyses. The following discussion covers the more detailed audit reports.

The report should begin with an executive summary that provides the owners/managers of the audited facility with a brief synopsis of the total savings available and the highlights of each EMO. The report should then describe the facility that has been audited, and provide information on the operation of the facility that relates to its energy costs. The energy bills should be presented, with tables and plots showing the costs and consumption. Following the energy cost analysis, the recommended EMOs should be presented, along with the calculations for the costs and benefits, and the cost-effectiveness criterion.

Regardless of the audience for the audit report, it should be written in a clear, concise and easy-to understand format and style. An executive summary should be tailored to non-technical personnel, and technical jargon should be minimized. The reader who understands the report is more likely to implement the recommended EMOs. An outline for a complete energy audit report is shown in Figure 2-6 below.

Energy Audit Report Format

Executive Summary
>   A brief summary of the recommendations and cost savings

Table of Contents

Introduction
>   Purpose of the energy audit
>   Need for a continuing energy cost control program

Facility Description
>   Product or service, and materials flow
>   Size, construction, facility layout, and hours of operation
>   Equipment list, with specifications

Energy Bill Analysis
>   Utility rate structures
>   Tables and graphs of energy consumptions and costs
>   Discussion of energy costs and energy bills

Energy Management Opportunities
>   Listing of potential EMOs
>   Cost and savings analysis
>   Economic evaluation

Energy Action Plan
>   Recommended EMOs and an implementation schedule
>   Designation of an energy monitor and ongoing program

Conclusion
>   Additional comments not otherwise covered

Figure 2-6 Outline of Energy Audit Report

2.2.8 The Energy Action Plan

An important part of the energy audit report is the recommended action plan for the facility. Some companies will have an energy audit conducted by their electric utility or by an independent consulting firm, and will then make changes to reduce their energy bills. They may not spend any further effort in the energy cost control area until several years in the future when another energy audit is conducted. In contrast to this is the company which establishes a permanent energy cost control program, and assigns one person—or a team of people—to continually monitor and improve the energy efficiency and energy productivity of the company. Similar to a Total Quality Management program where a company seeks to continually improve the quality of its products, services and operation, an energy cost control program seeks continual improvement in the amount of product produced for a given expenditure for energy.

The energy action plan lists the EMOs which should be implemented first, and suggests an overall implementation schedule. Often, one or more of the recommended EMOs provides an immediate or very short payback period, so savings from that EMO—or those EMOs—can be used to generate capital to pay for implementing the other EMOs. In addition, the action plan also suggests that a company designate one person as the energy monitor or energy manager for the facility if it has not already done so. This person can look at the monthly energy bills and see whether any unusual costs are occurring, and can verify that the energy savings from EMOs is really being seen. Finally, this person can continue to look for other ways the company can save on energy costs, and can be seen as evidence that the company is interested in a future program of energy cost control.

2.3 IMPLEMENTING THE AUDIT RECOMMENDATIONS

After the energy consumption data has been collected and analyzed, the energy-related systems have been carefully examined, the ideas for improvement have been collected, and management commitment has been obtained, the next steps are to obtain company support for the program, to choose goals, and to initiate action.

2.3.1 The Energy Action Team

Now that the preliminary audits have uncovered some energy management measures that can save significant amounts of money or can substantially improve production, funding for the changes and employee support are two additional critical ingredients for success. These can best be obtained with the help of a committee, preferably called something like the energy action team. The functions of this committee are given in Table 2-1.

Table 2-1. Functions of the Energy Action Committee

1. Create support within the company for energy management.
2. Generate new ideas.
3. Evaluate suggestions
4. Set goals.
5. Implement the most promising ideas.

No program will work within a company without employee support, particularly such a program as energy management which seems to promise employee discomfort at no visible increase in production. Therefore, one function of the energy action committee is to give representation to every important political group within the company. For this purpose, the committee must include people from unions, management, and every major group that could hinder the implementation of an energy management plan. The committee must also include at least one person with financial knowledge of the company, a person in charge of the daily operation of the facility, and line personnel in each area of the facility that will be affected by energy management. In a hospital, for example, the committee would have to include a registered nurse, a physician, someone from hospital administration, and at least one person directly involved in the operation of the building. In a university, the committee should include a budget officer, at least one department chairperson, a faculty member, a senior secretary, someone from buildings and grounds, and one or more students.

In addition to providing representation, a broadly based committee provides a forum for the evaluation of suggestions. The committee should decide on evaluation criteria as soon as possible after it is organized. These criteria should include first cost, estimated payback period or (for projects with a payback period longer than 2 years) the constant-dollar return on investment (see Chapter 4), the effects on production, the effects on acceptance of the entire program, and any mitigating effect on problems of energy curtailment.

The committee has the additional duty to be a source of ideas. These ideas can be stimulated by the detailed energy audit which clearly shows problems and areas for improvement. The energy manager should be aware, however, that most maintenance personnel become quickly defensive and that their cooperation, and hopefully their support, may be important. The specific tasks of this committee are to set goals, implement changes, and monitor results.

2.3.1.1 Goals

At least three different kinds of goals can be identified. First, *performance goals*, such as a reduction of 10 percent in Btu/unit product, can be chosen. Such goals should be modest at first so that they can be accomplished—in general, 10-30 percent reduction in energy usage for companies with little energy management experience and 8-15 percent for companies with more. These goals can be accompanied by goals for the reduction of projected energy costs by a similar amount. The more experienced the company is in energy management, the fewer easy saving possibilities exist; thus lower goals are more realistic in that case.

A second type of goal that can be established is an accounting goal. The ultimate objective in an energy accounting system is to be able to allocate the cost of energy to a product in the same way that other direct costs are allocated, and this objective guides the establishment of preliminary energy accounting goals. A preliminary goal would therefore be to determine the amount of electricity and the contribution to the electrical peak from each of the major departments within the company. This will probably require some additional metering, but the authors have found that such metering pays for itself in energy saving (induced by a better knowledge of the energy consumption patterns) in six months or less.

The third type of goal is that of employee participation. Even if an energy management program has the backing of the management, it will still fail without the support and participation of the employees. Ways to measure this include the number of suggestions per month; the dollar value of improvements adopted as a result of employee suggestions, per month; and the number of lights left on or machines left running unnecessarily, on a spot inspection. Work sampling has been used to estimate the percentage of time that people are working at various tasks—it can be used equally well on machines.

2.3.1.2 Implementing Recommendations

In addition to providing and evaluating ideas, setting goals, and establishing employee support, the energy action committee has the duty of implementing the most promising ideas that have emerged from the energy evaluation process. Members of the committee have the responsibility to see that people are assigned to each project, that timetables are established, that money is assigned, and that progress reporting procedures are set up and followed. It is then the committee responsibility to follow up on the progress of each project; this monitoring process is described in detail in the next section.

2.3.1.3 Monitoring

Energy management is not complete without monitoring and its associated feedback, and neither is the energy audit process. In an energy audit, monitoring discloses what measures contributed toward the company goals, what measures were counterproductive, and whether the goals themselves were too low or too high.

Monitoring consists of collecting and interpreting data. The data to collect are defined by the objectives chosen by the energy action committee. At the very least, the electrical and gas bills and those of other relevant energy sources must be examined and their data graphed each month. Monthly graphs should include: the total energy used of each type (kWh of electricity, therms [10^5 Btu] of gas, etc.); the peaks, if they determine part of the cost of electricity or gas; and any other factors that contribute to the bills. At the same time, other output-related measures, such as Btu/ ton, should also be calculated, recorded, and graphed.

The monitoring data should provide direct feedback to those most able to implement the changes. Often this requires that recording instruments be installed in a number of departments in addition to the meters required by the utility company. The additional expense is justified by increased employee awareness of the timing and amounts of energy consumed, and usually this awareness leads to a reduction in energy costs. Metering at each department also enables management to determine where the energy is consumed and, possibly, what is causing the energy consumption. Such metering also helps each department manager to understand and control the consumption of his or her own department.

Monitoring should result in more action. Find what is good, and copy it elsewhere. Find what is bad, and avoid it elsewhere. If the goals are too high, lower them. If the goals are too low, raise them. Wherever the difference between the planned objectives and the achievements is great, initiate an analysis to determine the reasons and then develop new objectives, initiate new action, and monitor the new results. In this way, the analysis, action, and monitoring process repeats itself.

2.4 SUMMARY

In this chapter, we have explained the mechanics of performing and implementing the energy audit as well as the reasoning behind the energy audit process. The energy audit process should be a dynamic feedback loop. The process starts with the analysis of past data. Then each energy system is examined for savings potential. Recommendations for energy cost saving actions are made and an energy action committee is formed. Next, attractive energy cost savings projects are implemented. The last step is the monitoring program which necessarily leads back to the first step of analysis, thus renewing the cycle.

REFERENCES

1. *1993 ASHRAE Handbook on Fundamentals*, American Society of Heating, Refrigerating and Air Conditioning Engineers, Inc., Atlanta, GA, May, 1993.
2. Simplified Energy Analysis Using the Modified Bin Method, American Society of Heating, Refrigerating and Air Conditioning Engineers, Inc., Atlanta, GA, 1984.
3. *IES Lighting Handbook*, Illuminating Engineering Society, New York, NY, 1993.
4. *Total Energy Management*, A handbook prepared by the National Electrical Contractors Association and the National Electrical Manufacturers Association, Washington, DC, Third Edition, 1986.

BIBLIOGRAPHY

Baumeister, Theodore, Eugene A. Avallone, and Theodore Baumeister, III, *Marks' Standard Handbook for Mechanical Engineers*, 8th edition, McGraw-Hill, New York, 1978.

Energy Conservation Guide for Industry and Commerce, National Bureau of Standards Handbook 115 and Supplement, 1978. Available through U.S. Government Printing Office, Washington, DC.

Facility Design and Planning Engineering Weather Data, Departments of the Air Force, the Army, and the Navy, 1978.

Instructions For Energy Auditors, Volumes I and II, U.S. Department of Energy, DOE/CS-0041/12&13, September, 1978. Available through National Technical Information Service, Springfield, VA.

Kennedy, W.J., Jr., "How to Conduct an Energy Audit," in Proceedings of the 1978 Fall AIIE Conference, Institute of Industrial Engineers, Atlanta, GA, 1978.

Kennedy, W.J., Jr., "An Introduction to HVAC Systems," in Proceedings of the 1980 Fall AIIE Conference, Institute of Industrial Engineers, Atlanta, GA, 1980.

McQuiston, Faye C., and Jerald D. Parker, *Heating, Ventilating, and Air-Conditioning*, Wiley, New York, 1977.

Peterson, Carl, *Energy Survey and Conservation Manual*, Granite School District, Salt Lake City, 1980.

Turner, Wayne C., "The Impact of Energy on Material Handling," Proceedings of the 1980 Conference on Material Handling, Materials Handling Institute, Pittsburgh, PA, 1980.

Turner, Wayne C., Editor, *Energy Management Handbook*, Second Edition, Fairmont Press, Atlanta, GA, 1993.

Threshold Limit Values for Chemical Substances and Physical Agents and Biological Exposure Indices, 1990-91 American Conference of Governmental Industrial Hygienists.

Thumann, Albert, *Handbook of Energy Audits*, Third Edition, Fairmont Press, Atlanta, GA, 1987.

Variable Air Volume Systems Manual, 1980 edition, Manual No. 74-5461, Honeywell, Inc., Minneapolis, MN, 1980.

Ventilation for Acceptable Indoor Air Quality, ASHRAE 62-1989, American Society of Heating, Refrigerating and Air-Conditioning Engineers, Inc., 1989.

Witte, Larry C., Philip S. Schmidt, and David R. Brown, *Industrial Energy Management and Utilization*, Hemisphere Publishing Corporation, Washington, DC, 1988.

1992 Handbook—HVAC Systems and Equipment, American Society of Heating, Refrigerating, and Air-Conditioning Engineers, Inc., Atlanta, GA, 1992.

1991 Handbook—HVAC Applications, American Society of Heating, Refrigerating, and Air-Conditioning Engineers, Inc., Atlanta, GA, 1991.

Chapter 3

Understanding Energy Bills

3.0 INTRODUCTION

Although temporary increases in energy supplies may cause short-term rate decreases or rate stability, the long-term prospect is for energy costs to continue to increase. This is particularly true in the case of electric power where the costs are associated as much with the cost of building the new generation facilities as with the cost of producing the electricity. The impact of an increase in energy costs can be easily seen by examining the rate schedules for the various fuel sources, yet few managers take the time to peruse and understand their utility's billing procedures.

Why? The reasons are many, but the main ones seem to be the following:

- Rate schedules are sometimes very complicated. They are difficult to understand, and the explanations developed by state utility regulatory boards and by the utilities themselves often confuse the customer rather than clarify the bill.

- Energy is too often treated as an overhead item. Even though energy is frequently a substantial component of a product's cost, the cost of the energy is almost always included as an overhead item rather than a direct cost. This makes the energy cost more difficult to account for and control. Consequently, management does not give energy costs the attention they deserve.

This must change and, in fact, is changing. More and more managers are trying to understand their rate schedules, and sometimes they even participate in utility rate increase hearings.

Managers should know what electric rate schedule they are under and how much they are charged for the various components of their electric bill: demand, consumption, power factor, sales tax, etc. They should also know the details of their other energy rate structures. This chapter covers rate schedules for the major energy sources utilized in this country. While the majority of the discussion focuses on electricity, attention is also given to gas, fuel oil, coal, and steam/chilled water.

3.1 ELECTRIC RATE STRUCTURES

3.1.1 Utility Costs

Perhaps the best way to understand electric utility billing is to examine the costs faced by the utility. The major utility cost categories are the following:

- *Physical Plant.* This is often the single biggest cost category. Because electric power plants have become larger and more technologically sophisticated with more pollution control requirements, the cost of building and operating an electric power generation facility continues to increase. Furthermore, the utility is required to have sufficient capacity to supply the peak needs of its customers while maintaining some equipment in reserve in case of equipment failures. Otherwise brownouts or even blackouts may occur. This added capacity can be provided with expensive new generating facilities. Alternatively, instead of building new facilities, many utilities are urging their customers to reduce their peak demand so that the existing facilities will provide sufficient capacity.

- *Transmission lines.* Another major cost category is the cost of transmission lines to carry the electricity from where it is produced to the general area where it is needed. Electricity is transmitted at relatively high voltages—often 500 to 1,000 kilovolts—to minimize resistance (I^2R) losses. This loss can be large or small depending on the transmission distances involved.

- *Substations.* Once the electricity reaches the general area where it is needed, the voltages must be reduced to the lower levels which can be safely distributed to customers. This is done with step-down transformers at substations. A few customers may receive voltage at transmission levels, but the vast majority do not.

- *Distribution systems.* After the voltage is reduced at a substation, the electricity is delivered to the individual customers through a local distribution system typically at a voltage level around 12 kilovolts. Most residential customers are supplied electricity at 120 and 240 volts, single-phase. In addition to these two voltages, commercial customers often take 230-volt, three-phase service. Some larger customers must also have 480-volt, three-phase service in order to power their large motors, ovens, and process equipment. The desired voltages are provided through the use of appropriate step-down transformers at the customer's specific location. Components of the distribution system which contribute to the utility's costs include utility poles, lines, transformers, and capacitors.

- *Meters.* Meters form the interface between the utility company and customer. Although the meter costs are relatively small, they are considered a separate item by the utility and are usually included in the part of the bill called the customer charge. The cost of a meter can range from under $50 for a residential customer to $1500 or more for an industrial customer requiring information on consumption, demand and power factor.

- *Administrative.* Administrative costs include salaries for executives, middle management, technical and office staff, as well as for maintenance staff. Office space and office equipment, taxes, insurance, and maintenance equipment and vehicles are also part of the administrative costs.

- *Energy.* Once the generation, transmission and distribution systems are in place, some form of primary energy must be purchased to fuel the boilers and generate the electricity. In the case of hydroelectric plants, the turbine generators are run by water power and the primary energy costs are small. Fossil fuel electric plants have experienced dramatic fluctuations in fuel costs depending on how national and world events alter the availability of oil, gas, and coal. The cost of fuel for nuclear power plants is reasonable, but the costs of disposing of the radioactive spent fuel rods, while still unknown, are expected to be relatively large.

- *Interest on debt.* This cost category can be quite large. For example, the interest on debt for a large power plant costing $500 million to $1 billion is substantial. Utilities commonly sell bonds to generate capital, and these bonds represent debt that the company must pay interest on.

- *Profit.* Finally the utility must generate enough additional revenue above costs to provide a reasonable profit to stockholders. The profit level for private utility companies is determined by the state utility regulatory commission and is called the *rate of return.* Public-owned utilities such as municipal utilities or rural electric cooperatives usually set their own rates and their profit goes back to their customers in the form of reduced municipal taxes or customer rebates.

Once you understand what costs contribute to an electric bill, the next step is to learn how these costs are allocated to the various customers. The *billing procedure*, also called the *rate schedule*, should be designed to reflect the true costs of generating the electric power. If the customers understand the problems faced by the utilities, they can help the utilities minimize these costs. Recent rate schedules and proposed new ones capture the true costs of generation much better than has been done in the past, but more changes are still needed.

3.1.2 Regulatory Agencies

Private electric and gas utilities are chartered and regulated by individual states, and are also subject to some federal regulation. The state utility regulatory agencies are most often called Public Utility Commissions or Public Service Commissions. Private utilities are called *Investor Owned Utilities* (IOUs), and their retail rates for residential, commercial and industrial customers are subject to review and approval by the state utility regulatory agencies.

Utility rates are set in two steps: first, the revenue requirements to cover costs plus profit is determined; second, rates are designed and set to recover these costs or revenue requirements [1]. The state regulatory agencies set a *rate of return* for utilities. The rate of return is the level of profit a utility is allowed to make on its investment in producing and selling energy. In developing rates, the costs of serving different classes of customers must be determined and allocated to the customer classes. Rates are then structured to recover these costs from the appropriate customer class. Such rate designs are called *cost-based rates.* Often, these costs are *average,* or *embedded costs,* and do not consider the *marginal costs* associated with providing electricity at different times of the day and different seasons of the year. Rate design is subject to many competing viewpoints, and there are many different objectives possible in rate setting.

When an IOU requests a rate increase, the state regulatory agency holds a public hearing to review the proposed rate increase, and to take testimony from the utility staff, consulting engineers, customers and the public at large. The utility presents its case for why it needs a rate increase, and explains what its additional costs are. If these costs are judged "prudent" by the state regulatory agency and approved, the utility is allowed to recover the costs, plus adding some of that cost to its rate base—which is the accumulated capital cost of facilities purchased or installed to serve the customers and on which the utility can earn its rate of return.

Many large utility customers participate actively in the rate hearings for their utility. Some state regulatory agencies are very interested in comments from utility customers regarding quality of service, reliability, lengths of outages, and other utility service factors. State regulatory agencies vary greatly in their attitude toward utility rate increases. Some states favor the utilities and consider their interests to be first priority, while other states consider the interests of the customers and the public as paramount.

Two other major categories of utilities exist: *public* or *municipal utilities* owned and operated by cities and local government entities; and *Rural Electric Cooperatives* (RECs) established under the Rural Electrification Administration and operated by customer Boards of Directors. State regulatory agencies generally do not exercise the same degree of control over public utilities and RECs, since these utilities have citizens and customers controlling the rates and making the operating decisions, whereas the IOUs have stockholders making those decisions. Municipals and RECs also hold public hearings or public meetings whenever rate increases are contemplated. Customers who have an interest in participating in these meetings are usually encouraged to do so.

Interstate transactions involving the wholesale sale or purchase of electricity between utilities in different states are subject to regulation by the Federal Energy Regulatory Commission (FERC) in Washington, DC. FERC also regulates the designation of some cogenerators and renewable electric energy suppliers as Qualifying Facilities (QFs) or Small Power Producers (SPPs). Few retail customers outside of those engaged in self-generation would have any reason to participate in the regulatory process at this level. FERC also licenses non-federal hydroelectric facilities.

3.1.3 Customer Classes and Rate Schedules

An electric utility must serve several classes of customers. These classes vary in complexity of energy use, amount of consumption, and priority of need. The typical customer class categories are residential, commercial and industrial. Some utilities combine commercial and industrial customers into one class while other utilities divide the industrial class into heavy industrial and light industrial customers.

The state regulatory agencies and utilities develop different rate schedules for each customer class. Electric rate structures vary greatly from utility to utility, but they all have a series of common features. The most common components of rate schedules are described below, but not all of these components are included in the rate schedule for every customer class.

- *Administrative/Customer Charge:* This fee covers the utility's fixed cost of serving the customer including such costs as providing a meter, reading the meter, sending a bill, etc. This charge is a flat monthly fee per customer regardless of the number of kWh of electricity consumed.

- *Energy charge:* This charge covers the actual amount of electricity consumed measured in kilowatt-hours. The energy charge is based on an average cost, or base rate, for the fuel (natural gas, fuel oil, coal, etc.) consumed to produce each kWh of electricity. The energy charge also includes a charge for the utility's operating and maintenance expenses.

  Many utilities charge a constant rate for all energy used, and this is called a *flat rate* structure. A *declining block* approach may also be used. A declining block schedule charges one price for the first block of energy (kWh) used and less for the next increment(s) of energy as more energy is used. Another approach is the *increasing block* rate where more is charged per increment as the consumption level increases. Although this approach would tend to discourage electric energy waste, it does not meet the *cost-based rate* criterion and is therefore not widely used.

- *Fuel cost adjustment:* If the utility has to pay more than its expected cost for primary fuel, the increased cost is "passed on" to the customer through use of a prescribed formula for a fuel adjustment cost. In times of rapidly increasing fuel prices, the fuel adjustment cost can be a substantial proportion of the bill. This concept was adopted when fuel costs were escalating faster than utility commissions could grant rate increases. However, utilities can also use the fuel adjustment cost to reduce rates when fuel costs are lower than the cost included in the base rate.

- *Demand charge:* The demand charge is used to allocate the cost of the capital facilities which provide the electric service. The demand charge may be "hidden" in the energy charge or it may be a separate charge; for example it may be expressed as $6.25 per kW per month for all kW above 10 kW. For large customers, the demand charge is generally based on their kilowatt demand load. For smaller users such as residential and small commercial customers this charge is usually averaged into the energy charge. The demand charge is explained in more detail in Section 3.1.6.2.

  Understanding the difference between *electric demand,* or power in kilowatts (kW), and *electric energy,* or consumption in kilowatt-hours (kWh), will help you understand how electric bills are computed. A helpful analogy is to think of an automobile where the speedometer measures the rate of travel in miles per hour, and the odometer measures the total miles traveled. In this instance, speed is analogous to electric power, and miles traveled is analogous to total energy consumed. In analytical terms, *power is the rate of use of electric energy,* and conversely, *energy is the time integral of the power.* Finally, the value of the power or demand a utility uses to compute an electric bill is obtained from a peak power measurement that is averaged over a short period of time. Typical averaging times used by various electric utilities are 15 minutes, 30 minutes and one hour. The averaging time prevents unreasonable charges from occurring because of very short, transient peaks in power consumption. Demand is measured by a demand meter.

- *Demand ratchet:* An industrial or commercial rate structure may also have a demand ratchet component. This component allows the utility to adequately charge a customer for creating a large kilowatt demand in only a few months of the year. Under the demand ratchet, a customer will not necessarily be charged for the actual demand for a given month. Instead the customer will be charged a percentage of the largest kW value during the last 11 months, or the current month's demand, whichever is higher.

- *Power factor:* If a large customer has a poor power factor, the utility may impose another charge, assessed as a function of that power factor. Power factor is discussed in detail in section 3.1.6.4.

  All of these factors are considered when a utility sets its *base rates—* the rates the utility must charge to recover its general cost of doing business. The term "base rates" should not be confused with the term "rate base" which was previously defined. The base rates contain an energy charge that is estimated to cover the average cost of fuel in the future. The fuel adjustment charge keeps the utility from losing money when the price of their purchased primary fuel is higher than was estimated in their base rates.

Figure 3-1 presents a generalized breakdown of these rate components by customer class.

In addition, there are also a number of other features of electric rates incorporated in the *rate structure* which includes the relationship and form of prices within particular customer classes. The rate structure is set to maintain equity between and within customer classes, ensuring that there is no discrimination against or preferential treatment of any particular customer group. Some of the factors considered in the rate structure are: season of use; time of use; quantity of energy used—and whether increased consumption is encouraged, discouraged, or considered neutral; and social aspects such as the desire for a "lifeline" rate for low-income or elderly customers. A number of these factors are illustrated in the description of the specific rate structures shown in examples for particular customer classes.

3.1.4 Residential Rate Schedules

As shown in Figure 3-1, there are many residential users, but each is a relatively small consumer. A typical residential bill includes an administrative/customer charge, an energy charge which is large enough to cover both the actual energy charge and an implicit demand charge, and a fuel adjustment charge. Residential rates do not usually include an explicit demand charge because the individual demand is relatively inconsequential and expensive to meter.

3.1.4.1 Standard residential rate schedule

Figure 3-2 presents a typical monthly rate schedule for a residential customer.

3.1.4.2 Low-use residential rate schedule

A typical low-use residential service rate is shown in Figure 3-3. This schedule, which is an attempt to meet the needs of those on fixed incomes, is used for customers whose monthly consumption never exceeds 500 kWh. In addition, it cannot exceed 400 kWh more than twice a year. This rate is sometimes referred to as a *lifeline rate*.

Typical schedule bills for:

Customer Class	Comments	Consumption (kWh)	Demand (kW)	Power factor (kvar)
1. Residential	Small user but large numbers of them	✓		
2. Commercial	Small to moderate user; relatively large numbers	✓		
3. Small industrial	Small to moderate user; fewer customers	✓	✓	
4. Large industrial	Large user with low priority; typically, only a few customers in this class, but they consume a large percentage of the electricity produced.	✓	✓	✓

Figure 3-1. Generalized breakdown of electric rate schedule components.

Customer charge: $8.00/month

Energy Charge: All kWh @ 6.972 ¢/kWh

Fuel adjustment: (A formula is provided by the utility to
                    calculate the fuel adjustment charge each
                    month. It is rather complex and will not be
                    covered here.)

Figure 3-2 Typical residential rate schedule.

Customer charge: $5.45/month

Energy charge: 5.865¢/kWh

Fuel adjustment: (A formula is provided for calculating this
                    charge.)

Figure 3-3 Low-use residential schedule.
(Courtesy of Oklahoma Gas and Electric Company)

3.1.4.3 Residential rate schedules to control peak uses
    Although individual residential demand is small, collectively residential users place a peak demand burden on the utility system because the majority of them use their electricity at the same times of the day during the same months of the year. Some utilities charge more for energy during peaking months in an attempt to solve this problem. Many utilities have an optional *time-of-day* or *time-of-use* rate which is supposed to help alleviate the daily peaking problem by charging customers more for electric use during these peak periods. A number of utilities also have a load management program to control customers' appliances.
    Figure 3-4 provides examples of a Florida utility's residential demand profile over a given 24-hour period during the weekdays. Figure 3-4 (a) shows the residential winter peak demand profile. This utility experiences one large peak around 9:00 a.m. and another somewhat smaller peak near 7:00 p.m. The first peak occurs when people get up in the

morning and start using electricity. They all turn up their electric heat, cook breakfast, take a shower, and dry their hair at about the same time on weekday mornings. Then in the evening, they all come home from work, start cooking dinner, turn the heat back up (or use it more because nights are colder than days) and turn on the TV set at about the same time. Figure 3-4 (b) shows the residential summer peak demand profile for the same utility.

Figure 3-4 (a)

Figure 3-4 (b)

- **Seasonal use rate schedule.** Figure 3-5 presents a residential rate schedule where the season of use is a factor in the rate structure. This utility has chosen to attack its residential peaking problem by charging more for electricity consumed in the summer months when the highest peaks occur.

Customer charge: $6.50/month

Energy Charge: On-peak season (June through October)
                       All kWh @ 7.728¢/kWh

                   Off-peak season (November through May)
                       First 600 kWh @ 7.728¢/kWh
                       All additional kWh @ 3.898¢/kWh

Fuel adjustment: (Calculated by a formula provided by the utility.)

Figure 3-5 Seasonal use residential rate schedule.
(Courtesy of Oklahoma Gas and Electric Company)

During the summer peak season this utility uses a constant charge or flat rate for all energy (7.728 cents/kWh) regardless of the amount consumed. In the off-peak season, however, the utility uses a declining block approach and charges a higher rate for the first 600 kWh of energy than it does for the remaining kilowatt-hour use.

- **Time-of-day or time-of-use pricing.** To handle the daily peaking problem, some utilities charge more for energy consumed during peak times. This requires the utility to install relatively sophisticated meters. It also requires some customer habit changes. Time-of-use pricing for residential customers is not very popular today; however, most utilities are required by their state regulatory agencies to provide a time-of-use rate for customers who desire one, so most utilities have some form of time-of-use pricing.

A sample time-of-day rate for residential customers served by Florida Power Corporation is shown in Figure 3-6.

- **Peak shaving.** Some utilities offer a discount to residential customers if the utility can hook up a remote control unit to cycle large electricity

Customer charge: $16.00/month

Energy charge:
  On-peak energy   10.857¢/kWh
  Off-peak energy   0.580¢/kWh

On-peak hours:
  November through March:
    Monday through Friday   6:00 a.m. to 10:00 a.m.
                                          .   6:00 p.m. to 10:00 p.m.

April through October:
  Monday through Friday   12:00 noon to 9:00 p.m.

Off-peak hours: All other hours

Figure 3-6 Sample time-of-day electric rate.
(Courtesy Florida Power Corporation, St. Petersburg, FL)

using appliances in the home (usually electric heaters, air conditioners and water heaters). This utility load control program is also called *load management*. This way the utility can cycle large appliance loads on and off periodically to help reduce demand. Since the cycling is performed over short periods of time, most customers experience little to no discomfort. This approach is rapidly gaining in popularity.

A sample load management rate for residential customers served by the Clay Electric Cooperative in Keystone Heights, FL, is shown in Figure 3-7. This rate provides a rebate to customers who agree to allow the utility to turn off their electric water heaters or air conditioners for short periods of time during peak hours. Note that the Clay Electric rate also includes an inclining block feature.

3.1.5 General Service Rate Schedules

A *general service rate schedule* is used for commercial and small industrial users. This is a simple schedule usually involving only consumption (kWh) charges and customer charges. Sometimes, demand (kW) charges are used; this requires a demand meter. (See Section 3.1.3 for a more detailed discussion of demand charges.)

Customer charge: $9.00/month

Energy charge: First 1000 kWh @ $0.0825/kWh
                 Over 1000 kWh @ $0.0930/kWh

Load management credit per month: Credit will be applied to the bill of all customers with load management switches who use 500 kWh or more per month as follows:

Electric water heater controlled January-December $4.00
Electric central heating controlled October-March
    for 5 to 7.5 minutes of each 25-minute period $3.00
Electric central air conditioner controlled
    April-September for 5 to 7.5 minutes of each
    25-minute period $3.00

Electric central heating controlled October-March
    for 12.5 minutes of each 25-minute period $8.00
Electric central air conditioner controlled
    April-September for 5 to 7.5 minutes of each
    25-minute period $8.00

Figure 3-7. Sample load management rate for residential service.
(Courtesy of Clay Electric Cooperative, Keystone Heights, FL)

The energy charge for this customer class is often substantially higher than for residential users for various noneconomic reasons. Some of these reasons include the fact that many businesses have widely varying loads depending on the health of the economy, and many businesses close after only a few months of operation—sometimes leaving large unpaid bills. In addition, some regulatory agencies feel that residential customers should have lower rates since they cannot pass on electric costs to someone else. For example, one rate schedule charges almost 8 cents/kWh for commercial users during peak season but only a little more than 5 cents/kWh for residential users during the same season.

3.1.6 Small Industrial Rate Schedules

A *small industrial rate schedule* is usually available for small industrial users and large commercial users. The service to these customers often becomes more complex because of the nature of the equipment used in the industry, and their consumption tends to be higher. Consequently, the billing becomes more sophisticated. Usually, the same cost categories occur as in the simpler schedules, but other categories have been added. Some of these are outlined below.

3.1.6.1 Voltage level.

One degree of complexity is introduced according to what *voltage level* the customer needs. If the customer is willing to accept the electricity at transmission voltage levels (usually 50,000 volts or higher) and do the necessary transforming to usable levels on-site, then the utility saves considerable expense and can charge less. If the customer needs the service at a lower voltage, then the utility must install transformers and maintain them. In that case, the cost of service goes up and so does the bill.

The voltage level charge can be handled in the rate schedule in several ways. One is for the utility to offer a percentage discount on the electric bill if the customer owns its own primary transformer and accepts service at a higher voltage than it needs to run its equipment. Another is to increase the energy charge as the voltage level decreases. (This method is shown in the example in Figure 3-8.) Installing their own transformers is often a significant cost-cutting opportunity for industrial users and should be explored. Maintaining transformers is a relatively simple (though potentially dangerous) task, but the customer may also need to install standby transformers to avoid costly shutdowns.

3.1.6.2 Demand billing.

Understanding industrial rate structures means understanding the concept of *demand billing*. Consider Figure 3-8 where energy demands on a utility are plotted against time for two hypothetical companies. Since the instantaneous demand (kW) is plotted over time, the integration of this curve (i.e., the area under the curve) is the total energy (kWh) consumed (see shaded area). Company B and Company A have the same average demand, so the total energy consumed by B equals that of A. Company B's peak demand and its average demand are the same, but Company A has a seasonal peak that is twice as high as its average demand. Because the kWh consumed by each are equal, their bills for consumption will be equal, but this seems unfair. Company B has a very flat demand structure so the utility can gear up for that level of service with high-efficiency equipment. Company A, however, requires the utility to supply about twice the capacity that company B needs but only for one short period of time during the year. This means the utility must maintain and gear up equipment which will only be needed for a short period of time. This is quite expensive, and some mechanism must be used by the utility to recover these additional costs.

Figure 3-8 Demand profiles for two hypothetical industrial firms.

To properly charge for this disproportionate use of facilities and to encourage company A to reduce its peak demand, an electric utility will usually charge industrial users for the peak demand incurred during a billing cycle, normally a month. Often a customer can achieve substantial cost reductions simply by reducing peak demand and still consuming the same amount of electricity. A good example of this would be to move the use of an electric furnace from peaking times to nonpeaking times (maybe second or third shifts). This means the same energy could be used at less cost since the demand is reduced. A peak shaving (demand control) example will be discussed in section 3-7.

3.1.6.3 Ratchet Clause

Many utility rate structures have a ratchet clause associated with their demand rate. To understand the purpose of the ratchet clauses, one must realize that if the utility must supply power to meet a peak load in July, it must keep that equipment on hand and maintain it for the next peak load which may not occur for another year. To charge for this cost, and to encourage customers to level their demand over the remaining months, many utilities have a ratchet clause.

A ratchet clause usually says that the billed demand for any month is a percentage (usually greater than 50%) of the highest maximum demand of the previous 11 months or the actual demand, whichever is greater. The demand is normally corrected for the power factor. For a company with a large seasonal peaking nature, this can be a real problem. A peak can be set in July during a heavy air conditioning period that the company in effect pays for a full year. The impact of ratchet clauses can be significant, but often a company never realizes this has occurred.

3.1.6.4 Power factor.

Power factor is a complex subject to explain, but it can be a vitally important element in a company's electrical bill. One company the authors worked with had a power factor of 51 percent. With their billing schedule, this meant they were paying a penalty of 56.9 percent on demand billing. With the addition of power factor correction capacitors, this penalty could have been avoided or minimized.

The power factor is important because it imposes costs on a utility that are not recovered with demand and energy charges. Industrial customers are more likely to be charged for a poor power factor. They create greater power factor problems for a utility because of the equipment they use. They are also more likely to be able to correct the problem.

To understand the power factor, you must understand electric currents. The current required by induction motors, transformers, fluorescent lights, induction heating furnaces, resistance welders, etc., is made up of three types of current:

1. *Power-producing current* (working current or current producing real power). This is the current which is converted by the equipment into useful work, such as turning a lathe, making a weld, or pumping water. The unit of measurement of the real power produced from working current is the kilowatt (kW).

2. *Magnetizing current* (wattless or reactive current). This is the current which is required to produce the flux necessary for the operation of induction devices. Without magnetizing current, energy could not flow through the core of a transformer or across the air gap of an induction motor. The unit of measurement of the reactive power associated with magnetizing current is the kilovar (kvar) or kilovolt-amperes reactive.

3. *Total current* (current producing apparent power or total power). This is the current that is read on an ammeter in the circuit. It is made up of the vector sum of the magnetizing current and the power-producing current. The unit of measurement of apparent power associated with this total current is the kilovoltampere (kva). Most alternating current (ac) powered loads require both kilowatts and kilovars to perform useful work.

Power factor is the ratio of actual (real) power being used in a circuit, expressed in watts or kilowatts, to the apparent power drawn from the power line, expressed in voltamperes or kilovolt-amperes. The relationship of kW, kvar, and kva in an electrical system can be illustrated by scaling vectors to represent the magnitude of each quantity, with the vector for kvar at a right angle to that for kW (Figure 3-9). When these components are added *vectorially*, the resultant is the kva vector. The angle between the kW and kva vectors is known as the *phase angle*. The cosine of this angle is the power factor and equals kW/kva.

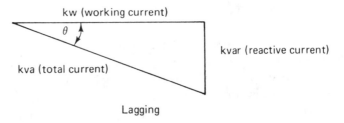

$$\theta = \text{phase angle} = \text{measure of net amount of inductive reactance in circuit}$$

$$\cos \theta = PF = \text{ratio of } \textit{real power} \text{ to } \textit{apparent power}$$

$$kva = \frac{kw}{\cos \theta} = \frac{kw}{PF} = \sqrt{(kw)^2 + (kvar)^2}$$

Figure 3-9 Diagram of ac component vectors

Unless some way of billing for a low power factor is incorporated into a rate schedule, a company with a low power factor would be billed the same as a company with a high power factor. Most utilities do build in a power factor penalty for industrial users. However, the way of billing varies widely. Some of the more common ways include:

- Billing demand is measured in kva instead of kW. A look at the triangle in Figure 3-9 shows that as the power factor is improved, kva is reduced, providing a motivation for power factor improvement.

- Billing demand is modified by a measure of the power factor. Some utilities will increase billed demand one percent for each one percent the power factor is below a designated base. Others will modify demand as follows:

$$Billed\ Demand = Actual\ \dot{Demand} \times \frac{Base\ Power\ Factor}{Actual\ Power\ Factor}$$

This way, if the actual power factor is lower than the base power factor, the billed demand is increased. If the actual power factor is higher than the base power factor, some utilities will allow the fraction to stay, thereby providing a reward instead of a penalty. Some will run the calculation only if actual power factor is below base power factor.

• The demand or consumption billing schedule is changed according to the power factor. Some utilities will change the schedule for both demand and consumption according to the power factor.

• A charge per kvar is used. Some companies will charge for each kvar used above a set minimum. This is direct billing for the power factor.

In addition, since a regular kW meter does not recognize the reactive power, some other measuring instrument must be used to determine the reactive power or the power factor. A kva meter can be supplied by the utility, or the utility might decide to only periodically check the power factor at a facility. In this case a utility would send a crew to the facility to measure the power factor for a short period of time, and then remove the test meter.

3.1.6.5 The rate schedule.

The previous few sections were necessary in order to be able to present a rate schedule itself in understandable terms. All these complex terms and relationships make it difficult for many managers to understand their bills. You, however, are now ready to analyze a typical rate schedule. Consider Figure 3-10.

Figure 3-10. Typical small industrial rate schedule.
(Courtesy of Oklahoma Gas and Electric Company)

<u>Effective in:</u> All territories served

<u>Availability:</u> Power and light service. Alternating current. Service will be rendered at one location at one voltage. No resale, breakdown, auxiliary, or supplementary service permitted.

<u>Rate:</u>
   A. <u>Transmission service (service level 1):</u>
     <u>Customer charge:</u> $637.00/bill/month

     <u>Demand charge applicable to all kW/month of billing demand:</u>
       On-peak season: $10.59/kW
       Off-peak season: $3.84/kW

     <u>Energy charge:</u>
       First two million kWh 3.257¢/kWh
       All kWh over two million 2.915¢/kWh

   B. <u>Distribution service (service level 2):</u>
     <u>Customer charge:</u> $637.00/bill/month

     <u>Demand charge applicable to all kW/month of billing demand:</u>
       On-peak season: $11.99/kW
       Off-peak season: $4.36/kW

     <u>Energy charge:</u>
       First two million kWh 3.297¢/kWh
       All kWh over two million 2.951¢/kWh

   C. <u>Distribution service (service levels 3 and 4):</u>
     <u>Customer charge:</u> $269.00/bill/month

     <u>Demand charge applicable to all kW/month of billing demand:</u>
       On-peak season: $12.22/kW
       Off-peak season: $4.45/kW

Figure 3-10. (*Continued*)

Energy charge:
    First two million kWh 3.431¢/kWh
    All kWh over two million 3.010¢/kWh

D. Secondary service (service level 5):
    Customer charge: $151.00/bill/month

Demand charge applicable to all kW/month of
billing demand:
    On-peak season: $13.27/kW
    Off-peak season: $4.82/kW

Energy charge:
    First two million kWh 3.528¢/kWh
    All kWh over two million 3.113¢/kWh

Definition of season:
    On-peak season   Revenue months of June-October of any year.
    Off-peak season: Revenue months of November of any year
                        through May of the succeeding year.

Late payment charge: A late payment charge in an amount equal to one
and one-half per cent (1-1/2%) of the total balance for services and
charges remaining unpaid on the due date stated on the bill shall be added
to the amount due. The due date shall be twenty (20) days after the bill is
mailed.

Minimum bill: The minimum monthly bill shall be the Customer Charge
plus the applicable Capacity Charge as computed under the above schedule. The Company shall specify a larger minimum monthly bill, calculated
in accordance with the Company's Allowable Expenditure Formula in its
Terms and Conditions of Service on file with and approved by the Commission, when necessary to justify the investment required to provide
service.

Determination of maximum demand: The consumer's Maximum Demand
shall be the maximum rate at which energy is used for any period of
fifteen (15) consecutive minutes of the month for which the bill is rendered
as shown by the Company's demand meter. In the event a consumer
taking service under this rate, has a demand meter with an interval

Figure 3-10. (*Continued*)

greater than 15 minutes, the company shall have a reasonable time to change the metering device.

Determination of billing demand: The Billing Demand upon which the demand charge is based shall be the Maximum Demand as determined above corrected for the power factor, as set forth under the Power Factor Clause, provided that no billing demand shall be considered as less than 65% of the highest on-peak season maximum demand corrected for the power factor previously determined during the 12 months ending with the current month.

Power factor clause: The consumer shall at all times take and use power in such manner that the power factor shall be as nearly 100% as possible, but when the average power factor as determined by continuous measurement of lagging reactive kilovoltampere hours is less than 80%, the Billing Demand shall be determined by multiplying the Maximum Demand, shown by the demand meter for the billing period, by 80 and dividing the product thus obtained by the actual average power factor expressed in per cent. The company may, at its option, use for adjustment the power factor as determined by tests during periods of normal operation of the consumer's equipment instead of the average power factor.

Fuel cost adjustment: The rate as stated above is based on an average cost of $1.60/million Btu for the cost of fuel burned at the company's thermal generating plants. The monthly bill as calculated under the above rate shall be increased or decreased for each kWh consumed by an amount computed in accordance with the following formula:

$$F.A. = A * \frac{(B*C)-D}{10^6} + \frac{P}{S} + \frac{OC}{OS} - Y$$

where

F.A. = fuel cost adjustment factor (expressed in $/kWh) to be applied per kWh consumed

A   = weighted average Btu/kWh for net generation from the company's thermal plants during the second calendar month preceding the end of the billing period for which the kWh usage is billed

Figure 3-10. (*Continued*)

B   = amount by which the average cost of fuel per million Btu during the second calendar month preceding the end of the billing period for which the kWh usage is billed exceeds or is less than $1.60/million Btu; any credits, refunds, or allowances on previously purchased fuel, received by the company from any source, shall be deducted from the cost of fuel before calculating B each month

C   = ratio (expressed decimally) of the total net generation from all the company's thermal plants during the second calendar month preceding the end of the billing period for which the kWh usage is billed to the total net generation from all the company's plants including hydro generation owned by the company, or kW produced by hydro generation and purchased by the company during the same period

D   = the amount of fuel cost per million Btu embedded in the base rate is $2.30

P   = the capacity and energy cost of electricity purchased by the Company, excluding any cost associated with "OC," during the second calendar month preceding the current billing month, excluding any capacity purchased in said month and recovered pursuant to Standard Rate Schedule PCR-1.

S   = total kWh generated by the company plus total kWh purchased by the company which are associated with the cost included in "P" during the second calendar month preceding the end of the billing period for which kWh use is billed

OC = the difference between the cost of cogenerated power and company-generated power (Note that this factor has been simplified for purposes of this book.)

OS = the company's appropriate Oklahoma retail kWh sales during the twelfth billing month preceding the current billing month

Y   = a factor (expressed in $/kWh) to reflect 90% of the margin (profits) from the non-firm off-system sales of electricity to other utilities during the second calendar month preceding the end of the billing period for which the kWh usage is billed.

Figure 3-10. (*Concluded*)

Franchise payment: Pursuant to Order Number 110730 and Rule 54(a) of Order Number 104932 of the Corporation Commission of Oklahoma, franchise taxes or payments (based on a percent of gross revenue) in excess of 2% required by a franchise or other ordinance approved by the qualified electors of a municipality, to be paid by the company to the municipality, will be added pro rata as a percentage of charges for electric service, as a separate item, to the bills of all consumers receiving service from the company within the corporate limits of the municipality exacting the said tax or payment.

Transmission, distribution, or secondary service: For purposes of this rate, the following shall apply:

Transmission service (service level 1), shall mean service at any nominal standard voltage of the company above 50 kv where service is rendered through a direct tap to a company's transmission source.

Distribution service (service levels 2,3, and 4), shall mean service at any nominal standard voltage of the company between 2,000 volts and 50 kv, both inclusive, where service is rendered through a direct tap to a company's distribution line or through a company numbered substation.

Secondary service (service level 5), shall mean service at any nominal standard voltage of the company less than 2,000 volts or at voltages from 2 to 50 kv where service is rendered through a company-owned line transformer. If the company chooses to install its metering equipment on the load side of the consumer's transformers, the kWh billed shall be increased by the amount of the transformer losses calculated as follows:

1% of the total kva rating of the consumer's transformers * 730 hours

Term: Contracts under this schedule shall be for not less than 1 year, but longer contracts subject also to special minimum guarantees may be necessary in cases warranted by special circumstances or unusually large investments by the company. Such special minimum guarantees shall be calculated in accordance with the company's allowable expenditure formula and its terms and conditions of service on file with and approved by the commission.

Let's examine the different components in this rate structure of Figure 3-10.

- *Voltage level.* This utility has chosen to encourage company-owned primary transformers by offering a cheaper rate for both demand and consumption if the company accepts service at a higher voltage level. To analyze what it could save from primary transformer ownership, a company only needs to calculate the dollar savings from accepting service at a higher voltage level and compare that savings to the cost of the necessary transformers and annual maintenance thereof. Transformer losses must be absorbed by the company, and the company must provide a standby transformer or make other arrangements in case of a breakdown. (See Problem 3.5 at the end of this chapter for an example.)

- *Demand billing.* This utility has chosen to emphasize demand leveling by assessing a rather heavy charge for demand.* Furthermore, the utility has emphasized demand leveling during its summer peaking season.

- *Consumption.* This utility uses a declining block rate for very large users, but this essentially amounts to a flat charge per kilowatt-hour for most consumption levels.

- *Power factor.* The utility has chosen to charge for the power factor by modifying the demand charge. They have decided all customers should aim for a power factor of at least 80 percent and should be penalized for power factors of less than 80 percent. To do this, the peak demand is multiplied by a ratio of the base power factor (80%) to the actual power factor if the actual power factor is below 80%; there is no charge or reward if the power factor is above 80%:

$$Billed\ Demand = Actual\ Demand \times \frac{Base\ Power\ Factor}{Actual\ Power\ Factor}$$

where the base power factor = .80.

- *Ratchet clause.* The utility has a ratchet clause which says that the billed demand for any month is "65% of the highest on-peak season maximum demand corrected for the power factor" of the previous 12 months or the actual demand corrected for power factor whichever is greater.

---

*Actually, charges are regional. For the Southwest, this is a rather large demand charge. For the Northwest, it would be cheap.

- *Miscellaneous.* Other items appearing in the rate schedule include fuel cost adjustment, late payment charge, and minimum bill. The fuel cost adjustment is based on a formula and can be quite significant. Anytime the cost of energy is calculated, the fuel cost adjustment should be included.

- *Sales tax.* One item not mentioned in the sample schedule is sales tax. Many localities have sales taxes of 6-8% or more, so this can be a significant cost factor. The cost of electrical service should include this charge. One item of interest: Some states have laws stating that *energy used directly in production should not have sales tax charged to it.* This is important to any industry in such a state with energy going to production. Some submetering may be necessary, but the cost savings often justifies this. For example, electricity used in a process furnace should not be taxed, but electricity running the air conditioners would be taxed.

Example 3-1. As an example of rate schedule calculations, let's use the schedule in Figure 3-10 to calculate the September bill for the company whose electric use is shown below:

    Month: September 1993
    Actual demand: 250 kW
    Consumption: 54,000 kWh
    Previous high billed demand: 500 kW (July 1993)

    Power factor: 75%
    Service level: Secondary (PLS, service level 5)
    Sales tax: 6%
    Fuel adjustment: 1.15¢/kWh (This value is calculated by the utility
                    company according to the formula given in the
                    rate schedule.)

As a first step, the demand should be calculated:

Power factor correction:

    billed demand   = (actual demand) * (.80/PF)
                         = 250 kW * (.80/.75)
                         = 266.7 kW

minimum billed demand (ratchet clause)
$$= (500 \text{ kW}) * (.65)$$
$$= 325 \text{ kW}$$
billed demand $= \text{max. } (266.7 \text{ kW}, 325 \text{ kW})$

billed demand $= 325 \text{ kW}$

demand charge (on-peak season)
$$= (325 \text{ kW}) (\$13.27/\text{kW})$$
$$= \underline{\$4312.75}$$

Consumption charge:

(54,000 kWh)($.03528/kWh)	= $1905.12
(54,000 kWh)($.0115/kWh)(fuel adjustment)	= $ 621.00
total consumption charge	= $2526.12

Customer charge:

$151.00

Total charge before sales tax:

$4312.75 + $2526.12 + $151.00 = $6989.87

Sales tax:

$6989.87 × (.06) = $ 419.39

Total*:

$6989.87 + $ 419.39 = $7409.26

3.1.7 Large Industrial Rate Schedules
Most utilities have very few customers that would qualify for or desire to be on a large industrial rate schedule. Sometimes, however, one or two large industries will utilize a significant portion of a utility's total generating capacity. Their size makes the billing more complex; therefore,

---

*Ignoring franchise payment and late charges.

a well-conceived and well-designed rate schedule is necessary.

Typically a large industrial schedule will include the same components as a small industrial schedule. The difference occurs in the amount charged for each category. The customer charge, if there is one, tends to be higher. The minimum kW of demand tends to be much higher in cost/ kW, but all additional kW may be somewhat lower (per kW) than on small industrial schedules. Similarly, the charge per kWh for consumption can be somewhat less. The reason for this is economy of scale; it is cheaper for a utility to deliver a given amount of electrical energy to one large customer than the same amount of energy to many smaller customers.

Figure 3-11 is an example of a large industrial schedule.

Figure 3-11. Large industrial rate schedule.
(Courtesy of Oklahoma Gas and Electric Co.)

STANDARD RATE SCHEDULE   Rate Code No. 530

Large Power and Light (LPL)
(TITLE AND/OR NUMBER)

Availability: Available on an annual basis by written contract to any retail customer. This schedule is not available for resale, standby, breakdown, auxiliary, or supplemental service. It is optional with the customer whether service will be supplied under this rate or any other rate for which he is eligible. Once a rate is selected, however, service will continue to be supplied under that rate for a period of 12 months unless a material and permanent change in the customer's load occurs.

Service will be supplied from an existing transmission facility operating at a standard transmission voltage of 69 kv or higher by means of not more than one transformation to a standard distribution voltage of not less than 2.4 kv. Such transformation may be owned by the company or customer. Service may be supplied by means of an existing primary distribution facility of at least 24 kv when such facilities have sufficient capacity.

Service will be furnished in accordance with the company's rules, regulations, and conditions of service and the rules and regulations of the Oklahoma Corporation Commission.

Net rate: Capacity charge:
    $13,750.00:   net per month for the first 2500 kilowatts (kW) or less of billing demand

Figure 3-11. (*Continued*)

4.20: net per month per kilowatt (kW) required in excess of 2500 kW of billing demand

.50: net per month for each reactive kilovoltampere (kvar) required above 60% of the billing demand

Plus an energy charge:

2.700¢: net per kilowatt-hour (kWh) for the first 1 million kWh used per month

2.570¢: net per kilowatt-hour (kWh) for all additional use per month

Determination of monthly billing demand: The monthly billing demand shall be the greater of (a) 2500 kW, (b) the monthly maximum kilowatt (kW) requirement, or (c) eighty percent (80%) of the highest monthly maximum kilowatt (kW) requirement established during the previous 11 billing months. The monthly maximum reactive kilovoltampere (kvar) required are based on 30-min integration periods as measured by appropriate demand indicating or recording meters.

Determination of minimum monthly bill: The minimum monthly bill shall consist of the capacity charge. The monthly minimum bill shall be adjusted according to adjustments to billing and kvar charges. If the customer's load is highly fluctuating to the extent that it causes interference with standard quality service to other loads, the minimum monthly bill will be increased $.50/kva of transformer capacity necessary to correct such interference.

Terms of payment: Payment is due within 10 days of the date of mailing the bill. The due date will be shown on all bills. A late payment charge will be assessed for bills not paid by the due date. The late payment charge shall be computed at 1-1/2 % on the amount past due per billing period.

Adjustments to billing:

1. Fuel cost adjustment: The rate as stated above is based on an average cost of $2.00/million Btu for the cost of fuel burned at the company's thermal generating plants. The monthly bill as calculated under the above rate shall be increased or decreased for each kWh consumed by an amount computed in accordance with the following formula:

Figure 3-11. (*Continued*)

$$FA = A \times \frac{B}{10^6} \times C$$

where FA= fuel cost adjustment factor (expressed in dollars per kWh)
to be applied per kWh consumed

A= weighted average Btu/kWh for net generation from the
company's thermal plants during the second calendar
month preceding the end of the billing period for which
the kWh usage is billed

B= amount by which the average cost of fuel per million Btu
during the second calendar month preceding the end of the
billing period for which the kWh usage is billed exceeds or
is less than $2.00/million Btu; any credits, refunds, or al-
lowances on previously purchased fuel received by the
company from any source shall be deducted from the cost
of fuel before calculating B each month

C= ratio (expressed decimally) of the total net generation from
all the company's thermal plants during the second calendar month preceding the end of the billing period for
which the kWh usage is billed to the total net generation
from all the company's plants including hydrogeneration
owned by the company, or kWh produced by
hydrogeneration and purchased by the company, during
the same period

2. Tax adjustment: If there shall be imposed after the effective date of
this rate schedule, by federal, state, or other governmental authority,
any tax, other than income tax, payable by the company upon gross
revenue, or upon the production, transmission, or sale of electric
energy, a proportionate share of such additional tax or taxes shall be
added to the monthly bills payable by the customer to reimburse the
company for furnishing electric energy to the customer under this rate
schedule. Reduction likewise shall be made in bills payable by the
customer for any decrease in any such taxes.

Additionally, any occupation taxes, license taxes, franchise taxes, operating permit fees required for engaging in business with any municipality, or for use of its streets and ways, in excess of two percent (2%) of gross revenues from utility business done within such municipality, shall be added to the billing of customers residing within such municipality when voted by the people at a regularly called franchise election. Such adjustment to billing shall be stated as a separate item on the customer's bill.

Figure 3-11. (*Concluded*)

3.1.8 Cogeneration and Buy-Back Rates

Since enactment of the Public Utility Regulatory Policy Act of 1978 (PURPA), there has been significant renewed interest in on-site-generated power. This can be from cogeneration (on-site generation of thermal heat with concurrent, sequential generation of electricity), windmills, solar thermal, solar photovoltaics, or other sources. Generation of this energy for use only on site is often not cost effective due to variability of loads. Resale of excess electricity (when it is available) to the local utility, however, often makes a non-utility electric generation project economically feasible.

PURPA specified that cogenerators that met certain minimum conditions would be designated as Qualifying Facilities (QFs) and would be paid Avoided Costs by the purchasing utilities. To comply with these requirements of PURPA, utilities have developed buy-back rates for this excess electricity. Since the value of this energy may be either less than or greater than the cost to the utility of generating it, buy-back usually requires a separate meter and a separate rate schedule.*

Cogeneration can be an attractive energy cost-saving alternative for facilities that need both electric power and large amounts of steam or hot water. The combined production of electricity and thermal energy can result in fuel savings of 10-30 percent over the separate generation costs. Cogeneration will be discussed again in a later chapter.

3.1.9 Others

Many other rate schedules are being developed as the needs dictate. For example, some utilities have a rate schedule involving interruptible and curtailable loads. An interruptible load is one that can be turned off at

---

*Remember, the time when the industry generates an excess of electricity is probably not a peak time, so the utility really does not need the power as badly.

certain times of the day or year. A utility offers a lower rate as an incentive to companies willing to help decrease the system demand during peaking times of the day or year.

A *curtailable* load is one that the company may be willing to turn off if given sufficient notice. For example, the utility may hear of a weather forecast for extreme heat or extreme cold which would result in a severe peaking condition. It may then call its curtailable customers and ask that all the curtailable loads be turned off. Of course, the utility is willing to compensate the customers for this privilege too.

In both cases, the utility compensates its customers for these loads by offering a reduction in the bill. In the case of curtailable loads, the rate reduction occurs every month during the peaking season whether or not the utility actually calls for the turnoff. In the case of interruptible loads, the basic rate is much lower to start with.

3.2 NATURAL GAS

Natural gas rate schedules are similar in structure to electric rate schedules, but they are often much simpler. Natural gas companies also experience a peaking problem. Theirs is likely to occur on very cold winter days and/or when supply disruptions exist. Due to the unpredictable nature of these peak problems, gas utilities normally do not charge for peak demand. Instead, customers are placed into *interruptible priority classes*.

A customer with a high priority will not be curtailed or interrupted unless absolutely necessary. A customer with the lowest priority, however, will be curtailed or interrupted whenever a shortage exists. Normally some gas is supplied to keep customer's pipes from freezing and pilot lights burning. To encourage use of the low-priority schedules, utilities charge significantly less for this gas rate. Most gas utilities have three or four priority levels. Some utilities allow customers to choose their own rate schedule, while others strictly limit the choice.

Figure 3-12 presents a sample rate schedule for four priority levels. Here the industrial customer is limited in choice to priorities 3 and 4.

Some points are demonstrated in this collection of schedules. First, the energy costs decrease as the priority goes down, but the probability of a curtailment or interruption dramatically increases. Second, the winter residential rate has an increasing block component on the block of gas use over 10 Mcf/month. Only very large residential consumers would approach this block, so its intent is to discourage wanton utilization. Like electric rates, fuel cost adjustments do exist in gas rates. Sales taxes also apply to natural gas bills. Again, some states do not charge sales tax on gas used directly in production.

Natural gas rates differ significantly in different parts of the country. Gas is relatively cheap in the producing areas of Oklahoma, Texas and Louisiana. It is much more expensive in other areas where it must be transported over long distances through transmission pipes. For example, in Florida, gas is almost twice as expensive as shown in the rate structure of Figure 3-12. Gas supplied by Gainesville Regional Utilities (Gainesville, FL) is priced under a flat rate structure (i.e. it does not drop in price with increased use). It costs around $6.00/Mcf for residential use, and almost $5.00/Mcf for commercial use. Interruptible gas service for larger users costs about $4.00/Mcf.

3.3 FUEL OIL AND COAL

Fuel oils are a very popular fuel source in some parts of the country, but they are rarely used in others. Natural gas and fuel oil can generally be used for the same purpose so the availability and price of each generally determines which is used.

Fuel oils are classified as *distillates* or *residuals*. This classification refers to the refining or distillation process. Fuel oils 1 and 2 are distillates. No. 1 oil can be used as a domestic heating oil and diesel fuel. No. 2 oil is used by industry and in the home. The distillates are easier to handle and require no heat to maintain a low viscosity; therefore, they can be pumped or poured with ease.

Residual fuel oils include nos. 4, 5 and 6. Optimum combustion is more difficult to maintain with these oils due to variations in their characteristics that result from different crude oil origins and refining processes. No. 6 or residual bunker C is a very heavy residue left after the other oils have been refined. It has a very high viscosity and must be heated in cold environments to maintain a *pour point* (usually somewhere around 55°F).

The sulfur content of fuel oil normally ranges from .3 to 3.0 percent. Distillates have lower percentages than residuals unless the crude oil has a very high sulfur level. Sulfur content can be very important in meeting environmental standards and thus should be watched carefully.

Billing schedules for fuel oils vary widely among geographical areas of the country. The prices are set by market conditions (supply vs. demand), but within any geographical area they are fairly consistent. Within each fuel oil grade, there is a large number of sulfur grades, so shopping

Residential		Commercial	
Priority 1		Priority 2	
Winter		Winter	
First 1 ccf/mo	$5.12	First 1 ccf/mo	$6.79
Next 2.9 Mcf/mo	$5.347/Mcf	Next 2.9 Mcf/mo	$5.734/Mcf
Next 7 Mcf/mo	$3.530/Mcf	Next 7 Mcf/mo	$5.386/Mcf
Over 10 Mcf/mo	$3.725/Mcf	Next 90 Mcf/mo	$4.372/Mcf
		Next 1900 Mcf/mo	$4.127/Mcf
Summer		Next 6000 Mcf/mo	$3.808/Mcf
First 1 ccf	$5.12/Mcf	Over 8000 Mcf/mo	$3.762/Mcf
Next 2.9 Mcf/mo	$5.347/Mcf		
Over 3 Mcf/mo	$3.633/Mcf		
		Summer	
		First 1 ccf	$6.79
		Next 2.9 Mcf/mo	$5.734/Mcf
		Next 7 Mcf/mo	$5.386/Mcf
		Next 90 Mcf/mo	$4.372/Mcf
		Next 100 Mcf/mo	$4.127/Mcf
		Next 7800 Mcf/mo	$3.445/Mcf
		Over 8000 Mcf/mo	$3.399/Mcf

Industrial		Industrial	
Priority 3		Priority 4	
(Second Interruptible)		(First Interruptible)	
First 1 ccf	$19.04	First 4000 Mcf/mo or	
Next 2.9 Mcf/mo	$5.490/Mcf	fraction thereof	$12,814.00
Next 7 Mcf/mo	$5.386/Mcf	Next 4000 Mcf/mo	$3.168/Mcf
Next 90 Mcf/mo	$4.372/Mcf	Over 8000 Mcf/mo	$3.122/Mcf
Next 100 Mcf/mo	$4.127/Mcf		
Next 7800 Mcf/mo	$3.445/Mcf		
Over 8000 Mcf/mo	$3.399/Mcf		

Summer periods include the months from May through October.
Winter periods include the months from November through April.

Figure 3-12. Gas schedules for one utility.
(Courtesy Oklahoma Natural Gas Company)

around can sometimes pay off. Basically, the price is simply a flat charge per gallon, so the total cost is the number of gallons used times the price per gallon.

Like fuel oil, coal comes in varying grades and varying sulfur content. It is, in general, less expensive than fuel oil per Btu, but it does require higher capital investments for pollution control, coal receiving and handling equipment, storage, and preparation. Coal is priced on a per ton basis with provisions for or consideration of sulfur content and percent moisture.

Finally, coal does not burn as completely as other fuels. If combustion air is properly controlled, natural gas has almost no unburned combustibles, while fuel oil has only a small amount. Coal, however, is much more difficult to fully combust.

3.4 STEAM AND CHILLED WATER

In some areas of the country, customers can purchase steam and chilled water directly instead of buying the fuel and generating their own. This can occur where there are large-scale cogeneration plants (steam), refuse-fueled plants (steam), or simple economics of scale (steam and/or chilled water). In the case of both steam and chilled water, it is normal to charge for the energy itself (pounds of steam or ton-hours of chilled water) and the demand (pounds of steam per hour or tons of chilled water). A sample hypothetical rate schedule is shown in Figure 3-13.

These rates are often competitive with costs of self-generated steam and chilled water. Purchasing steam and chilled water conserves considerable amounts of capital and maintenance monies. In general, when steam or chilled water is already available, it is worthy of consideration. The primary disadvantage is that the user does not have control of the generating unit. However, that same disadvantage is also true of electricity for most facilities.

3.5 WATER AND WASTEWATER

The energy analyst also frequently looks at water and wastewater use and costs as part of the overall energy management task. These costs are often related to the energy costs at a facility, and are also amenable to cost control. Water use should be examined, and monthly bills should be analyzed similarly to energy bills to see if unusual patterns of consump-

Steam

    Steam demand charge:

    $1500.00/month for the first 2000 lb/h of demand or any portion thereof

    $550.00/month/1000 lb/h for the next 8000 lb/h of demand

    $475.00/month/1000 lb/h for all over 10,000 lb/h of demand

    Steam consumption charge:

    $3.50/1000 lb for the first 100,000 lb of steam per month

    $3.00/1000 lb for the next 400,000 lb of steam per month

    $2.75/1000 lb for the next 500,000 lb of steam per month

    $2.00/1000 lb for the next 1 million lb of steam per month

    Negotiable for all over 2 million lb of steam per month

Chilled water

    Chilled water demand charge:

    $2500.00/month for the first 100 tons of demand or any portion thereof

    $15.00/month/ton for the next 400 tons of demand

    $12.00/month/ton for the next 500 tons of demand

    $10.00/month/ton for the next 500 tons of demand

    $9.00/month/ton for all over 1500 tons of demand

    (One ton is defined as 12,000 Btu/h, and an hour is defined as any 60 consecutive min.)

    Chilled water consumption charge:

    $.069/ton • h for the first 10,000 ton • h/month

    $.06/ton • h for the next 40,000 ton • h/month

    $.055/ton • h for the next 50,000 ton • h/month

    $.053/ton • h for the next 100,000 ton • h/month

    $.051/ton • h for the next 100,000 ton • h/month

    $.049/ton • h for the next 200,000 ton • h/month

    $.046/ton • h for the next 500,000 ton • h/month

Base rates: Consumption rates subject only to escalation of charges listed in conditions of service and customer instructions

Figure 3-13. Hypothetical steam and chilled water rate schedule.

tion are occurring. Water treatment and re-use may be cost effective in areas where water costs are high.

Wastewater charges are usually based on some proportion of the metered water use since the wastewater solids are difficult to meter. This can needlessly result in substantial increases in the utility bill for processes which do not contribute to the wastewater stream (e.g., makeup water for cooling towers and other evaporative devices, irrigation, etc.). A water meter can be installed at the service main to measure the loads not returning water to the sewer system. This can reduce the wastewater charges by up to two-thirds.

3.6 MONTHLY ENERGY BILL ANALYSIS

Once the energy rate structures have been examined, management should now understand how the company is being charged for the energy it uses each month. This is an important piece of the overall process of energy management at a facility. The next step in the examination of energy costs should be to review the bills and determine the average, peak and off-peak costs of energy used during at least the past twelve months.

Energy bills should be broken down into components that can be controlled by the facility. These cost components can be listed individually in tables and then plotted. For example, electricity bills should be categorized by demand costs per kW per month, and energy costs per kWh. The following example illustrates this analysis for an industry in Florida.

Example 3-2. A company in central Florida that fabricates metal products receives electricity from its electric utility at the following general service demand rate structure.

Rate structure:
(Minimum demand of 20 kW/month to qualify for rate)
Customer cost   = $21.00 per month
Energy cost   = $0.04 per kWh
Demand cost   = $6.50 per kW per month
Taxes   = Total of 8%
Fuel adjustment   = A variable amount per kWh each month (which may be a cost or a credit depending on actual fuel costs to the utility).

The electric energy use and costs for that company for a year are summarized below:

Summary of Energy Usage and Costs

Month	kWh Used (kWh)	kWh Cost ($)	Demand (kW)	Demand Cost ($)	Total Cost ($)
Mar	44960	1581.35	213	1495.26	3076.61
Apr	47920	1859.68	213	1495.26	3354.94
May	56000	2318.11	231	1621.62	3939.73
Jun	56320	2423.28	222	1558.44	3981.72
Jul	45120	1908.16	222	1558.44	3466.60
Aug	54240	2410.49	231	1621.62	4032.11
Sept	50720	2260.88	222	1558.44	3819.32
Oct	52080	2312.19	231	1621.62	3933.81
Nov	44480	1954.01	213	1495.26	3449.27
Dec	38640	1715.60	213	1495.26	3210.86
Jan	36000	1591.01	204	1432.08	3023.09
Feb	42880	1908.37	204	1432.08	3340.45
Totals	569,360	$24,243.13	2,619	$18,385.38	$42,628.51
Monthly Averages	47,447	$2,020.26	218	$1,532.12	$3,552.38

The energy analyst must be sure to account for all the taxes, the fuel adjustment costs, the fixed charges, and any other costs so that the true cost of the controllable energy cost components can be determined. In the electric rate structure described above, the quoted costs for a kW of demand and a kWh of energy are not complete until all these additional costs are added. Although the rate structure says that there is a basic charge of $6.50 per kW per month, the actual cost including all taxes is $7.02 per kW per month. The average cost per kWh is most easily obtained by taking the data for the 12-month period and calculating the cost over that period of time. Using the numbers from the table, one can see that this company has an average energy cost of ($42,628.51)/(569,360 kWh) = $0.075 per kWh.

The utility cost data are used initially to analyze potential Energy Management Opportunities (EMOs) and will ultimately influence which EMOs are recommended. For the example above, an EMO that reduces peak demand would save the company $7.02 per kW per month. Therefore, the energy analyst should consider EMOs that would involve using certain equipment during the night shift when the peak load is significantly lower than the first shift peak load. EMOs that save both energy and demand on the first shift would save costs at a rate of $0.075 per kWh. Finally, EMOs that save electrical energy during the off-peak shift should be examined too, but they may not be as advantageous; they would only save at the rate of $0.043 per kWh because they are already using off-peak energy and there would not be any additional demand cost savings.

The energy consumption should be plotted as well as tabulated to show the patterns of consumption pictorially. The graphs often display some unusual feature of energy use, and may thus help highlight periods of very high use. These high-use periods can be further examined to determine whether some piece of equipment or some process was being used much more than normal. The energy auditor should make sure that any discrepancies in energy use are accounted for. Billing errors can also show up on these plots, although such errors are rare in the authors' experience.

Figures 3-14 and 3-15 show graphs of the annual kilowatt-hour and kilowatt billing for the data from the preceding example. An energy auditor examining these graphs should ask a number of questions. Because the months of May through October are warm months in Florida, the kilowatt-hour use during these months would be expected to be higher than during the winter months. However, July shows unexpectedly low usage. In this case the company took a one-week vacation during July, and the plant energy consumption dropped accordingly. In other cases, this kind of discrepancy should be investigated, and the cause determined. The variations between December, January and February again bear some checking. In this example, the plant also experienced shutdowns in December and January. Otherwise, the facility's kilowatt-hour use seems to have a fairly consistent pattern over the twelve-month period.

Kilowatt use also needs some examination. The 18 kW jump from April to May is probably the result of increased air-conditioning use. However, the 9 kW drop from May to June seems odd especially since kilowatt-hour use actually increased over that period. One might expect demand to drop in July commensurate with the drop in energy use, but as long as the plant operated at normal capacity on any day during the

Figure 3-14

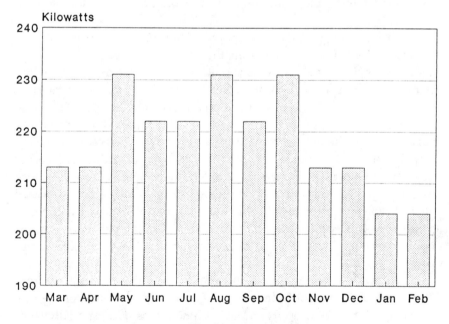

Figure 3-15

month of July, it would be likely to establish about the same peak demand as it did in June. Other causes of large variations for some facilities can be related to meter reading errors, equipment and control system malfunctions, and operational problems.

3.7 ACTIONS TO REDUCE ELECTRIC UTILITY COSTS

Typical actions to reduce kWh consumption involve replacing existing lights with more efficient types; replacing electric heating and cooling equipment with more efficient models; adding insulation to walls and ceilings; replacing motors with high efficiency models and using variable speed drives; recovering heat from air compressors, refrigeration units, or production processes to heat water for direct use or to pre-heat water for steam production; and replacing manufacturing or process equipment by more energy efficient models or processes.

Most of these actions will also result in demand reductions and produce savings through lower kW charges. Other actions that specifically reduce demand involve controlling and scheduling existing loads to reduce the peak kW value recorded on the demand meter. An energy management computer that controls demand is usually better than manual control or time-clock control. If several large motors, chillers, pumps, fans, furnaces or other high kW loads are in use at a facility, then electric costs can almost always be saved through demand limiting or control. All of these areas for savings will be examined in detail in the subsequent chapters.

Example 3-3. As an example of the savings that can be obtained by demand control, consider the use of four large machines at a production facility where each machine has a demand of 200 kW. The machines could be controlled by a computer which would limit the total demand to 400 kW at any one time. This company has chosen to limit the use of the machines by operational policy which states that no more than two machines should be turned on at any given time.

One morning at 8:00 am a new employee came in and turned on the two idle machines. At 8:30 am the plant foreman noticed that too many machines were running and quickly shut down the extra two machines. What did this employee's mistake cost the facility?

The immediate cost on the month's electric bill has two components. Using the demand rate from Example 3.2, the immediate cost is calculated as:

Demand cost increase   = 400 kW * $7.02/kW
                       = $2808

Energy cost increase   = 400 kW * 0.5 hr * $.043/kWh
                       = $8.60 for the energy.

If the utility rate structure includes a 70% demand ratchet, there would be an additional demand for the next 11 months of (.70 * 800 kW) – 400 kW = 160 kW. This would further increase the cost of the mistake as follows:

Ratchet cost increase   = 11 mo * $7.02/kW mo * 160 kW
                        = $12,355.20.
Total cost of mistake   = $2808 + $8.60 + $12,355.20
                        = $15,171.80

Under certain conditions a customer may be able to save money by shifting to another rate category [2]. Consider the example of a manufacturing facility which has a meter for the plant area and a separate meter for the office area. The plant is on a demand rate, but the office area is on a non-demand rate since it has a typical demand of around 19 kW for any month. Under the demand rate structure in section 3.6 customers are billed on the demand rate for one full year starting from any month in which their demand exceeds 20 kW for any 30 minute time window. If the office area could establish a very short peak demand of 20 kW or greater for one month it would automatically be shifted to the demand rate, and could likely benefit from the lower cost per kWh on that rate.

3.8 UTILITY INCENTIVES AND REBATES

Many utility rate structures include *incentives* and *rebates* for customers to replace old, energy inefficient equipment with newer, more energy efficient models. Utilities offer such incentives and rebates because it is cheaper for them to save the energy and capacity for new customers than it is to build new power plants or new gas pipelines to supply that additional load. In addition, stringent environmental standards in some areas makes it almost impossible for electric utilities to build and operate new facilities—particularly those burning coal. Helping customers install more energy efficient electrical and gas equipment allows utilities to delay the need for new facilities, and to reduce the emissions and fuel purchases for the units they do operate.

Direct incentives may be in the form of low interest loans that can be paid back monthly with energy savings resulting from the more efficient equipment. Incentives may also be in the form of lower rates for the electricity used to run higher efficiency lights and appliances, and more efficient process equipment. Other incentives include free audits from the utilities and free technical assistance in identifying and installing these energy efficiency improvements.

Indirect incentives also exist, and are often in the form of a special rate for service at a time when the utility is short of capacity, such as a time of day rate or an interruptible rate. The time of day rate offers a lower cost of electricity during the off-peak times, and often also during the off-season times. Interruptible rates allow large use customers to purchase electricity at very low rates with the restriction that their service can be interrupted on short notice. (See Section 3.1.9 for a discussion of interruptible and curtailable loads.)

Rebates are probably the most common method that utilities use to encourage customers to install high efficiency appliances and process equipment. Utilities sometimes offer a rebate tied to the physical device— such as $1.00 for each low-wattage fluorescent lamp used or $10.00 per horsepower for an efficient electric motor. Other rebates are offered for reductions in demand—such as $250 for each kW of demand that is eliminated. Metering or other verification techniques may be needed to insure that the proper kW reduction credit is given to the customer. The load management rate structure shown in Figure 3-7 is a form of rebate for residential customers.

Incentives or rebates can substantially improve the cost effectiveness of customer projects to replace old devices with new, high efficiency equipment. In some cases, the incentives or rebates may be great enough to completely pay for the difference in cost in putting in a high efficiency piece of equipment instead of the standard efficiency model. Additional discussion and examples of utility incentives and rebates, and how they affect equipment replacement decisions is provided in subsequent chapters.

3.9 SUMMARY

In this chapter we have analyzed rate schedules and costs for electricity in detail. We have also examined rates and costs for natural gas, coal, fuel oil, steam, and chilled water. A complete understanding of all the rate schedules is vital for an active and successful energy management program.

In the past, few managers have understood all of the components of these rate schedules, and very few have even seen their own rate schedules. The future successful manager will not only be familiar with the terms and the schedules themselves, but he or she will also likely work with utilities and rate commissions toward fair rate-setting policies.

REFERENCES

1. Resource: *An Encyclopedia of Utility Industry Terms, Pacific Gas and Electric Company*, San Francisco, CA, 1985.
2. Capehart, Barney L. and John F. Mahoney, "Minimizing Commercial Electric Bills Through Choice of Rate Structures," *Strategic Planning and Energy Management*, Vol. 9, No. 2, 1989.

BIBLIOGRAPHY

Berg, Sanford V., Editor, *Innovative Electric Rates*, Lexington Books, D.C Heath and Company, Lexington, MA, 1983.

Berg, Sanford V. and John Tschirhart, *Natural Monopoly Regulation: Principles and Practices*, Cambridge University Press, 1988.

Bonbright, James C., Albert L. Danielson, and David B. Kamerschen, *Principles of Public Utility Rates*, Public Utility Reports, Inc., Arlington, VA, 1988.

Crew, Michael A. and Paul R. Kleindorfer, *The Economics of Public Utility Regulation*, The MIT Press, Cambridge, MA, 1986.

Zajac, Edward E., *Fairness or Efficiency—An Introduction to Public Utility Pricing*, Ballinger Publishing Co., Cambridge, MA, 1978.

Hirsh, Richard F., *Technology and Transformation in the American Electric Utility Industry*, Cambridge University Press, 1989.

Howe, Keith M. and Eugene F. Rasmussen, *Public Utility Economics and Finance*, Prentice-Hall, Inc., Englewood Cliffs, NJ, 1982.

Chapter 4

Economic Evaluation

4.0 INTRODUCTION

Once a group of energy management opportunities—EMOs—has been identified, the energy manager must determine the cost-effectiveness of the EMOs in order to recommend them to management for implementation. The EMOs should be ranked, with the most cost-effective ones implemented first. There are many measures of cost-effectiveness, and many businesses and industries use their own methods or procedures to make the final decisions. The basic elements of cost-effectiveness analysis are discussed in this chapter and some of the useful techniques or measures of cost-effectiveness are presented.

In Section 4.2 we apply simple economic measures of cost effectiveness to evaluate alternative energy management opportunities. In Section 4.3 we present some of the classical ideas of engineering economy and include sections on cash flow analysis. Section 4.4 compares alternative patterns of cash flow and determines cost-effectiveness measures using the time value of money. In Section 4.5 we show how to take taxes into account. In Section 4.6 we show how to incorporate inflation into economic analyses and discuss the difference between current dollars and constant dollars as they apply to energy management analyses. Finally, several examples of economic evaluations of energy management opportunities are presented.

4.1 COSTS AND COST ESTIMATES

Costs and benefits must be identified and calculated in order to perform an economic analysis. Depending on the purpose of the analysis, costs can be categorized in several ways. For the purpose of comparing energy projects, we choose to divide them into costs at the start of a project, costs that take place throughout the duration of the project, and salvage costs.

4.1.1 Initial Costs

Initial costs are defined to be the costs that occur at the beginning of a project. These include the purchase price and cost of installation, as well as charges for any engineering work that must be done, any permits that must be obtained, and any renovation that must be performed before a project can be started. Methods of estimating these costs range from approximations based on past experience (rules of thumb) to detailed analysis in which as many factors as possible are considered. (See Reference 1 for a good treatment of this subject.) Initial costs must be reduced by any salvage value obtained from selling the item that the new one replaces. Any tax credits on the initial investment are also subtracted from the initial cost (see Section 4.5).

The accuracy of the estimate, and the cost of the effort used to derive the estimate, should depend on the consequences of making an error based on the estimate. If the estimate is used to determine whether the project is feasible, approximate methods are good enough. If the estimate is used to obtain financing for the project, the estimates should be the best possible.

4.1.2 Operation and Maintenance Costs

Operation and maintenance costs are those costs which occur on a regular basis, and which are often considered over an annual period. These costs include the cost of labor and material that can be directly attributed to repairing and maintaining equipment. Examples are annual cleaning of lamps and light fixtures, and twice a year inspection and tune-up of an oil-fired boiler. There may also be indirect costs such as those shared with other projects like supervision, janitorial supplies, etc. Energy costs—which are also a part of the operation costs—are frequently regarded as indirect costs in companies where the cost of the energy needed to produce one unit of output is not known. Where the energy cost per unit of product is known (such as in steel or aluminum plants, for example), these costs are often regarded as direct costs.

An accurate estimate of annual costs is difficult. When the costs involve building-related repair and maintenance, the Unit Price Standards Handbook of the U.S. Department of Defense is excellent [2]. This book gives labor-hour estimates by craft and specifies materials and special equipment needed for each task. Another source for this information is Sack [3]. Also, vendors usually have estimates of the time and skill levels required to maintain their products, although these are not always accurate.

4.1.3 Salvage Values

Most equipment items have a cash value at any given time after their purchase, and this salvage value must be a part of any cost-effectiveness analysis. Unfortunately the salvage value depends on many unknowns—the market for used equipment, how the equipment was used, how well it was maintained, and more. If the salvage value is a deciding factor in the comparison of two alternatives (it usually is not), it can be varied in the analysis to see whether to spend additional time and money to better define it. Trade magazines carry advertisements for used equipment, and they can be used to estimate the salvage value. The tax consequences of salvage values are discussed in Section 4.5.1.

4.1.4 Energy Costs

Energy costs are really part of operation costs, but since we are interested in energy management, we will specifically address them here in a separate section. These costs depend on the type of energy (gas, electricity, coal, steam, etc.) and on the billing schedule of the particular utility. The computation and analysis of energy costs was discussed in detail in Chapter 3. Energy costs are regularly occurring costs that are accumulated to provide an annual cost that is included with other operation and maintenance costs.

EMOs which reduce energy costs by reducing energy use or by reducing the peak demand will be evaluated with respect to how much energy cost savings occurs because of the particular investment in more energy efficient equipment or the changes in operating procedures. For example, if a company is considering buying a new high efficiency motor, it must compare the purchase cost of a standard motor with the purchase cost of the high efficiency model. The initial cost for the EMO will then be the cost difference or cost premium between the two motors. The difference in energy use by the two motors will provide the cost savings to pay back the added cost of the high efficiency motor.

4.2 SIMPLE ECONOMIC ANALYSIS METHODS

Whether a given investment in new equipment or process technology is cost effective is answered by an economic analysis which identifies the costs and the savings, and evaluates a pre-selected cost-effectiveness measure. There are a number of different cost-effectiveness measures, and they fall into two main categories: those that do not take into account the time value of money and those that do. First, we will look at two simple measures which do not take into account the time value of money.

4.2.1 Simple Payback Period

The most widely used economic decision measure to determine cost-effectiveness is the Simple Payback Period or SPP. This measure reflects the time required to recover an investment without considering the time value of money.

$$SPP = \frac{Initial\ Cost}{Annual\ Savings}$$

(4-1)

Example 4-1

An EMO that adds roof insulation to a facility at a cost of $1000 and produces annual savings of $500 from reduced heating and cooling costs would have a SPP of:

$$SPP = \$1000/\$500/yr = 2\ years.$$

Many businesses and industries use only the SPP measure, and further insist that the longest SPP they will accept is two years. As long as the SPP is short, ignoring the time value of money is not a critical flaw in this approach.

The SPP measure is the easiest of all cost-effectiveness measures to evaluate, and it is easily understood by workers and management. No assumptions are required for the cost of money, and no specific lifetime of the EMO is required. It is assumed however, that the lifetime of the EMO is longer than the time computed in the SPP.

4.2.2 Return on Investment

Another common cost-effectiveness measure is the Return On Investment or ROI. As defined here, this measure still does not include the time value of money, but it does require an estimate of the lifetime of the EMO. The ROI measure uses the lifetime of the investment in the EMO and depreciates its cost uniformly over its lifetime. The form of the ROI measure that we use here is:

$$ROI = \frac{(Annual\ Savings - (Cost/Lifetime))}{Initial\ Cost}$$

(4-2)

The ROI is considered as a percentage value per year.

Example 4-2

For the roof insulation EMO discussed above in Example 4-1, we might estimate the life of the insulation at 10 years, and then the ROI would be:

$$ROI = ((\$500/yr) - (\$1000/10 \text{ yrs}))/\$1000 = 40\%/year.$$

The level of an acceptable ROI is very dependent on the particular company or organization that is making the investment decision. Some companies set a minimum level of ROI—called the Hurdle Rate; as long as an EMO has an ROI that exceeds the corporate Hurdle Rate, it will be approved. Other companies consider what ROI might be obtained by investing in a new product or new process. They then require an energy project to produce a level of ROI that is greater than that for these other alternative investments before the energy project will be approved.

The ROI cost-effectiveness measure is not always defined as we have shown. Many other forms exist, and some companies have their own corporate standard for the calculation of the ROI measure, as discussed by Turvaville [4]. While the ROI measure does not explicitly consider the time value of money, the depreciation of cost over the lifetime of the measure does introduce an element of time dependence for the results. We define the term rate of return to include the time value of money, and this measure is described in Section 4.4.2.

4.3   ECONOMIC ANALYSIS USING THE TIME VALUE OF MONEY: DISCOUNTED CASH FLOW ANALYSIS

Energy management expenditures are generally justified in terms of avoided energy costs. The expenses usually come at the beginning of a project, with the savings benefits coming later, and the question arises of how to account for the savings in future years. This introduces the concept of the time value of money, and how to compare savings in one year with those in another year. The problem that arises from simply adding the future annual savings together is caused by the fact that a given sum of money at the beginning of a year should have increased by the end of the year since that amount of money could have been invested or used for some productive purpose to increase its value. Therefore, if you are trying to repay the cost of a project by adding the dollars saved at the end of the first year of a project to the dollars saved several years later, you are working with unlike quantities. But all of these quantities are cash, and there should be some way to treat them so that they are equivalent in value, and can be added. This method of treatment is called discounted cash flow analysis, and it is a fundamental approach that is necessary to understand in order to correctly account for energy costs and savings in different years.

The authors have not used the customary engineering economy tables because most of the users of this book have access to either small financial calculators or a microcomputer with financial programs. In addition, the use of tables often obscures the pattern of cash flows to the detriment of financial planning.

4.3.1 Discounted Cash Flow Analysis—Methodology

In any discipline there is an underlying methodology. The guiding steps of discounted cash flow analysis are the following:

Step 1:   Define the Alternatives. The problem is stated and the list of possible solutions is refined to those selected for additional analysis.

Step 2:   Estimate Relevant Costs. Each alternative is defined in terms of its cash flows including the amount, timing, and direction (benefit or cost) of each.

Step 3:   Analyze Viable Alternatives. The cash flows identified in step 2 are analyzed and compared using the discounted cash flow method described in this chapter. The most cost-effective alternatives can then be identified.

Step 4:   Perform Sensitivity Analyses. The cash flow analysis can be examined in order to determine the effects of uncertainties in the estimates of costs and discount rates to help provide the best solution.

With this underlying methodology, energy management projects can be evaluated for their cost-effectiveness using the time value of money. Several other considerations that are important in performing this analysis may come up. One is still a question of how accurate a cost estimate for the project is needed. Most of the time, a company is examining alternatives for performing a certain task, so the actual cost of the alterative is not so important as the price differences among the various alternatives. The differences in operating costs tell the savings expected for each alternative, so the operating costs need to be calculated with reasonable accuracy.

Two additional pieces of information are needed, and they are more subjective. One is the *lifetime* of the energy management opportunity, and the other is the *discount rate* for the business or industry. Estimates of lifetimes are often difficult to obtain, but can usually be found by contacting the manufacturers or suppliers of the equipment or item being considered. Customer service representatives from electric and gas utilities can often provide these lifetime estimates. The discount rate is often company-specific. It must be supplied by the company accounting department or corporate accounting level, and is usually known for evaluating investments in new product lines or new facilities.

Once the purchase costs, operating costs, savings benefits, lifetimes, and discount rates have been specified, several cost-effectiveness measures using the time value of money can then be produced. For analytical purposes capital costs are assumed to take place at the end of year 0 within the project life. Annual costs are usually treated as occurring at the end of the year in which they are incurred. A salvage value is treated as occurring at the end of the year in which the item is sold.

4.3.2 Data Collection and Presentation

Once cost data have been collected, the next step is to prepare the data in a form useful for analysis. All the cash flows in a given year must be labeled as to type, amount, and direction (benefit or cost). Direction is shown as positive for benefit and negative for cost.

A typical example is given in Figure 4-1. This approach has the advantage that cash flows arranged in this way can be converted into equivalent cash flows (see below) with the use of very simple formulas and without the use of engineering economy tables. In Figure 4-1, the First Cost, the Operating and Maintenance Costs, and the Energy Cost Savings are added together to get the Net Cash Flow for each year.

4.3.3 Equivalence of Cash Flows

One of the goals of this chapter was to find a method which correctly allows the addition of two cash flows occurring at different times. The general approach used is to reduce cash flows occurring at different times to a common basis through the use of a *discount rate*. One example of a discount rate is basic bank interest. If, for example, a person can get 10% annual interest on his or her money at a bank, then $1000 today is worth

Year	First Cost	Operating and Maintenance	Energy Cost Savings	Net Cash Flow
0	-$150,000			-$150,000
1		-$10,000	$45,000	$35,000
2		-$11,000	$50,000	$39,000
3		-$12,000	$56,000	$44,000
4		-$13,000	$62,000	$49,000
5	$30,000 (salvage value)	-$14,500	$69,000	$84,500

Figure 4-1 Example of net cash flows.

$1100 one year from today, and the two amounts can be considered *equivalent at a 10% discount rate.* If, in another example, the annual inflation rate is 12%, then $1000 today has the same purchasing power as $1120 one year from today, and these two amounts can be considered equivalent at a 12% discount rate. As is shown in Equation 4-3, an amount P today is equivalent to an amount F, n periods in the future, at a discount rate of (100*I)% per period.

$$F = P (1 + I)^n \qquad (4\text{-}3)$$

Another way of showing this is Equation 4-4 which defines the *equivalent present worth* P of a cash flow F, occurring n years in the future, where I is the annual discount rate:

$$P = \frac{F}{(1 + I)^n} \qquad (4\text{-}4)$$

Equation 4-4 is appropriate in the many situations where the return on an investment can be considered analogous to the return on a savings account in a bank, with the interest left in the bank. Suppose that $1000 is invested now, at an annual interest rate of 12 percent. This investment will grow as shown in Figure 4-2. Note that this growth can be expressed in terms of equation 4-4. Not all investments give a return as easily identified as bank interest, but the analogy is a valuable one.

[Note: If the return is compounded monthly with a nominal annual rate I, then the equivalent annual rate is $(1 + I/12)^{12}$. Under usual circumstances, annual compounding—using I as the return rate per year and n as the number of years—is adequate for computational purposes.]

End of Year	Value	
0	$1000 * (1+.12)^0$	$= \$1000$
1	$1000 * (1+.12)^1$	$= \$1120$
2	$1000 * (1+.12)^2$	$= 1120 * (1+.12)^1$
		$= 1254$
3	$1000 * (1+.12)^3$	$= 1254 * (1+.12)^1$
. . .	. . . .	
n	$1000 * (1+.12)^n$	

Figure 4-2 Growth of compound interest.

Equation 4-4 can be used to determine the equivalent present worth of cash flows from Figure 4-1, as shown in Figure 4-3. A discount rate of 12% is used.

Year	Net cash flow	Discount factor	Equivalent present worth
0	-$150,000	$1/(1+.12)^0$	-$150,000
1	35,000	$1/(1+.12)^1$	31,250
2	39,000	$1/(1+.12)^2$	31,090
3	44,000	$1/(1+.12)^3$	31,318
4	49,000	$1/(1+.12)^4$	31,140
5	84,500	$1/(1+.12)^5$	47,948
		Total:	$22,746

Figure 4-3. Calculating the equivalent present worth
of a cash flow stream.

Present worth equivalence is a common form of economic equivalence that can be used to evaluate the cost-effectiveness of an energy project or EMO. Another common form is *uniform annual worth* equivalence. The equivalent uniform annual worth of a cash flow stream is a series of equal cash flows occurring at the end of every year of the project that would create an equivalent present worth to the original cash flow stream. Examples of such equivalent uniform payments are mortgage payments, where the monthly payments cover the purchase price of a house as well as the annual interest payments; and car lease agreements, where the lease payments cover the purchase price of the car plus interest minus the estimated value of the car at the time it is returned.

Equation 4-5 relates a present worth P to its equivalent uniform annual worth A [5]:

$$A = P \times \frac{I \times (1+I)^n}{(1+I)^n - 1}$$

(4-5)

where   I = the discount rate per period
          n = number of periods

Equation 4-5 can be used to determine the equivalent annual worth of the cash flow stream of Figure 4-1 in either of two ways:

(1)   The individual cash flows can be converted into their equivalent present worth cash flows using Equation 4-4, then into equivalent annual worth cash flows using Equation 4-5, and, finally, summed, or

(2)   Equation 4-5 can be used directly on the equivalent present worth of Figure 4-3. Using the latter procedure, assuming a 12% discount rate, gives

$$A = 22,746 \times \frac{.12 \times 1.12^5}{1.12^5 - 1}$$

$$= \$6310$$

(4-6)

Either the present worth method or the equivalent uniform annual worth method may be used when comparing projects or equipment items with the same service lives. Comparing projects that have different service lives is a more complex problem. (For a more detailed explanation of this problem, see Reference 6.)

One approach is to define what is to be done between the end of the shorter project and the end of the longer project and add that cost to the cost of the shorter project. This method can be cumbersome, but it is the only easy way to deal with costs or benefits that are changing at different rates throughout the length of the project. If the projects are both serving a single function that will not be renewed or if it is known that the equipment will not be replaced at the end of its service life, then the service life for comparison is taken as the longer of the two service lives with all cash flows after the end of the shorter service life assumed to be zero. By using this longer service life, either an annual worth or a present worth comparison can be used.

One other approach is to consider each project to be repeated until a common ending time is reached. For example, a project with a life of ten years might be compared to a project of life five years. By repeating the five year lifetime project once, both projects will have a total lifetime of ten years—and now the two projects can be correctly compared.

4.4 COST-EFFECTIVENESS MEASURES USING THE TIME VALUE OF MONEY

We can now use the principles developed so far to perform a more advanced level of economic cost-effectiveness analysis which includes the time value of money. For a company that uses a discount rate above 7 percent, or a company that is willing to consider investment payback periods longer than two years, it is imperative to include the time value of money into the cost-effectiveness measures that are used to evaluate the desirability of an energy project. The effect of including the time value of money is to lengthen the time it takes to pay back the initial cost of the project with annual energy savings, or to reduce the rate of return from the energy project. Several common cost-effectiveness measures using the time value of money are discussed below.

4.4.1 Benefit-Cost Ratio For A Project

The simplest cost-effectiveness measure that includes the time value of money is the *Benefit-Cost Ratio* or *BCR*. In this method, the annual savings are discounted when they occur in future years, and are added together to find the present value of the annual savings over a specified period of time. The Benefit-Cost Ratio is then calculated by dividing the present value of the savings by the initial cost. A ratio greater than one means that the investment will more than repay itself, even when the discounted future savings are taken into account. The larger the Benefit-Cost Ratio is, the more attractive the project investment is.

Equation 4-5 can be solved for P, which will then be the present worth of the annual amounts of energy savings, for a given discount rate, and a given project life.

Example 4-3

Consider the insulation EMO discussed earlier in Example 4-1 which saved $500 in annual energy costs, cost $1000 to install, and had a lifetime of ten years. If we assume a discount rate of 12%, the present worth of ten years' annual savings of $500 is found as:

$$P = A * \frac{(1+I)^n - 1}{I*(1+I)^n}$$

$$= \$500 * \frac{(1+.12)^{10} - 1}{.12*(1+.12)^{10}}$$

$$= \$2825.10$$

$$BCR = P/\$1000 = 2.83$$

This is a very attractive Benefit-Cost Ratio, even when the time value of money is included.

4.4.2 Rate of Return For A Project

A second common cost-effectiveness measure using the time value of money is the *rate of return*—sometimes called the discounted rate of return (ROR) or internal rate of return (IRR). In this method, the cash flows for the project are reduced to their present values, and the cash flows are summed up to get a total result. When a value of the discount rate is found that produces a sum of costs and savings that equals zero, that discount rate is called the project rate of return. In other words, the project rate of return is the discount rate at which the present value of the project savings exactly equals the initial project cost. Since the annual savings are discounted, this method is also called the discounted rate of return method.

Consider the energy project shown with cash flows in Figure 4-3. Since the total equivalent present worth is positive—$22,746 at the assumed 12% discount rate, the original investment of $150,000 yields at least a 12% rate of return. To find the actual rate of return of this investment, replace the 12% by successively larger values until the total present worth is zero. For this project, this rate of return is found to be 17.28%. Since finding this discount rate is an iterative process, a computer or financial calculator provides an automated calculation process to get the rate of return in a quick and easy manner.

Example 4-4

Consider a particular energy management project that involves purchasing an induction heater costing $700,000, and has a salvage value at the end of year 5 estimated as $100,000. It will incur a maintenance cost of $60,000 each year beginning in year 2 (the first year is covered as part of the purchase agreement), and will return savings starting at $450,000 during year 3 and increasing 20% each year thereafter. No savings will occur in years 1 and 2 because of costs associated with terminating the contract for gas to fuel the old furnace. For purposes of the problem, assume that the discount rate used to calculate the present worth is 15% and that inflation is neglected. The cash flows of this project can be arranged and the calculations performed as shown in Figure 4-4. Since the net present worth of this project is positive at a 15% discount rate, the rate of return of this project is greater than 15%. If a company considers projects only with a rate of return of at least 15%, this project should be taken into consideration. It might be worthwhile, moreover, to recalculate these figures at higher rates of return to find out exactly how attractive this investment is.

Year	One-time costs	Maintenance	Return from project	Net Cash flow (NCF)	Discount factors	Discounted cash flow
0	−700	—	—	−700	1	−700.0
1	—	—	—	0	$1/1.15$	0
2	—	−60	—	−60	$1/1.15^2$	−45.4
3	—	−60	450	+390	$1/1.15^3$	+256.4
4	—	−60	540	+480	$1/1.15^4$	+274.4
5	100	−60	648	+688	$1/1.15^5$	+342.1
				Net present worth ($\times$ 1000):		$127.5

Figure 4-4 Present worth calculations without taxes or inflation.

4.5 DEPRECIATION AND TAXES

Depreciation and taxes have a significant effect on the economic analysis of energy projects, and must be considered for all large scale projects. Depreciation itself is not a cash flow, but depreciation methods do affect taxes, and those taxes are cash flows. Most companies are taxed by local, state, and federal governments. This taxation varies in what is taxed, what tax credits are given, and when tax amounts and credits occur. The variation is by location, by governmental unit, and by year, but the principles of taxation remain the same.

In general, a corporate tax rate is applied only to profit, so anything that decreases profit decreases the tax amount and thus creates a tax credit. Capital investments in equipment or other property that wears out in some way, and has a lifetime of more than one year, can be depreciated, and the depreciation can be treated as an operating expense and thus can create a tax credit. The details of depreciation and tax laws change from year to year and place to place. Each time a new federal administration takes office, it seems that significant changes to the tax codes are discussed. We use the 1986 Tax Code to work the examples in this book. Therefore, before performing an after-tax economic analysis, it is essential to determine (from an accountant or a tax lawyer) what taxes are applicable to the particular situation under study. Then these taxes must be taken into account.

4.5.1 Major tax changes in 1981 and 1982

Federal tax laws and depreciation schedules for corporations are subject to frequent and substantial change by the U.S. Congress. The National Energy Act of 1978 made some sweeping changes in tax laws, and instituted a number of tax credits for investments in energy saving devices and technologies. For example, special tax credits were designed to increase the desirability of purchasing solar equipment. In the early 1980s two major federal tax laws were enacted which made major changes in how earlier depreciation and corporate taxes were calculated. These were the Economic Recovery Tax Act of 1981 and the Tax Equity and Fiscal Responsibility Act of 1982. These laws removed the tax credits for solar water heaters and other energy conservation investments.

Under the Tax Equity and Fiscal Responsibility Act of 1982, equipment was divided into four tax life categories: 3-year, 5-year, 10-year, and 15-year. Most machine tools and many energy management capital expenses have a 3-year tax life; most other capital expenses are in the 5-year tax category. The depreciation schedule for equipment purchased between 1981 and 1986 is called the Accelerated Cost Recovery System (ACRS), and percentages each year for each category can be found in most engineering economy books [5]. In calculations, these percentages are applied to the original capital expense *without regard to the salvage value*. Furthermore, the tax life category for the equipment is that specified by the law; the actual life anticipated is irrelevant. The cash consequences of this depreciation schedule are credits equal to the corporate tax rate multiplied by the depreciation for the given year. (Although depreciation is not a cash flow, it is regarded by the government as an expense; it therefore lowers the taxable profit and creates a tax credit.)

The 1981 and 1982 tax changes continued the provision of an investment tax credit which was designed to stimulate investment in tangible property by providing a ten percent tax credit. This investment tax credit was removed in 1985. However, when the government decides to stimulate industry investments in new processes and new technology, some form of the investment tax credit is likely to re-appear.

4.5.2 More tax changes in 1986

White, Agee and Case in reference 5 describe the latest tax code changes by saying that "The Tax Reform Act of 1986 virtually rewrote the book on depreciation and income taxes." A new depreciation schedule called the Modified Accelerated Cost Recovery System (MACRS) was put into place for depreciating all tangible property placed into service after 1986. Regular MACRS defined eight property classes which included the original four from ACRS, and in particular, added a seven-year property class. Regular MACRS contains five methods for depreciating property, ranging from a double declining balance method to a straight line depreciation method. A portion of a depreciation table using MACRS is shown in Table 4.1.

Property class

Year	3-year	5-year	7-year
1	33.33%	20%	14.29%
2	44.45%	32%	24.49%
3	14.81%	19.2%	17.49%
4	7.41%	11.52%	12.49%
5		11.52%	8.93%
6		5.76%	8.93%
7			8.93%
8			4.46%

Table 4.1 Partial table for MACRS depreciation values

4.5.3 Project evaluation including taxes

The general procedure for determining the cost-effectiveness of projects after taxes has the following steps:

1. Estimate the before-tax cash flows for each year of the project life, and list the net cash flow for each year.

2. Calculate the tax savings or cost associated with each cash flow, and list the net amount of tax consequence (positive or negative) for each year.

3. Add the results from step 1 and step 2, and list this amount for each year. This is the net after-tax cash flow.

4. For each year, find the discount factor necessary to equate cash flows for that year to present value cash flows, at a prescribed rate of return. The formula was presented earlier:

$$P = \frac{F}{(1 + I)^n}$$

where P is the present worth equivalent, F the net after-tax cash flow, I the rate of return, and n the number of years the payment is to be discounted.

5. Multiply the net after-tax cash flows by the appropriate discount factors, and add the results.

Example 4-5

This procedure for project evaluation including taxes is illustrated in Figure 4-7 for the purchase of an induction heater described in Example 4-4. Suppose (1) that the induction heater falls into the 5-year depreciation category, (2) that the company is subject to taxes which total 34%, and (3) that these taxes are all based on the net profit. Then the cash flows associated with the $700,000 investment are given by $700,000 \times .34 \times P_i$, where P_i is the depreciation percentage for year i. Here $P_1 = .20$, $P_2 = .32$, etc. using the MACRS values from Table 4-1. From Example 4-4, maintenance costs are $60,000 per year starting in year two; and the project returns $450,000 in year three, $540,000 in year four, and $648,000 in year five.

The one-time cost of 59,700 in year five is equal to the salvage value of $100,000 minus the undepreciated first cost of 5.76% of $700,000 since we are only interested in a five year project, and MACRS still has one year of depreciation left in year six. Since there is a net present worth for this project of over $21,000 after taxes, it is a good candidate project. Since the investment shows a positive present worth at the after-tax rate of return of 15%, the investment is yielding at least that great a rate of return. The actual ROR or IRR is 15.9% after taxes. Thus, most companies would find this project to be an economically attractive one.

See table on page 136.

4.5.4 Future tax law changes

The only certainty about tax law is that it will change. Taxes are used by government to advance some particular philosophy of government/ business interaction. As administrations change, so do tax laws. We are using the 1986 Tax Code to provide examples here, but the energy manager should stay up-to-date on current tax provisions and understand how they relate to any potential investments.

4.6 INFLATION

Inflation should be taken into account in most economic analyses. Otherwise, all cash flows are treated as if they were inflating at the same rate, and the results of the analysis may be distorted. In this section, the procedures of the previous sections are modified to take inflation into account.

4.6.1 Terminology

The only new definitions needed are those of current dollars and constant dollars. *Current dollars* are the dollars that are actually spent at a given time. These dollars have a purchasing power that decreases with inflation. In contrast, *constant dollars* are dollars with a constant purchasing power. These dollars represent purchasing power at a particular time and are generally labeled by that time, for example, *1980 dollars*. With a fixed-rate mortgage it is possible to buy a house for a specific amount and to pay off the mortgage at a constant amount per month, thus paying off a

Year	One-time costs	NCF* before taxes	Taxes	Cash flow from deprec.	NCF after taxes	Discount factors	Discounted after-tax cash flow
0	-700.0	-700.0	0.0	0.0	-700.0	1.0000	-700.0
1	0.0	0.0	0.0	47.6	47.6	.8696	41.4
2	0.0	-60.0	20.4	76.2	36.6	.7561	27.7
3	0.0	390.0	-132.6	45.7	303.1	.6575	199.3
4	0.0	480.0	-163.2	27.4	344.2	.5718	196.8
5	59.7	647.7	-220.2	27.4	514.6	.4972	255.9

Present worth ($\times$ 1000):   21.1

*Net cash flow

Figure 4-7. Present worth calculations after taxes, without inflation.

loan in an amount that decreased in value each month. Therefore, the mortgage holders are paying back an amount originally calculated in constant dollars and doing so in current dollars. So long as the lenders can continue to borrow money at lower rates than the rates they charge for the money they are lending, this system is viable. When the minimum rate at which the lenders can borrow money rises to the level of the mortgage rates, however, the lending institutions are in trouble. This kind of problem led to some of the major financial difficulties and reforms of the early 1980s.

The relationship between a current dollar amount D_{curr} and an equivalent constant dollar amount D_{cons} is given by

$$D_{cons} = \frac{D_{curr}}{(1 + I)^n}$$

(4-7)

where n is the number of years between the year when the money is to be spent and the reference year for D_{cons} and where I is the applicable inflation rate. Note that Equation 4-7 is equivalent to Equation 4-4 except that I now represents the inflation rate instead of the annual discount rate.

4.6.2 Incorporating Inflation
Into Economic Studies

The method described in section 4.5.3 can be extended to problems involving inflation as follows:

Step 1. Calculate the current dollar before-tax cash flows as before, this time taking into account any predicted inflationary effects. In particular, you will need to obtain some estimate for the inflation rates of utilities, labor, and equipment and to include these rates explicitly in the estimation of current dollar costs.

Step 2. For each year, add the cash flows to obtain net current dollar cash flows before taxes.

Step 3.   Apply relevant tax factors and depreciation to determine after-tax cash flows, still in current dollars.

Step 4.   Use Equation 4-7, with the projected inflation rate in the consumer price index for the value of I, in order to calculate discount factors for each year. These factors, when multiplied by current dollar figures, can yield their constant dollar equivalents. If the company doing the analysis uses a current-dollar rate of return, however, this percentage should be used as I, and the sum of the cash flows discounted by I will be the present worth equivalent of this cash flow. In that case, step 5 is not needed.

Step 5.   Calculate for each year the discount factors corresponding to the chosen constant-dollar rate of return. For each year, multiply the discount factor calculated in step 4 by the new discount factor; the result is a set of factors that yield the discounted constant-dollar present worth for any after-tax cash flow.

Step 6.   Using the after-tax net cash flows and the discount factors from step 5, compute the discounted after-tax constant-dollar net cash flow.

4.6.3 Project evaluation with taxes and inflation

Example 4-6

As an example of the application of these steps, consider the following situation: A proposed energy management measure can be initiated now with full operation expected to start 1 year from now. At present prices, the electrical savings will be $30,000/year and the labor cost will be $10,000/year. The equipment will cost $50,000 and is in the 3-year tax life category. The equipment is expected to last 5 years. The present salvage value of comparable 5-year-old equipment is $10,000 and is increasing at 6%/year. The planning horizon is 5 years. Projected inflation rates are as follows: electricity, 20%; labor, 4%; general inflation rate, 4%. The applicable tax rate is 34%, the prescribed after-tax constant-dollar rate of return is 15%. Is the investment worthwhile? The calculations are shown in Figures 4-8, 4-9, and 4-10. Since the investment shows a positive present worth at the after-tax constant dollar rate of return of 15%, the investment is yielding at least that great a rate of return.

Year	Electrical Savings	Labor Costs	NCF
1	0	0	0
2	$36,000	–$10,400	$25,600
	$(= 30,000 \times 1.2)$	$(= 10,000 \times 1.04)$	
3	$43,200	–$10,816	$32,384
	$(= 30,000 \times 1.2^2)$	$(= 10,000 \times 1.04^2)$	
4	$51,840	–$11,249	$40,591
5	$62,208	–$11,699	$50,509

Figure 4-8. Calculation of current-dollar net cash flow
(excluding one-time costs).

Year	Current to constant dollars	Constant to present worth dollars	Summary discount rate
0	$1.0000 = 1/1.04^0$	$1.0000 = 1/1.15^0$	1.0000
1	$.9615 = 1/1.04^1$	$.8696 = 1/1.15^1$	.8361
2	$.9246 = 1/1.04^2$	$.7561 = 1/1.15^2$	.6991
3	.8890	.6575	.5845
4	.8548	.5718	.4887
5	.8219	.4972	.4086

Figure 4-9. Calculation of summary discount rate.

Year	One-time costs	NCF before taxes	Taxes	Cash flow from deprec.	NCF after taxes	Discount factors	Discounted after-tax cash flow
0	-50.0	-50.0	0.0	0.0	-50.0	1.0000	-50.0
1	-	0.0	0.0	16.7	16.7	.8361	13.9
2	-	25.6	-8.7	22.2	39.1	.6991	27.3
3	-	32.4	-11.0	7.4	28.8	.5845	16.8
4	-	40.6	-13.8	3.7	30.5	.4887	14.9
5	13.4	50.5	-17.2	0.0	33.3	.4086	13.6
	$(=10 \times 1.06^5)$					Present worth ($\times 1000$):	36.6

Figure 4-10 Present worth calculations after taxes, with inflation.

4.7 ENERGY FINANCING OPTIONS

Many facilities find it difficult to finance energy management opportunities even though the measures are clearly cost effective. However, there are a number of methods available for paying for the energy efficiency improvements, and the energy manager should know about them and be able to present the financing options to management at the same time as he or she presents the EMOs. The following is a brief description of some of the energy financing options and their costs.

4.7.1 In-house Capital

In-house capital or internal financing is the most common way that companies finance their energy efficiency measures. Under this method, the company uses funds available within the organization to finance a project or purchase the equipment under consideration. The cost of using these funds is what they could have produced if invested somewhere else or in another project. If a company is short of in-house capital, it should consider other financing methods for implementing cost-effective energy efficiency measures.

4.7.2 Utility Rebates/Incentives

Some of the electric utility companies offer rebates or other incentives to help their customers reduce the initial cost of energy efficiency improvements. These were discussed in Chapter Three and are summarized again here.

- **Rebates**—The rebate has been the most common utility incentive in recent years. The utility reimburses the company for a portion of the cost of implementing the energy efficiency measure. Rebates may be based on load reduction ($ per kW), or based on a fixed dollar amount for each energy-efficient product purchased ($ per item).

- **Direct Utility Assistance**—Some utilities pay the installing contractor part or all of the cost of the energy efficiency measure. Some other utilities provide the energy efficiency products or services directly to the customer through utility personnel or through contractors selected by the utility.

- **Low Interest Loans**—Some utilities offer low-interest financing for energy conservation projects. The loan payments may be added to the customer's utility bill.

4.7.3 Debt financing/loans

* **Commercial lending institutions:** If internal funds are not available or are limited to fund only a small percentage of the energy efficiency projects, one option to consider is borrowing money from a lending institution. If the loan payments are less than the monthly savings from the EMO, the company will still come out ahead. The cost of this financing method is the interest rate that the bank charges for lending the money. This cost must be included in the implementation cost analysis.

* **Government loans:** Some states make loans available to small businesses to help them reduce their energy consumption. For example, the State of Florida has a loan program called the Florida Energy Loan Program (FELP). To qualify for a FELP loan, a company must have 200 employees or less, and must have made a profit for the past two consecutive years. Eligible conservation measures include replacing equipment with more energy efficient models; renovation of the business (air conditioning ducts, relocation of thermostats, etc.) for the purpose of energy conservation; and building retrofits including replacement of windows and heating, ventilation, and air conditioning units. The term of the loan is for a maximum of 7 years. The monthly payment is approximately 75% of the average monthly energy savings. The interest rate varies: it is 5% for loans up to $5,000; 7% for $5,001 to $10,000; and 9% for $11,000 to $25,000.

4.7.4 Leases

"Leasing" is a familiar financing option. In energy project financing, the monthly lease payment will usually be less than the monthly avoided energy cost or energy savings; this will result in a positive cash flow situation. Sometimes, the value of the equipment and its installation is amortized over the term of the lease and the lease has a purchase option. In most cases, the term of the lease is shorter than the life of the equipment with the purchase option being the fair market value at the end of the lease.

* **Capital leases** are essentially installment purchases. Little or no initial capital outlay is required to purchase the equipment. The company is considered the owner of the equipment and may take deductions for depreciation and for the interest portion of payments to the lessor. Capital leasing is "on-balance-sheet" financing (recorded on the company's balance sheet as both a liability and an asset). Capital leases are offered by banks, leasing companies, installation contractors, suppliers and some utilities.

- **Operating leases** are another option. The lessor owns the equipment and leases it to the company for a monthly fee during the contract period. The lessor claims the tax benefits associated with the depreciation of the equipment. At the end of the contract term, the company can elect to purchase the equipment at fair market value (or for a predetermined amount), renegotiate the lease, or have the equipment removed. Shared savings financing offers many of the advantages of operating leases.

4.7.5   Performance Contracts
(Shared Savings, ESCO, Guaranteed Savings)

For companies who do not have funds available in-house, this energy financing option can be attractive. The basis for this type of financing is quite simple; the avoided energy cost resulting from the increased energy efficiency of the new equipment pays for the equipment. However, the company must be sure the cost-benefit analysis has been done correctly, and must check to see that the new equipment is being used properly.

Performance contracting can be simple if the project involves something like a straight-forward lighting retrofit where the cost savings can easily be calculated. In most cases, the cost and risk of the project is largely taken by the contractor, but the user must relinquish a certain amount of control. Also there are many of the factors to be considered which will require in-house expertise. Nevertheless, performance contracting is an option that should be considered carefully because of the benefits it can bring if done right.

- **Shared savings:** This option is particularly useful to businesses that do not have the capital to pursue needed energy projects. Like all energy financing options, this option is based on using the energy cost savings in the existing energy budget to pay for the project. The shared savings company is usually responsible for the engineering study, energy and economic evaluation, purchase and installation, and maintenance. The customer pays a percentage (50%) of the savings to the shared savings company for a specified period of time (5-10 years).

- **Performance contract for energy services**: This is a variation of a shared savings program. The energy service company (ESCO) contracts with the customer to provide certain energy services such as providing specific levels of heating, cooling, illumination, and use of equipment for other purposes. The ESCO then takes over the payment of the utility bills that are directly related to the energy services provided.

- **Insurance guarantee of energy savings:** This approach is promoted by some equipment manufacturers and contractors. It consists of an agreement to ensure that the periodic energy cost savings from the energy-efficient equipment will exceed an established minimum dollar value. Often, the minimum guaranteed savings value is set equal to the financing payment value for the same period in order to ensure a positive cash flow during the financing term. The company's risk is minimized with this option, but the cost increases because it includes the cost of an indirect insurance premium.

4.8 SUMMARY

This chapter has discussed the economic analysis method needed in order to evaluate projects which include cash flow components inflating at different rates. Taxes have also been included, as they must be in most economic analyses. This method will be used throughout the rest of this book.

REFERENCES

1. Phillip F. Ostwald, Cost *Estimating for Engineering and Management, Second Edition*, Prentice-Hall, Englewood Cliffs, N.J., 1984.
2. *Unit Price Standards Handbook*, Stock No. 008-047-00221-1, Washington, D.C., Nov. 1977 (also known as NAVFAC P-716, ARMY TB 420-33, and AIR FORCE 85-56).
3. Thomas F. Sack, *A Complete Guide to Building and Plant Maintenance*, Prentice-Hall, Englewood Cliffs, N.J., 1971.
4. L. Jackson Turvaville, "Current ROI Techniques Practiced by U.S. Companies," Proceedings of the 1990 International Industrial Engineering Conference, San Francisco, CA, May, 1990.
5. John A. White, Marvin H. Agee, and Kenneth E. Case, *Principles of Engineering Economic Analysis, Third Edition*, Wiley, New York, 1989.
6. Donald G. Newnan, *Engineering Economic Analysis, Third Edition*, Engineering Press, San Jose, Calif. 1988.

BIBLIOGRAPHY

Eugene L. Grant, W. Grant Ireson, and Richard S. Leavenworth, *Principles of Engineering Economy, Eighth Edition*, John Wiley and Sons, New York, NY. 1990.

Kenneth K. Humphreys, *Basic Cost Engineering, Second Edition*, Marcel Dekker, New York, NY, 1987.

Gideon Samid, *Computer-Organized Cost Engineering*, Marcel Dekker, New York, NY, 1990.

G.J. Thuesen and W.J. Fabrycky, *Engineering Economy, Eighth Edition*, Prentice-Hall, New York, NY, 1993.

Chapter 5
Lighting*

5.0. INTRODUCTION

The lighting system provides many opportunities for cost-effective energy savings with little or no inconvenience. In many cases, lighting can be improved and operation costs can be reduced at the same time. Lighting improvements are excellent investments in most commercial businesses because lighting accounts for a large part of the energy bill— ranging from 30-70% of the total energy cost. Lighting energy use represents only 5-25% of the total energy in industrial facilities, but it is usually cost-effective to address because lighting improvements are often easier to make than many process upgrades.

While there are significant energy-use and power-demand reductions available from lighting retrofits, the minimum lighting level standards of the Illuminating Engineering Society (IES) should be followed to insure worker productivity and safety. Inadequate lighting levels can decrease productivity, and they can also lead to a perception of poor indoor air quality.

Used as a starting place for an energy management program, lighting can attract immediate employee attention and participation, since everyone has ideas about lighting. Lighting is also seen as a barometer of the attitude of top managers toward energy management: if the office of the president of a company is an example of efficient lighting, then employees will see that energy management is taken seriously. A lighting retrofit program can be a win-win proposition for the business owner and the employees as it can improve morale, safety, and productivity while reducing life-cycle costs. This chapter provides a brief description of lighting systems, their characteristics, and retrofit options.

---

*This chapter was substantially revised and updated by Mr. Mark Spiller of Gainesville (Florida) Regional Utilities.

5.1 COMPONENTS OF THE LIGHTING SYSTEM

A lighting system consists of light sources (lamps), luminaires (or fixtures), and ballasts. Each component will affect the performance, energy use and annual operating cost of the lighting system. This section discusses each of these components, and provides the basic information on lighting technology needed to successfully accomplish lighting Energy Management Opportunities (EMOs).

5.1.1 Lamp Characteristics

Lamps are rated a number of different ways and each characteristic is a factor to consider in the lamp selection process. The basic ratings include: luminous efficacy (lumens/watt); color temperature (Kelvins); color rendering index (CRI); cost ($); rated life (operation hours); and labor required for relamping. Lamps should carry recognizable name brands and should be purchased from a reputable vendor. Some off-brand lamps, particularly those from some foreign countries, have low light output and short lives.

5.1.1.1 Luminous Efficacy

The luminous efficacy of a lamp is an estimate of the light output (lumens) divided by the electrical power input (watts) under test conditions. Lamps operating outside their design envelope may suffer reduced efficacy. For example, 34-watt energy-saving fluorescent lamps should not be used in environments with temperatures below 60°F; at lower temperatures, they are prone to flickering, their light output is low, and they have short lives. Table 5.1 presents lamp data on many commonly used lamps. More data is available from lamp manufacturers and the IES Lighting Handbook [1].

5.1.1.2 Color Temperature

The color temperature of a lamp describes the appearance of the light generated compared to the perceived color of a blackbody radiator at that temperature on the absolute temperature scale (i.e., Kelvin scale). For example, a daylight fluorescent lamp rated at 6300 Kelvins appears bluish, while a warm-white fluorescent lamp rated at 3000 Kelvins appears yellowish. Figure 5-1 shows the color temperatures of commonly used fluorescent lamps.

The energy manager should be sensitive to lamp color when recommending lighting changes and should not recommend changing lamp color unless nearly everyone is in favor of the proposed change. Understanding the color needs of the facility is important too. Some merchan-

Table 5.1. Light Source Characteristics [ref 2, Table 13.4]

	Incandescent, Including Tungsten Halogen	Fluorescent	High-Intensity Discharge			
			Mercury Vapor (Self-Ballasted)	Metal Halide	High-Pressure Sodium Improved Color	Low-Pressure Sodium
Wattages (lamp only)	15-1500	15-219	40-1000	175-1000	70-1000	35-180
Life[a] (hr)	750-12,000	7500-24,000	16,000-15,000	1500-15,000	24,000 (10,000)	18,000
Efficacy[a] (lumens/W) lamp only	15-25	55-100	50-60 (20-25)	80-100	75-140 (67-112)	Up to 180
Lumen maintenance	Fair to excellent	Fair to excellent	Very good (good)	Good	Excellent	Excellent
Color rendition	Excellent	Good to excellent	Poor to excellent	Very good	Fair (very good)	Poor
Light direction control	Very good to excellent	Fair	Very good	Very good	Very good	Fair
Source size	Compact	Extended	Compact	Compact	Compact	Extended
Relight time	Immediate	Immediate	3-10 min	10-20 min	Less than 1 min	Immediate
Comparative fixture cost	Low: simple fixtures	Moderate	Higher than incandescent and fluorescent	Generally higher than mercury	High	High
Comparative operating cost	High: short life and low efficiency	Lower than incandescent	Lower than incandescent	Lower than mercury	Lowest of HID types	Low
Auxiliary equipment needed	Not needed	Needed: medium cost	Needed: high cost	Needed: High cost	Needed: High cost	Needed: high cost

[a]Life and efficacy ratings subject to revision. Check manufacturers' data for latest information.

Figure 5-1. Fluorescent Lamp Color Temperatures [ref 1, Figure 5-17]

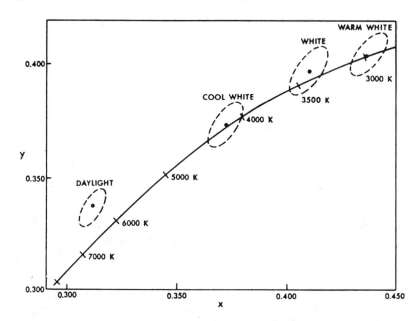

dise looks better at a particular color temperature. Example: meat in grocery store looks more appealing under warm lamps which accentuate the reds hues. The same meat in a meat-packing facility can be illuminated with lamps of a higher color temperature because visual appeal is not a factor.

5.1.1.3 Color Rendering Index

The color rendering index (CRI) is a relative indication of how well colors can be distinguished under the light produced by a lamp with a particular color temperature. The index runs from 0 to 1; where a high CRI indicates good color rendering. Many commonly-used lamps have poor CRIs (e.g., the CRI of a typical warm-white fluorescent lamp is 0.42, the CRI of a typical cool white fluorescent lamp is 0.67).

Rare earth elements in the phosphors of high-efficiency lamps increase light conversion efficacy and color rendition. Light sources such as metal halide lamps have better color rendering abilities than high pressure sodium lamps. The need for accurate color rendering depends on the particular task. Matching colors in a garment factory or photo laboratory will require a much higher CRI lamp than assembling large machine parts.

The color rendering abilities of commonly used lamps are described in Table 5.2.

Table 5.2 Color Rendition of Various Lamp Types

Incandescent, high-intensity discharge lamps

	Filament[a]	Clear Mercury	White Mercury	Deluxe White[b] Mercury	Multi-Vapor[b]	Lucalox[a]
Efficacy lm/w	Low	Medium	Medium	Medium	High	High
Lamp appearance effect on neutral surfaces	Yellowish white	Greenish blue-white	Greenish white	Purplish white	Greenish white	Yellowish
Effect on "atmosphere"	Warm	Very cool, greenish	Moderately cool, greenish	Warm, purplish	Moderately cool, greenish	Warm, yellowish
Colors strengthened	Red Orange Yellow	Yellow Green Blue	Yellow Green Blue	Red Yellow Blue	Yellow Green Blue	Yellow Orange Green
Colors grayed	Blue	Red, orange	Red, orange	Green	Red	Red, blue
Effect on complexions	Ruddiest	Greenish	Very pale	Ruddy	Grayed	Yellowish
Remarks	Good color rendering	Very poor color rendering	Moderate color rendering	Color acceptance similar to CW fluorescent	Color acceptance similar to CW fluorescent	Color acceptance approaches that of WW fluorescent

(*Continued*)

Table 5.2 Color Rendition of Various Lamp Types (*Continued*)

Fluorescent lamps

	Cool[b] White	Deluxe[b] Cool White	Warm[a] White	Deluxe[a] Warm White	Daylight	White	Soft White/Natural
Efficacy lm/w	High	Medium	High	Medium	Medium-high	High	Medium
Lamp appearance effect on neutral surfaces	White	White	Yellowish white	Yellowish white	Bluish white	Pale yellowish white	Purplish white
Effect on "atmosphere"	Neutral to moderately cool	Neutral to moderately cool	Warm	Warm	Very cool	Moderately warm	Warm pinkish
Colors strengthened	Orange Yellow Blue	All nearly equal	Orange Yellow	Red Orange Yellow Green	Green Blue	orange Yellow	Red Orange
Colors grayed	Red	None appreciably	Red, green blue	Blue	Red, orange	Red, green blue	Green, blue
Effect on complexions	Pale pink	Most natural	Sallow	Ruddy	Grayed	Pale	Ruddy pink
Remarks	Blends with natural daylight; good color acceptance	Best overall color rendition; simulates natural daylight	Blends with incandescent light; poor color acceptance	Good color rendition; simulates incandescent light	Usually replaceable with CW	Usually replaceable with CW or WW	Tinted source; usually replaceable with CWX or WWX

[a]Greater preference at lower levels.
[b]Greater preference at higher levels.
Source: General Electric Technical Pamphlet TP-I 19[3]. Reprinted with permission of General Electric Co.

5.1.2 Lamp Types

Lamps come in a variety of types and have a wide range of characteristics. Choosing the appropriate lamp type depends on the lighting task. Often, the least expensive lamp to buy is not the least expensive to operate. The energy manager should be familiar with the different options available for providing the desired lighting levels.

5.1.2.1 Incandescent Lamps

Incandescent lamps render colors well, are inexpensive to purchase, easily dimmed, small, and controllable which is useful for product display. However, they have relatively short lifespans, low efficacy and are susceptible to failure from heat and vibration. Incandescent lamps rated for long life or rough service have a correspondingly low efficacy. Some energy-saving lamps have a poorly supported filament and should not be used in environments with vibration or other mechanical stresses.

Most incandescent lamps tend to darken with age as tungsten is lost from the filament and deposited on the lamp walls. Figure 5-2 shows most of the commonly used lamps.

A Lamps, These lamps are low cost and commonly used in sizes of 20-1500 Watts. They project light out in all directions. In old industrial plants, look for large A-lamps in pendant fixtures where the lamps are left on most of the time. These are good candidates for replacement with HID lamps.

Reflector (R) Lamps. These lamps are usually more expensive than A-lamps and offer better control of the direction in which light is cast due to a reflective paint on the lamp wall. They have a focal point in back of the lamp, which results in the light from the lamp being dispersed broadly by the reflective surface of the lamp.

Ellipsoidal Reflector (ER) Lamps. These lamps cost about the same as R-lamps, but they are longer and have a focal point in front of the lamp. This location of the focal point results in the light being more concentrated as it leaves the lamp, and thus the beam is narrower than from an R lamp.

Quartz-Halogen Lamps. These lamps have a short life and low efficacy. They can be a good choice for areas which need lighting on an irregular basis. These lamps should not be cleaned.

Halogen Lamps. These lamps have a higher efficacy and cost than the lamps listed above and are available in many of the same lamp configurations.

Figure 5-2. Incandescent Bulb Shapes and Designations [ref 1, p. 8-5]

5.1.2.2 Fluorescent Lamps

Fluorescent lamps have high efficacy, long life, and low surface luminance; they are cool and are available in a variety of colors. Figure 5-3 shows many of the commonly used fluorescent lamps.

Figure 5-3. Fluorescent Lamp Shapes and Designations
[ref 8, Figure 2-14]

Typical Fluorescent Lamps. Fluorescent lamps are available in standard, high output (HO), and very high output (VHO) configurations. The HO and VHO lamps are useful for low-temperature environments and areas where a lot of light is needed with minimal lamp space.

Energy-Saving Lamps. Energy-saving fluorescent lamps which can replace the standard lamps can reduce power demand and energy use by about 15%. They will also decrease light levels about 3-10%. These lamps can only be used with ballasts designed and rated for energy-saving lamps and should not be used in areas in which the temperature falls below 60°F. Fixtures subject to direct discharge from air-conditioning vents are not good candidates for energy-saving fluorescent lamps. Figure 5-4 lists many of the energy-saving lamps available.

T-Measures for Lamps. The T-measure for a fluorescent lamp is the measure of the diameter of the lamp in eighths of an inch. Thus, a T12 lamp is twelve-eighths of an inch (or 1 1/2 inches) in diameter, while a T10 lamp is 1 1/4 inches in diameter.

T10 Lamps. T10 lamps typically contain phosphors which produce high efficacy and color rendition. They will operate on most ballasts designed for T12 lamps.

T8 Lamps with Electronic Ballasts. T8 lamps produce an efficacy of up to 100 lumens/watt, the highest efficacy of any fluorescent lamp. They will

Figure 5-4. Energy Saving Fluorescent Lamps [ref 1, Figure 8-115]

Lamp description	Lamp watts	Lamp watts replaced	Lamp current (A)	Lamp volts (V)	Lamp life† (h)	Nominal length (mm)	Nominal length (in.)	Base (end caps)	3000K RE70	3500K RE70	4100K RE70	3000K RE80	3500K RE80	4100K RE80	5000K RE80
Rapid start															
F17T8	17	—	0.265	70	20,000	610	24	Med. Bipin	1325	1325	1325	1375	1375	1375	—
F25T8	25	—	0.265	100	20,000	914	36	Med. Bipin	2125	2125	2125	2200	2200	2200	—
F32T8	32	—	0.265	137	20,000	1219	48	Med. Bipin	2850	2850	2850	2975	2975	2975	2700b
F40T8	40	—	0.265	172	20,000	1524	60	Med. Bipin	3600	3600	3600	3725	3725	3725	—
F40T12/U/3	36	40	—	84	12,000	610	24	Med. Bipin	2800b	2800b	2800b	—	—	—	—
F40T12/U/6	34	40	0.45	64	16,000	610	24	Med. Bipin	2090	2090	2025a	—	—	—	—
F30T12	25	30	0.453	73	18,000	914	36	Med. Bipin	—	—	—	—	—	—	—
F40T12	34 – 36 ‖	40	0.46	—	20,000	1219	48	Med. Bipin	2800	2800	2800	2880	2880	2880	—
F48T12/HO	55	60	—	126	12,000	1219	48	Reces. DC	3850a	4075	3850a	4400b	8500a	8600c	—
F96T12/HO	95	110	0.83	137	12,000	2438	96	Reces. DC	8430	8430	8430	8620	—	—	—
F96T12/1500	195	215	1.58	64	12,000	2438	96	Reces. DC	—	—	—	—	—	—	—
F48PG17	95	110	1.53	144	12,000	1219	48	Reces. DC	—	—	—	—	—	—	—
F96PG17	185	215	1.57	—	12,000	2438	96	Reces. DC	—	—	—	—	—	—	—
Preheat start															
F40T12	34	40	0.45	84	15,000	1219	48	Med. Bipin	—	—	—	—	—	—	—
F90T17	86	90	—	—	9,000	1524	60	Mog. Bipin	—	—	—	—	—	—	—
Instant start (Slimline)															
F48T12	30 – 32	38 – 40	—	—	9,000	1219	48	Single pin	2610	2610	2610	2700b	—	—	—
F96T8	40 – 41	50 – 51	—	—	7,500	2438	96	Single pin	—	—	—	—	—	—	—
F96T8b	56	—	0.26	267	15,000	2438	96	Single pin	—	—	—	5800	5800	5800	—
F96T12	60	75	0.44	153	12,000	2438	96	Single pin	5675	5675	5675	5850	5850	5850	—

Lamp description	Lamp watts‡	Lamp watts replaced	Nominal Lumens‡§											
			4150K CRI 60+ Cool white	3000K CRI 50+ Warm white	3470K CRI 60+ White	6380K CRI 70+ Daylight	5000K CRI 90+ C50	4160K CRI 80+ Deluxe cool white	2990K CRI 70+ Delux warm white (soft white)	4200K CRI 40+ Lite-white	3200K CRI 80+ Optima 32™·d	3570K CRI 80+ Natural	7500K CRI 90+ C75	5500K CRI 90+ Vita-lite·d
Rapid start														
F17T8	17	—	—	—	—	—	—	—	—	—	—	—	—	—
F25T8	25	—	—	—	—	—	—	—	—	—	—	—	—	—
F32T8	32	—	—	—	—	—	—	—	—	—	—	—	—	—
F40T8	40	—	—	—	—	—	—	—	—	—	—	—	—	—
F40T12/U/3	36	40	2350[a]	2425[a]	—	—	—	—	—	2500[a]	—	—	—	—
F40T12/U/6	34	40	2480	2530	2550	—	—	—	—	2620	—	—	—	—
F30T12	30	30	1975	2025	—	—	—	—	—	—	—	—	—	—
F40T12	34–36[b]	40	2670	2730	2700	2310	2010	1930	1925	2800[a]	2010	—	—	—
F48T12HO	55	60	—	—	—	—	—	—	—	3900[a]	—	—	—	—
F96T12HO	95	110	8020	8030	8000	6750[c]	—	5750	—	8400	—	—	—	6000[d]
F96T12/1500	188	215	13430	—	—	—	—	—	—	13,880	—	—	—	—
F48PG17	95	110	5700[a]	—	—	—	—	—	—	—	—	—	—	—
F96PG17	185	215	13,500[a]	—	—	—	—	—	—	14,100[a]	—	—	—	—
Preheat start														
F40T12	34	40	2700	—	—	—	—	—	—	—	—	—	—	—
F90T17	86	90	5765	—	—	—	—	—	—	—	—	—	—	—
Instant start (Slimline)														
F40T12	30.7	40	2475[a]	2575[a]	—	—	—	—	—	—	—	—	—	—
F96T8	40	50	3450[a]	—	—	—	—	—	—	2525	—	—	—	—
F96T12	60	75	5430	5570	5370	4730	4050	3950	3900	5670	4200	—	—	4015

* The life and light output ratings of fluorescent lamps are based on their use with ballasts that provide proper operating characteristics. Ballasts that do not provide proper electrical values may substantially reduce lamp life, light output or both.

† Rated life under specified test conditions at 3 hours per start. At longer burning intervals per start, longer life can be expected.

‡ "RE" indicates "RARE EARTH" type phosphors. This nomenclature has been developed by NEMA to define a system of color rendering information. RE 70 designates a CRI range of 70 - 79, RE 80 a range of 80 - 89, and RE 90 ≥ 90.

§ At 100 hours. When lamp is made by more than one manufacturer, light output is the average of all manufacturers submitting data.

‖ Also in 32 watt cathode-cutout but with reduced life.

[a] General Electric.
[b] Sylvania.
[c] Phillips.
[d] Duro-Test.

not generally operate on standard ballasts rated for T12 lamps. Because they are smaller than the T12 lamps, it is more difficult to replace them with the wrong lamps when they fail. They also use less of the toxic materials found in larger fluorescent lamps. T12 fixtures can be retrofitted with T8 lamps by using socket adapters and replacing the old ballasts with the T8 compatible ballasts.

Compact Fluorescent Lamps (CFLs). These twin-tube (TT) and double twin-tube (DTT) lamps are designed to replace many frequently used incandescent bulbs. These lamps can be used to reduce energy use and power demand by over 70%. For example, a 13-Watt fluorescent TT lamp can be used to replace a 60-Watt incandescent lamp. The light produced is similar in appearance to that of an incandescent lamp (i.e., color temperature of 2700 Kelvins for both). Since these lamps produce less heat, space cooling costs are also reduced. Many TT lamps/ballasts have a low power factor of about 0.5-0.6. A high power factor ballast with low harmonic distortion should be specified if a large number of these lamps is used.

Compact fluorescent lamps can be installed as a screw-in or hard-wired conversion kit and have a lifetime of at least 10,000 hours. Frequent cycles of short operation hours will significantly reduce lamp life. One-piece screw-in compact fluorescent lamps have higher life-cycle costs than two-piece screw-in models or the hard-wire models because the useful ballast life ends when the lamp burns out. Using the hard-wire kits will eliminate the possibility of an uninformed maintenance worker throwing away a two-piece screw-in conversion kit when the lamp fails. A number of facilities such as apartment buildings and motels find that screw-in compact fluorescent lamps are often stolen because they are easy to remove and are fairly expensive. Figure 5-5 lists the characteristics of some of the compact fluorescent lamps available.

5.1.2.3 High Intensity Discharge (HID) Lamps

These lamps are relatively expensive initially but offer low life-cycle costs due to long life and high efficacy. In general, the larger the HID lamp the higher the efficacy. The HID lamp efficacy can be affected by the lamp position, and some of these lamps have a significant color shift and loss of efficacy near the end of their rated life. Figure 5-6 shows some of the commonly used HID lamps.

Mercury Vapor Lamps. Mercury vapor lamps were the first HID lamps. They can offer good color rendering and low-to-moderate efficacy. Self-ballasted mercury vapor lamps are a direct replacement for large incan-

Figure 5-5. Compact Fluorescent Lamps [ref 1, Figure 8-117]

Generic designation NEMA	Lamp watts	Bulb type	Base	Rated life[†] (h)	Maximum overall length (mm)	Maximum overall length (in.)	Lamp current (A)	Lamp voltage (V)	Approx. initial lumens[‡]	Lumens per watt	Color[†] temperature and/or CRI
Twin tube 2700K, 3500K, 4100K, 5000K§ CRI 80 +											
CFT5W/G23	5	T-4	G23	7500	105	$4\frac{1}{4}$	0.180	38	250	50	82
CFT7W/G23	7	T-4	G23	10,000	135	$5\frac{7}{16}$	0.180	45	400	57	82
CFT9W/G23	9	T-4	G23	10,000	167	$6\frac{9}{16}$	0.180	59	600	67	82
CFT13W/GX23	13	T-4	GX23	10,000	191	$7\frac{1}{2}$	0.285	60	888	68	82
CFT5W/2G7	5	T-4	2G7	10,000	85	$3\frac{11}{32}$	0.180	35	250	50	82
CFT7W/2G7	7	T-4	2G7	10,000	115	$4\frac{17}{32}$	0.180	45	400	57	82
CFT7W/2G7	9	T-4	2G7	10,000	145	$5\frac{23}{32}$	0.180	59	600	67	82
CFT13W/2GX7	13	T-4	2GX7	10,000	175	$6\frac{29}{32}$	0.285	59	900	69	82
FT18W/2G11[2]	18	T-5	2G11	12,000	229	9	0.375	60	1250	69	82
FT18W/2G11RS	18	T-5	2G11	20,000	267	$10\frac{1}{2}$	0.250	76	1250	69	82
FT24W/2G11	24 – 27	T-5	2G11	12,000	328	$12\frac{29}{32}$	0.340	91	1800	69	82
FT36W/2G11	36 – 39	T-5	2G11	12,000	422	$16\frac{5}{8}$	0.430	111	2900	76	82
FT40W/2G11	40	T-5	2G11	20,000	574	$22\frac{19}{32}$	0.270	169	3150	79	82
FT50W/2G11	50	T-5	2G11	14,000	574	$22\frac{19}{32}$	0.43	147	4000	80	—
Quad 2700K, 3500K, 4100K, 5000K; CRI 80 +											
CFQ9W/G23	9	T-4	G23-2	10,000	111	$4\frac{3}{8}$	0.180	15	575	21	82
CFQ13W/GX23	13	T-4	GX23-2	10,000	125	$4\frac{29}{32}$	0.285	15	860	59	82
CFQ10W/G24d	10	T-4	G24d-1	10,000	118	$4\frac{5}{8}$	0.140	64	600	60	82
CFQ13W/G24d	13	T-4	G24d-1	10,000	152	6	0.170	91	900	69	82
CFQ18W/G24d	18	T-4	G24d-2	10,000	175	$6\frac{29}{32}$	0.220	100	1250	69	82
CFQ28W/G24d	26	T-4	G24d-3	10,000	196	$7\frac{23}{32}$	0.315	105	1800	69	82
CFQ15W/GX32d	15	T-5	GX32d-1	10,000	140	$5\frac{1}{2}$	—	60	900	60	—
CFQ22W/GX32d	20	T-5	GX32d-2	10,000	151	$5\frac{15}{16}$	—	53	1200	60	—
CFQ26W/GX32d	27	T-5	GX32d-3	10,000	173	$6\frac{13}{16}$	—	54	1600	59	—
CFQ10W/G24q	10	T-4	G24q-1	10,000	117	$4\frac{5}{8}$	0.190	64	600	60	82
CFQ13W/G24q	13	T-4	G24q-1	10,000	152	6	0.170	77	900	69	82
CFQ18W/G24q	18	T-4	G24q-2	10,000	173	$6\frac{13}{16}$	0.110	80	1250	69	82
CFQ26W/G24q	26	T-4	G24q-3	10,000	194	$7\frac{5}{8}$	0.158	80	1800	69	82

(Continued)

Figure 5-5. Compact Fluorescent Lamps [ref 1, Figure 8-117] (Continued)

Typical incandescent lamp substitutes (Compact fluorescent lamps, internal ballast)*

Generic designation	Ballast type	Incandescent equivalent (W)	Lamp watts	Bulb type	Base	Related life† (h)	Maximum overall length** (mm)	(in.)	Lamp current (A)	Lamp voltage (V)	Approx initial lumens‡ (Candle-power)	lumens per watt	Color* temperature and / or CRI
7 W³	Electronic	25	7	T-4	Med. Screw	10,000	140	$5\frac{1}{2}$	0.140	120	400	57	—
11 W³	Electronic	40	11	T-4	Med. Screw	10,000	140	$5\frac{1}{2}$	0.170	120	600	55	—
11 W reflector³	Electronic	50W R30	11	R-35	Med. Screw	10,000	148	$5\frac{13}{16}$	0.170	120	(315)	—	—
11 W globe³	Electronic	30	11	G-32	Med. Screw	10,000	168	$6\frac{9}{16}$	0.170	120	450	41	—
15 W¹,²,³	Electronic	60	15	T-4	Med. Screw	10,000	172	$6\frac{25}{32}$	0.240	120	900	60	81
15 W reflector³	Electronic	75W R40	15	RSB	Med. Screw	10,000	183	$7\frac{7}{32}$	0.240	120	(1335)	—	—
15 W globe³	Electronic	50	15	G-38	Med. Screw	10,000	188	$7\frac{13}{32}$	0.240	120	700	47	—
15 W globe²	Magnetic	40 - 60	18	G-30	Med. Screw	9000	160	$6\frac{5}{16}$	—	120	700	47	82
17 W decorative diffuser¹	Electronic	60	17	T-24	Med. Screw	10,000	149	$5\frac{7}{8}$	0.265	120	950	57	—
18 W decorative diffuse²	Electronic	75	18	T-24	Med. Screw	10,000	183	$7\frac{7}{32}$	0.240	120	1100	61	—
18 W reflector	Electronic	75W R40	18	R-40	Med. Screw	10,000	142	$5\frac{19}{32}$	0.240	120	800	44	—
18 W⁴	Electronic	75	18	T-4	Med. Screw	10,000	175	$6\frac{7}{8}$	—	120	1100	61	—
20 W¹,²,³	Electronic	75	20	T-4	Med. Screw	10,000	203	8	0.265††	120	1200	60	—
22 W circular lamp²	Magnetic	75	22	T-9	Med. Screw	10,000	130	$5\frac{1}{8}$	—	120	1200	54	52
23 W¹,³	Electronic	90	23	T-4	Med. Screw	10,000	176	$6\frac{5}{16}$	0.325	120	1550	67	—
26 W²	Electronic	90	26	T-5	Med. Screw	10,000	203	8	0.430	120	1500	58	—

(Continued)

Figure 5-5. Compact Fluorescent Lamps [ref 1, Figure 8-117] (Continued)

Square 2700 K and 3500 K²

F102D/827/4P	10	2D	GR10Q 4Pl	8000	94	$3\frac{20}{32}$	0.18	72	650	65	82
F162D/827/4P	16	2D	GR10Q 4Pl	8000	140	$5\frac{1}{2}$	0.195	103	1050	66	82
F212D/827/4P	21	2D	GR10Q 4Pl	8000	140	$5\frac{1}{2}$	0.26	101	1350	64	82
F282D/827/4P	28	2D	GR10Q 4Pl	10,000	203	8	0.32	107	2050	73	82
F382D/827/4P§§, || ||	38	2D	GR10Q 4Pl	10,000	203	8	0.43	110	2850	79	82

Note. Many Compact Fluorescent Lamps come equipped with normal (formally called low) power-factor ballasts, either integral or as auxiliaries. To minimize line losses and to maximize energy savings and lamp efficiency specifiers should require high power factor ballasts and should insist that luminaires, utilizing these lamps, be so equipped.
* These values are the averages of several manufacturers data.
† At 3 hours per start.
‡ Values for reflector lamps are in beam candlepower.
§ Not all color temperatures available from each manufacturer.
|| Also available in red, blue, green.
Screw base adaptors with integral ballasts are available for retrofitting twin and quad tube lamps into incandescent fixtures.
** Some lamps are available with shorter MOL's.
†† LLD's not yet available for these lamps
‡‡ Lamps are available with higher power factors and lower line currents.
§§ Add 2 watts for cathodes when operating on rapid-start circuits.
|| || Rapid start life is estimated as 12000 hours (a 3 or more hours per start.
¹ Philips
² General Electric
³ Osram
⁴ Sylvania

Figure 5-6. HID Lamp Shapes and Designations [ref 8, Figure 2-16]

descent lamps but have only 30-50% of the efficacy of typical mercury vapor lamps. Mercury vapor lamps are good candidates for replacement with more efficient light sources such as metal halide, high-pressure sodium, and low-pressure sodium.

Metal Halide lamps. Many of these lamps produce bright white light. They are used in applications which require good color rendering and high lighting levels such as sports facilities. They do not have as long a life as most other HID lamps. Some metal halide lamps can be installed in fixtures designed to operate mercury vapor lamps.

High Pressure Sodium (HPS) Lamps. HPS lamps offer high efficacy, long life, and a relatively small light source which is easily controlled. The light produced does not render colors well but is useful for tasks in which color rendering is not critical. These lamps are usually the best choice for warehouses, factories, exterior floodlighting, and streetlighting.

5.1.2.4 Low-Pressure Sodium (LPS) Lamps

LPS lamps offer the highest efficacy of any light source (i.e., up to 180 lumens/Watt). They have a long life but are fairly large compared to HID lamps. Their size requires a larger, more complex reflector design to efficiently utilize the light produced. The color of the light is nearly a monochromatic yellow under which very little color discrimination is possible. LPS lamps are common in European street lighting systems.

LPS is useful for interior security lighting. The distinctive yellow light tells law enforcement personnel that the premises should be unoccupied. Thieves and vandals can be mistaken for employees working late when better color-rendering light sources such as fluorescent lamps are used for night-time security.

5.1.3 Ballasts

Light from discharge-type lamps (e.g., fluorescent, mercury vapor, metal halide, high-pressure sodium, and low-pressure sodium) is produced indirectly by a cathode exciting a gas in which an electrical arc forms which then emits light. In a fluorescent lamp, mercury vapor emits ultra-violet radiation which strikes the sides of the lamp wall where phosphors convert it to visible light. A ballast is required to start and operate all discharge lamps.

Each lamp/ballast combination should have test results which state the ballast factor (BF). The lumen rating for a lamp is based on a particular lamp/ballast combination with a ballast factor of 1.0. The ballast factor indicates the light output of a particular system relative to a standard test ballast on which the lamp lumen ratings are based. For example, if a four-foot F32T8 has a rated lamp light output of 3000 lumens, and it is used with a ballast which has a ballast factor of 1.1 for the lamp/ballast combination, the system produces about 3300 lumens.

The ballast label usually gives the electric current drawn for particular lamp types, states whether the ballast does not contain polychlorinated biphenyls (PCB), and shows a wiring diagram. Many older ballasts, especially those manufactured before 1975, utilize PCB oil which contains some of the most toxic chemicals known. Leaking PCB-laden ballasts should be handled as a hazardous waste material during disposal. Ballasts which do not contain PCB can currently be landfilled in most states.

National standards require that all new ballasts have a minimum efficiency and contain no PCBs. Ballasts are usually labeled low-heat (LH), very-low heat (VLH), super-low heat (SLH), energy-saving, or electronic. Not all electronic ballasts on the market will operate at higher efficiency than magnetic ballasts. Claims of "full light output" can be distracting. Use the ballast factor to determine the actual efficiency of the ballast. Many of these new ballasts have much longer operating lives than the ballasts they replace. Some low-efficiency replacement ballasts are still available but should not be used.

Unlike magnetic core-and-coil ballasts, some electronic ballasts can be dimmed, but they also can produce significant levels of harmonic distortion. When recommending electronic ballasts, specify ballasts with a total harmonic distortion (THD) less than 20%.

Strobing and flickering lamps strain the ballast. Strobing occurs as the ballast attempts to restart and operate the lamp(s). The problem is sometimes due to a loose or corroded connection, but most often the lamp has come to the end of its useful life. As lamps age, their light output decreases and they become more difficult to start and operate. If checking the connections and lamps does not reveal the cause of the flickering, then ballast replacement may be necessary.

5.1.4 Luminaires (Fixtures): Lenses, Diffusers, and Reflectors

The luminaire is the complete lighting fixture. It consists of a housing socket, the light source (lamps) and the components which distribute the light such as the lens, the diffuser and the reflector.

The coefficient of utilization (CU) of a lighting fixture is the ratio of the light leaving the fixture to the light produced by the lamps. Light is absorbed and converted to heat by surfaces, by lamp-to-lamp interactions, and by the lenses. Tables of common CU values can be found in the IES Lighting Handbook [1].

<u>5.1.4.1 Types of Fixtures</u>

Some common types of lighting fixtures are listed below. Retrofit options are noted where appropriate.

Jars and Globes: These fixtures typically have low CU values. Although they generally use incandescent lamps, compact fluorescent lamps can also be used in jar and globe fixtures.

Wall Surface-Mounted: These fixtures have moderate CU values.

Pendant: These fixtures utilize a variety of lamp types. They often use bare lamps, but are sometimes fitted with globes, lenses, or diffusers. Large incandescent lamps (e.g., 500-1500 Watt) are frequently used, but more efficient sources are available. These fixtures often hang from the ceilings of production and warehouse areas. They can sometimes be lowered to increase the light levels at the work surfaces if the resulting light distribution is acceptable.

Track and Recessed: These fixtures (sometimes referred to as downlights) baffle some light when A-lamps or deeply-recessed R-lamps are used. Most of the baffled light is converted to heat. Ellipsoidal reflector lamps (ER) are designed to cast more light out of recessed fixtures. If an adjustable-depth can-type fixture without a specular reflector looks bright inside, there is potential for using an ER lamp. However, a protruding ER lamp is not a good application.

Although there are some excellent downlights available which have been designed for CFLs, in general a compact fluorescent screw-in lamp should not be used in a track or recessed fixture unless lighting levels can be reduced. A CFL casts nearly all of its light out to the sides instead of downward where it is needed in a downlight.

Area Lights: These fixtures are commonly known as barn lights. They cast light in nearly all directions. Much of the light is lost to the sky and overhead trees, or it trespasses off the property. Cutoff luminaires which use reflectors to direct the light where it is needed can produce the same light levels for the area of interest with about a third of the power.

Streetlights: Some types such as barnlights and cobraheads cast light upward and away from the area of interest. Cutoff luminaires offer a good alternative here too.

Figure 5-7 demonstrates the application of well-controlled lighting fixtures to reduce outdoor lighting costs an poorly applied fixtures.

Exit Signs: These signs typically use two 20-Watt or two 25-Watt incandescent lamps which can be replaced by a 7-Watt compact fluorescent TT lamp. This will reduce operating costs by more than 75%. These TT lamps can last up to 20,000 hours when run continuously.

Floodlights: Look for low-efficacy sources such as mercury vapor lamps. High pressure sodium floodlights offer excellent savings.

5.1.4.2 Lenses, Diffusers and Reflectors

Adding lenses or reflectors to a lighting fixture changes the light distribution pattern (i.e., photometrics). Replacing the lens or diffuser can be helpful in situations where the system was poorly designed or the use of the space has changed. For example, many office areas which were lighted appropriately for traditional paperwork now have glare on video terminals because people are using computers instead of typewriters or pens. Polarized lenses can reduce the glare but they also reduce the amount of light leaving the fixture. Figure 5-8 shows how relationships can change when the workstation is altered.

Glare from the specular surfaces reduces visibility. This veiling glare can be reduced with careful fixture positioning and proper selection of diffusers. Glare on video display terminals can often be alleviated by tipping the monitor down or putting a wedge under the back. Figure 5-9 shows how glare can reach the eyes of a worker.

Most fixtures have built-in reflectors. A fixture with a baked white enamel finish has a reflectivity of about 0.88. The practice of removing half the lamps and adding a reflector to a fixture will reduce the amount of

Figure 5-7. Effective Outdoor Lighting [ref 7, Figure 11]

The top two illustrations indicate effective ways of achieving desired lighting for the area involved (indicated by the dotted line). The bottom two approaches waste energy and the "light trespass" effect of the "spill light" can irritate neighbors.

Figure 5-8. Prevailing Lighting Relationships [ref 7, Figure 7]

Prevailing relationships between workers, their tasks, and their environments are altered significantly when traditional "white paper tasks" (on left) are changed to VDT-based tasks (on right).

Figure 5-9. Veiling Reflections [ref 7, Figure 3]

Veiling reflections commonly occur when the light source is directly above and in front of the viewer, in the "offending zone."

light leaving a fixture. Therefore, this is only practical when the area is over-lighted.

The performance of lenses, diffusers, and reflectors depends on environmental factors such as dirt accumulation, oxidation, glare, discoloration due to UV light exposure, and vandalism. Surface finishes react with airborne chemicals and lose some light transmission capability as they age. Choosing materials appropriate for the work environment and cleaning the fixtures regularly will prolong their useful lives. Outdoor lighting presents some special problems. Vandals often delight in shooting at exterior lighting fixtures. Polycarbonate lenses resist damage due to firearms and other projectiles. However, they also discolor from prolonged exposure to UV light produced by the lamp or the sun. Acrylic lenses typically last longer than even UV-inhibited polycarbonate lenses. Table 5-3 gives a comparison of the performance of acrylic and polycarbonate lenses over time.

5.2 DETERMINING LIGHTING NEEDS

A variety of techniques are available for estimating the lighting levels in a given space. The average illuminance method described in the Illuminating Engineering Society IES Lighting Handbook [1] incorporates the major variables affecting light utilization: amount of light produced by lamps, amount of light exiting the fixture, mounting height and spacing of fixture, fixture photometrics, lumen dirt depreciation, lamp lumen depreciation, ballast factor, and room surface finish characteristics. A worksheet to calculate lighting levels is also found in the IES Lighting Handbook.[1]

The IES has developed standards for appropriate lighting levels for typical applications. Lighting levels are generally expressed in terms of illuminance, which is measured in footcandles. Table 5.4 shows the illuminance category for a number of commercial and industrial applications.

Once the appropriate illuminance category has been identified, Table 5.5 can be used to determine the range of illuminance values needed to achieve desirable lighting levels. In using these tables, the IES recommends that the lower values be used for occupants whose age is under 40 and/or where the room reflectance is greater than 70% and that the higher values be used for occupants more than 55 years old and/or where the room reflectance is less than 30%. For occupants between 40 and 55 years of age and where the reflectance is 30-70% or where a young occupant is combined with low reflectance or an older person is in a high-reflectance environment, the intermediate values should be used. In addition, the

Table 5-3. Comparison of Acrylic and Polycarbonate Lenses
[ref 7, Table 4]

	Acrylic	High-Impact Acrylic	Polycarbonate
Light Transmission	92%	90%	88%
Aging-Light Stability	10-15 yrs	10-15 yrs	3-4 yrs
Impact Strength	1	10X	30X but degrades rapidly
Haze	Under 3%	Under 3%	Under 3% but degrades
Scratch Resistance	Very Good	Good	Good
Burning Character (U.L. Class)	Slow	Slow	Self-Ext.
Smoke Generation	Slight	Slight	High
Resistance to Heat	200°F	180°F	250°F
Type Smoke	Nontoxic	Nontoxic	Toxic
Relative Cost	1X	2X	4X

Table 5.4. Lighting Recommendations For Specific Tasks

Area-activity	Illuminance category
Bakeries	D
Classrooms	D to E
Conference rooms	D
Drafting rooms	E to F
Hotel lobbies	C to D
Home kitchens	D to E
Inspection, simple	D
Inspection, difficult	F
Inspection, exacting	H
Machine shops	D to H
Material handling	C to D
Storage, inactive	B
Storage, rough, bulky items	C
Storage, small items	D
	Footcandles
Building entrances	1-5
Bulletin boards, bright surroundings, dark surfaces	100
Bulletin boards, dark surroundings, bright surfaces	20
Boiler areas	2-5
Parking areas	1-2

Table 5.5. Illuminance Categories and Illuminance Values

Type of activity	Illuminance category	Ranges of illuminances		Reference work plane
		Lux	Footcandles	
Public spaces with dark surroundings	A	20-30-50	2-3-5	General lighting throughout spaces
Simple orientation for short temporary visits	B	50-75-100	5-7.5-10	
Working spaces where visual tasks are only occasionally performed	C	100-150-200	10-15-20	
Performance of visual tasks of high contrast or large size	D	200-300-500	20-30-50	
Performance of visual tasks of medium contrast or small size	E	500-750-1000	50-75-100	Illuminance on task
Performance of visual tasks of low contrast or very small size	F	1000-1500-2000	100-150-200	
Performance of visual tasks of low contrast and very small size over a prolonged period	G	2000-3000-5000	200-300-500	
Performance of very prolonged and exacting visual tasks	H	5000-7500-10000	500-750-1000	Illuminance on task, obtained by a combination of general and local (supplementary lighting)
Performance of very special visual tasks of extremely low contrast and small size	I	10000-15000-20000	1000-1500-2000	

need for speed and accuracy influences the amount of light needed, with higher speed and accuracy demanding more light.

Lighting levels can sometimes be reduced if there is sufficient contrast between the work and its surroundings. However, too much contrast between the work and the ambient environment will fatigue the eyes. For example, occupants of dark paneled offices who work with pencil on brightly-illuminated white paper will often experience eye fatigue.

The illuminance values considered acceptable have changed through the years in response to emerging technology. Recommended light levels were relatively low in the incandescent lamp era due to restrictions resulting from heat production. From 1945-1960 more efficient light sources were developed and energy costs plummeted, leading to a tripling of the recommended values [3]. Rising energy and power demand costs in the 1970's spurred the industry to reduce the recommended lighting levels to the minimum necessary to provide adequate illumination.

5.3 MAINTAINING THE LIGHTING SYSTEM

In addition to a proper choice of light sources, ballasts, and luminaires, the efficiency of a lighting system depends on maintenance policies. Maintenance includes both cleaning and relamping.

5.3.1 Luminaire Maintenance

The performance of lenses, diffusers, and reflectors depends on environmental factors such as dirt accumulation, oxidation, vandalism, and degradation due to ultra-violet (UV) light exposure. Typical fluorescent lamp performance under various temperature conditions is shown in Figure 5-10.

Lamps, fixtures, reflectors, lenses, and diffusers collect dust and insects. Dust accumulation on lighting fixtures and on surfaces adjacent to lighting fixtures reduces light utilization by up to 40 percent and increases heat production. Periodic cleaning of the fixtures will maintain higher and more uniform light levels. All lamps should be cool before cleaning. Gloves should be worn when cleaning any mirror-like reflective part of a luminaire. Quartz lamps should not be cleaned.

Outdoor lighting has some special maintenance problems. Poorly-sealed gaskets allow insects to clog the lenses of outdoor lighting fixtures. Dead insects can completely block out light. Overgrown vegetation can also reduce lighting levels from outside fixtures. Regular trimming of

Figure 5-10. Lamp Performance vs. Temperature

shrubs and trees will help to fully utilize these light sources. Small trees planted under or near lighting fixtures can quickly grow to block the light source.

5.3.2 Establishing the Lighting System Maintenance Schedule

Establishing a good maintenance schedule for a lighting system takes three steps. First, you must determine the maintenance characteristics of the luminaires in your facility. Table 5-6 lists the maintenance categories for a variety of luminaires.

The next step is to determine what dirt conditions the luminaires are likely to experience. Table 5-7 shows dirt conditions for representative areas. Once you know both the maintenance category from Table 5-5 and the appropriate dirt conditions for the facility from Table 5-7, then use the graphs in Figure 5-11 to set the luminaire maintenance schedule. These graphs show the effect that dirt accumulation has upon lighting levels over a period of months.

These graphs are for average conditions. Actual conditions may warrant a more frequent cleaning schedule than the graphs indicate. Using open fixtures and diffusers often reduces the potential for light loss since there are fewer surfaces for settling, and particles and dust can be

Table 5.6. Maintenance Categories for Luminaire Types

Maintenance category	Top enclosure	Bottom enclosure
I	1. None	1. None
II	1. None 2. Transparent with 15% or more uplight through apertures 3. Translucent with 15% or more uplight through apertures 4. Opaque with 15% or more uplight through apertures	1. None 2. Louvers or baffles
III	1. Transparent with less than 15% uplight through apertures 2. Translucent with less than 15% uplight through apertures 3. Opaque with less than 15% uplight through apertures	1. None
IV	1. Transparent unapertured 2. Translucent unapertured 3. Opaque unapertured	1. None 2. Louvers
V	1. Transparent unapertured 2. Translucent unapertured 3. Opaque unapertured	1. Transparent unapertured 2. Translucent unapertured
VI	1. None 2. Transparent unapertured 3. Translucent unapertured 4. Opaque unapertured	1. Transparent unapertured 2. Translucent unapertured 3. Opaque unapertured

Table 5-7. Degrees of Dirt Conditions

	Very clean	Clean	Medium	Dirty	Very dirty
Generated dirt	None	Very little	Noticeable but not heavy	Accumulates rapidly	Constant accumulation
Ambient dirt	None (or none enters area)	Some (almost none enters)	Some enters area	Large amount enters area	Almost none excluded
Removal or filtration	Excellent	Better than average	Poorer than average	Only fans or blowers if any	None
Adhesion	None	Slight	Enough to be visible after some months	High—probably due to oil, humidity, or static	High
Examples	High-grade offices, not near production; laboratories; clean rooms	Offices in older buildings or near production; light assembly; inspection	Mill offices; paper processing; light machining	Heat treating; high-speed printing; rubber processing	Similar to "Dirty" but luminaires within immediate area of contamination

Source: Courtesy of the Illuminating Engineering Society of North America.

Figure 5-11. Luminaire Dirt Depreciation Factors

[Fraction of full light output is the scale on the y-axis.] (Courtesy of the Illuminating Engineering Society [1])

more easily removed from these fixtures. However, an environment with potentially explosive dust; such as flour, corn, coal, etc., has to have sealed explosion-proof fixtures. Typical lamp lumen depreciation, luminaire dirt depreciation, and room surface dirt depreciation curves for fluorescent lamps are shown in Figure 5-12.

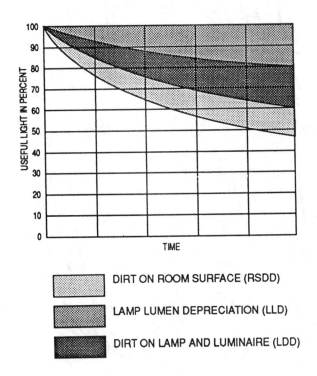

DIRT ON ROOM SURFACE (RSDD)

LAMP LUMEN DEPRECIATION (LLD)

DIRT ON LAMP AND LUMINAIRE (LDD)

Figure 5-12. Lumen Maintenance Curve
For Fluorescent Lighting Systems [ref 8, Figure 2-20]

5.3.3 Relamping strategies.

The usual strategy for replacing lamps in many facilities is to wait until a lamp burns out and then replace it (called spot relamping). This relamping strategy is not necessarily the best one for a facility to follow because it does not consider such factors as labor costs or lumen depreciation. It is often more economical to replace all of the fluorescent and HID lamps in a facility at one time (called group relamping). Spot relamping is more labor intensive and results in less efficient lighting than group relamping. However, spot relamping can be more practical for lamps with a short life such as incandescent lamps.

The reason that spot relamping results in a loss of lighting efficiency is that the amount of light that comes from a lamp declines with the age of the lamp. Performance curves, such as those in Figures 5-13, 5-14, and 5-15 show how the light output is reduced as a function of ordinary usage. Note that the life of a lamp is measured in hours of use rather than installed hours. In addition to degraded performance of individual lights, the total lighting system performance is decreased as individual lamps burn out. Typical mortality curves, showing the percent of lamps in service as a function of time, are given in Figures 5-16, 5-17, and 5-18.

As these figures show, lamp mortality at 85-100% of rated life is about twenty-five times that of lamps aged 0-70% of rated life. Therefore, fixtures should be group relamped when lamps are between 70-80% of the rated lamp life.

The average useful lamp life is shown in Figure 5-20.

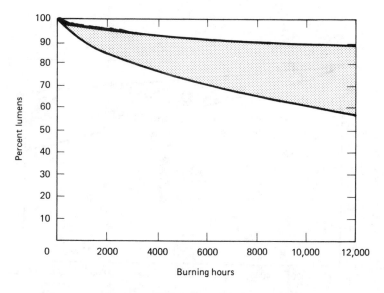

Figure 5-13. Typical lumen maintenance curve for fluorescent lamps. (From General Electric Technical Pamphlet TP-105. Courtesy of General Electric Co.)

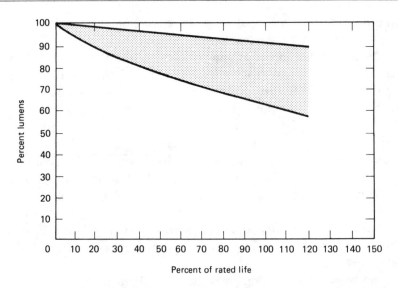

Figure 5-14. Typical lumen maintenance curve for filament lamps.
(From General Electric Technical Pamphlet TP-105.
Courtesy of General Electric Co.)

Figure 5-15. Lumen Maintenance Curve for HID Lamps
(From General Electric Technical Pamphlet TP-105.
Courtesy of General Electric Co.)

Figure 5-16. Typical mortality curve for fluorescent lamps.
(From General Electric Technical Pamphlet TP-105.
Courtesy of General Electric Co.)

Figure 5-17. Typical mortality curve for filament lamps.
(From General Electric Technical Pamphlet TP-105.
Courtesy of General Electric Co.)

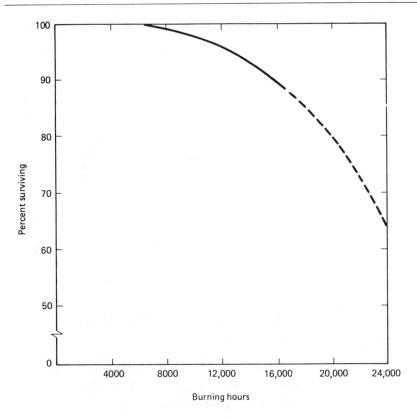

Figure 5-18. Typical mortality curve for HID (mercury vapor) lamps.
(From General Electric Technical Pamphlet TP-105.
Courtesy of General Electric Co.)

Since most F40T12 four-foot fluorescent lamps are rated for 20,000 hours, they should be replaced after 16,000 hours of operation. Lamp life ratings indicate the point at which half of the lamps are likely to have failed. Fluorescent lamp lifetimes are rated at three hours per start, HID lamps are rated at five hours per start. Operating the lamps for shorter periods will reduce lamp life.

Group relamping can:

1. **Reduce Labor Costs:** Spot relamping can require up to 30 minutes to move furnishings or equipment, set up, replace the lamp, and put equipment away (e.g., ladder, lamps, tools). In group relamping, a fixture can usually be relamped and cleaned in 5 minutes. The loss-of-use cost of replacing a few lamps before they have burned out is

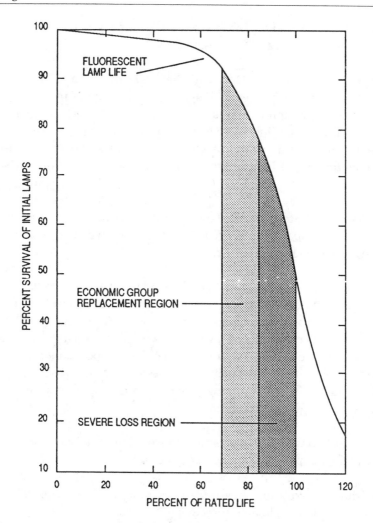

Figure 5-19. Lumen Maintenance Curve For Fluorescent Lighting
Systems [ref 8, Figure 2-20]

generally less than the increased cost of the labor to individually
replace burned out lamps.

2. **Reduce Lamp Costs:** Purchasing a large number of lamps at one time
    allows for high volume discounts. Fewer purchases result in less time
    spent ordering, receiving, and stocking lamps.

3. **Allow Lamp Maintenance to be Scheduled:** Relamping can be sched-
    uled for slow periods for the maintenance staff. Scheduled relamping

also allows a regular schedule to be set for regular inspection and cleaning of lamps and fixtures.

4. **Maintain Higher and More Uniform Lighting Levels:** The light output of a lamp decreases with age. Group relamping insures that all lamps have a high light output. Fewer lamp failures, less flickering, and reduced swirling and color shift produce a safer, more comfortable work environment. If the lighting system was over-designed to allow for loss of light levels, group relamping may allow for some delamping.

5. **Reduce Inventory Needs:** Fewer lamps must be stored in inventory since fewer spot failures occur in group relamped fixtures when the relamping interval is set correctly.

6. **Insure the Correct Lamp Use:** Spot relamping often results in the installation of a variety of lamp types with inconsistent light output levels, lifespans, and colors. This can occur when the inventory of spare lamps runs out and the person who purchases a quick replacement either does not know the correct lamp type, cannot locate the correct type, or does not realize that the cheapest lamp may not be the best value. Group relamping provides an opportunity to install the newest energy-efficient lamps.

7. **Extend Ballast Life:** Ballasts have to work harder to start and operate strobing lamps which are near the end of their life.

8. **Reduce Interruptions in Work Area:** Group relamping prevents most of the unplanned lamp replacements.

General Electric [4] has provided the following cost formulas for determining relamping costs:

Spot Replacement Costs
$$C = L + S \tag{5-1}$$
Group Relamping Costs

$$C = \frac{L + G}{I} \tag{5-2}$$

where: C = total replacement cost per lamp
          L = net price per lamp
          S = spot replacement labor cost per lamp
          G = group replacement labor cost per lamp
          I = group relamping interval (% of rated lamp life)

Example 5-1: An office building contains a number of small (400 ft^2) rooms, each of which has four two-lamp fluorescent fixtures. Every time a maintenance person changes lamps, they must bring a ladder into the room and clear away furniture. It takes the person 15 minutes to replace one lamp. It takes 25 minutes to replace all the lamps in a room and clean the luminaires if all the work is done at one time. The lamps cost $0.85 each, and labor costs are $10/hour. The lamps are used about 2000 hours/ year. The average lamp life is 20,000 hours. Determine whether group relamping with I = 0.8 is preferable to spot relamping for this building.

Solution: Find the cost for spot relamping and group relamping using equations 5-1 and 5-2, and select the method with lowest cost.

         L = $0.85 per lamp
         S =   (time to replace lamp in hours) × (cost of labor per hour)
            =   15/60 × $10 = $2.50
         G =   (time to replace lamp in hours) × (cost of labor per hour)
            =   3/60 × $10 = $0.50.
         I =   0.8
         and thus
         C$_{spot}$ = 0.85 + 2.50
                         = $3.35/lamp
         C$_{group}$ = (0.85 + 0.50)/0.80
                         = $1.69/lamp

Therefore, the decision should be made to use group relamping. The reduced labor cost more than offsets the cost of the lost lamp life; in fact, the savings from group relamping is almost twice the full cost of each lamp.

5.4 THE LIGHTING SURVEY

To perform an objective evaluation of the lighting system, the energy auditor must gather the following data: lighting needs/objectives, hours/days that lighting is required, lighting levels, type of lamps, age of lamps, age of the lighting installation, ambient environment of lighting fixtures (e.g., dust exposure, air temperature, etc.), room surface characteristics, type of ballasts, condition of fixtures, and relamping practices. Building plans are not useful unless the facility was built as planned and few modifications have been made.

Forms are useful for recording the specific lighting data needed. Chapter Two, Figure 2-5 provided a sample lighting data collection form. Figure 5-20 shows a sample data collection form for recording the lighting system condition, while Figure 5-21 illustrates a sample form for recording lighting needs.

Two basic surveys should be conducted to look for savings opportunities: one to see how the facility operates while in production and another to determine the lighting practices when the facility is dormant or shut down for the night.

5.4.1 Interviews with employees

Interviews with the managers and workers help the energy auditor to evaluate the relamping and maintenance practices, determine problems with the lighting system, ascertain the employees' satisfaction level, find out when light is needed, and uncover the potential for cost savings. The first question should be—Are you happy with your lighting? Major retrofits such as fixture replacement and color changes affect everyone in the work environment, so the opinions of all participants should be considered.

5.4.2 Measuring Light Levels

A light meter is needed to measure illuminance levels. An inexpensive analog light meter is practical and rugged for screening lighting levels and determining relative values. A self-calibrating digital light meter (photometer) is very useful to get fast, accurate, and repeatable measurements. Using a lightmeter, the auditor should record light intensity readings for each area and task of the facility.

Taking notes on the types of tasks performed in each area will help the auditor select alternative lighting technologies that might be more energy efficient. Other items to note are the areas that may be infrequently used and are potential candidates for controlling the lighting with occupancy sensors, or the areas where daylighting may be feasible.

Figure 5-20. Lighting System Condition Form

Location _____   Light type (HPS, FL, etc.) _____

Lamps _____

| Lighting level (fc) | No. | Watts | | % burning | Condition of system* | |
		Each	Total		Luminaires	Reflective surfaces
_____	—	—	—	—	————	————
_____	—	—	—	—	————	————
_____	—	—	—	—	————	————
_____	—	—	—	—	————	————
_____	—					

*From Table 5-7

Figure 5-21. Lighting needs form.

Location	Hours when light is needed	Importance of color rendition	Task lighting possible?	Light levels required (fc)
————	————	————	————	————
————	————	————	————	————
————	————	————	————	————
————	————	————	————	————
————	————	————	————	————

5.5 REGULATORY/SAFETY ISSUES

The lighting industry is encountering increasing safety and environmental concerns. Some of the materials used in lighting fixtures are, or will soon be, labeled hazardous for disposal.

For example, older ballasts and capacitors may contain polychlorinated biphenyl (PCB) oil which should be sent to a facility certified for handling hazardous wastes. Federal law 40CFR Part 761 requires proper disposal of leaking ballasts containing PCB oil. Fluorescent lamps contain mercury vapor (Hg), antimony (Sb), cadmium (Cd) and other toxic chemicals. The new T8 lamps (e.g., 1" diameter) use less phosphor material and mercury gas than the conventional T12 lamps (e.g., 1-1/2" diameter). Using T8 lamps should reduce disposal costs and environmental impacts.

5.5.1 Safety Issues

Lamps are fragile and break easily. Broken fluorescent lamps are difficult to transport and recycle. Areas subject to vibration or other mechanical stress should be illuminated with durable lamps or with fixtures which have adequate containment for broken lamps within the fixture housing. Delamping inexpensive fluorescent lighting fixtures can also be hazardous if the lamp pins come in contact with the fixture housing.

Insulation placed on top of lighting fixtures recessed in the ceiling may pose a fire hazard unless the fixture is rated for insulation contact. The insulation can increase the operating temperature of the lamps and ballasts of fluorescent or high intensity discharge (HID) fixtures, which will reduce the lifespan of all the system components.

High intensity discharge lamps have arc tubes which operate at high temperatures. Pieces of hot arc tube can fall from the fixture if the lamp wall is fractured. Some manufacturers recommend using lenses or fixture housings capable of containing incendiary materials.

5.5.2 Energy Policy Act of 1992

As part of the Federal Energy Policy Act of 1992, Congress placed restrictions on the production and sale of inefficient fluorescent and incandescent lamps [10]. These restrictions will require facility managers to take a hard look at the emerging high-efficiency lighting technologies when many of the lamps currently in use become difficult or impossible to find.

Standard cool-white and warm-white F40T12 lamps, the fluorescent lamps most often seen in commercial lighting fixtures, are specifically targeted by the Act. The manufacture and sale of low-efficacy F96T12 lamps and many reflector-type and PAR-type incandescent lamps are also restricted.

Table 5-8 from the Energy Policy Act of 1992 lists the minimum lamp efficacy and color rendering index requirements for fluorescent and incandescent lamps. Manufacturers must comply with these restrictions between 18 and 36 months after the date of the Act—which was October 24, 1992.

While lamps can be exempted from these restrictions for specialty applications, this law means that facility managers will be able to reduce life-cycle operation costs for lighting more easily. Since there are some gaps in suitable lamp replacements, particularly for incandescent downlights, the law will also spur research and development of new lighting technology. The Act will be evaluated and amended, if necessary, around the year 2000.

5.6 IDENTIFYING POTENTIAL EMOS

Lighting is used primarily for workplace illumination, for safety, and for decoration. In each of these uses, the same three questions can be asked: (1) How much light is needed? (2) How must the light be controlled? (3) How can lighting be provided most efficiently? When examining an existing system of lighting, the answers to these questions can be used to decrease lighting cost and improve lighting efficiency.

Lighting improvements provide cost savings in a number of ways:

Table 5-8. Criteria for Manufacturing Lamps

Fluorescent Lamps

Lamp Type	Nominal Lamp Wattage	Minimum CRI	Minimum Average Lamp Efficacy (LPW)	Effective Date (Months)
4-foot medium bi-pin	>35W	69	75.0	36
	<35W	45	75.0	36
2-foot U-shaped	>35W	69	68.0	36
	<35W	45	64.0	36
8-foot slimline	>65W	69	80.0	18
	<65W	45	80.0	18
8-foot high output	>100W	69	80.0	18
	<100W	45	80.0	18

Incandescent Reflector Lamps

Nominal Lamp Wattage	Minimum Average Lamp Efficacy (LPW)	Effective Date (Months)
40-50	10.5	36
51-66	11.0	36
67-85	12.5	36
86-115	14.0	36
116-155	14.5	36
156-205	15.0	36

reduced energy use and power demand; reduced heat production; lower life-cycle lamp costs; reduced need for maintenance; and increases in safety and productivity. In some instances, improving lighting quality increases worker productivity, and can result in greater profitability for a facility since the benefits of even a small change in worker productivity can vastly outweigh the relamping and maintenance costs. Figure 5-22 presents the results of a study of productivity in an office building when the lighting levels were first cut in half, then restored.

In examining the lighting system, the energy auditor should ask three questions: first, whether the light is needed; second, whether the correct amount of light is being used; and third, what is the most cost-effective lighting technology to use to supply the correct amount and quality of light. To locate the energy management opportunities, the auditor should specifically:

• Identify and characterize the visual tasks, and determine the contrast of the work to the surrounding surfaces.

• Look for the potential to use daylighting and task-specific lighting to displace high ambient, artificial lighting levels.

Productivity

At 100fc

At 50fc (a productivity loss of 28%)

Immediately after restoration of 100fc

Figure 5-22. Productivity Loss from Light Reduction [ref 6, Figure 2]

• Determine the appropriate lighting levels and the quality of light needed.

• Select alternative lighting systems to meet the needs, and analyze the cost-effectiveness of each alternative.

• Select the best alternative to implement.

The remainder of this section provides some specific recommendations for areas that can result in cost-effective improvements in lighting systems. The advice of a qualified lighting consultant should be solicited before undertaking any major lighting retrofit to ensure the task is completed with the best available technology and the lowest life-cycle costs.

5.6.1 Delamping

Major savings can be obtained by removing some of the lamps that are producing excessive levels of illumination. The lighting levels in many facilities have been over-designed to allow for poor maintenance and relamping practices. The first place to check for excessive light levels is in corridors and at work stations. The range of footcandles for each task was given in Table 5-1. Other places where lighting should be examined carefully include warehouses and storage areas.

There is often a good potential for delamping fixtures when planned group relamping is practiced. For example, a fluorescent light fixture with four new F40T12 40-watt lamps may provide over twice the actual illumination specified by IES standards for a particular task. In such a case, half of the lamps can often be removed while still providing the area with sufficient light levels and light distribution patterns. Hallways with 4-lamp fixtures frequently offer a good opportunity for removing half the lamps while maintaining adequate light levels.

If two lamps are removed from a four-lamp fixture, it is usually better to remove either the inboard or outboard set. Which set depends on the fixture design. Measuring the light levels after removing each set of lamps in turn will reveal which set should be removed. The best light distribution is typically achieved when the lamps are centered in the fixture. There are low-cost kits available for repositioning lamps. Although some reflector sales people advertise their product as capable of maintaining the same fixture light output with half the lamps, much of the increase in light levels is due to centering the lamps in the fixture.

There is usually one ballast for each two fluorescent lamps. The ballast for rapid-start lamps will continue to consume some power even when the lamps are removed. Disconnecting the ballast and capping the power leads will eliminate this power draw and provide a readily accessible replacement ballast already mounted in the fixture. The ballasts in instant-start (IS) fixtures automatically disconnect when the lamps are removed. Leaving burned-out lamps in IS fixtures reduces ballast life.

Example 5-2: The packing and shipping area in a plastic jar production facility is lighted with 50 fluorescent fixtures that each have four F40T12 34-watt lamps. It operates two shifts per day for 250 days a year. Light level measurements show that the average illumination level is about 110 footcandles. Is delamping warranted, and if so, how much can be saved if electricity costs 8 cents per kWh?

Solution: The IES illumination level standard (Tables 5-4 and 5-5) for packing and shipping—which are material handling tasks—is 50 footcandles as an upper limit for tasks involving large items. Thus, half of the lamps can be removed if the resulting light distribution pattern is still acceptable. Assume that each light fixture has two ballasts, and that each ballast serves two lamps. The two lamps and ballast removed will save 80 watts for the two lamps, and an additional 15%—or 12 watts—for the ballast. The total energy cost savings from this delamping can be calculated as follows:

$$\text{Cost savings} = (92 \text{ watts/fixture}) \times (50 \text{ fixtures}) \times (16 \text{ hours/day}) \times$$
$$(250 \text{ days/year}) \times (1 \text{ kWh/1000 wh}) \times (\$.08/\text{kWh})$$
$$= \underline{\$1472/\text{year}}$$

5.6.2 Task Lighting

Ambient lighting levels can be reduced when adequate task lighting is supplied for the work. Ambient lighting levels of 25 footcandles or less are frequently sufficient if the individual work areas have sufficient light from task-dedicated lighting fixtures. Figure 5-23 demonstrates the placement of supplementary luminaires.

5.6.3 Relamping

Replacing existing lamps, ballasts and luminaires with newer, more energy-efficient models offers the potential for significant savings on lighting system costs in many facilities. Lamp replacement is often a

Figure 5-23. Supplementary Luminaires [ref 1, Figure 9-11]. Examples of placement of supplementary luminaires: a. Luminaire located to prevent veiling reflections and reflected glare; reflected light does not coincide with angle of view. b. Reflected light coincides with angle of view. c. Low-angle lighting to emphasize surface irregularities. d. Large-area surface source and pattern are reflected toward the eye. e. Transillumination from diffuse source.

simple procedure if the new lamp works with the existing ballast and fixture. For example, replacing existing F40T12 40-watt lamps with F40T12 34-watt lamps in appropriate areas offers an easy and cost-effective lighting system improvement. Replacing existing F40T12 40-watt lamps with F32T8 32-watt lamps is a little more complicated as it requires a change of ballasts and lamp sockets in addition to the lamp change. This may still be very cost-effective, but it costs more for equipment and labor, and must be analyzed to see if it is truly cost-effective for a particular facility.

As another example, a 2'×4' lighting troffer (fixture) using four F40T12, 34-watt energy-saving lamps can be retrofitted with three F40T10 40-watt high-efficacy lamps or three F32T8 lamps and an electronic ballast. These alternatives will maintain or increase light levels and reduce energy use by about 14% and 35%, respectively.

There are many potential lamp substitutions possible, and it is important to know what kind of substitutions are reasonable. Table 5-9 shows some of the lamp substitutions which can produce lower life-cycle costs. Many other lamps are also suitable for replacing commonly used lamps.

Table 5-10 shows more possible lamp substitutions including some compact fluorescent lamps.

Table 5-9. Lamp Substitutions [ref 5, Table 5.8]

Present lamp	Substitute	Light level	Energy saved (w/%)
60-w In	30-w RI	100%	30/50
(1000 h)	50-w RI	200%	10/16
60-w In	50-w RI	200 + %	10/16
(2500 h)	55-w PAR/FL	200 + %	5/8
75-w In	55-w PAR/FL	150%	20/27
100-w In	75-w IR	125%	25/25
(750 h)	75-w PAR/FL	200 + %	25/25
	75-w ER	200 + %	25/25
100-w In	75-w PAR/FL	100 + %	25/25
(2500 h)			
150-w PAR/FL	100-w PAR/FL	70%	50/33
150-w R/FL	100-w PAR/FL	150%	50/33
200-w In	150-w PAR/FL	200 + %	50/33
(750 h)	(2000 h)		
30-w F	30-w EEL	87%	5/16
40-w F	40-w EEL	89%	6/15
96-w F	96-w EEL	91%	15/20
96-w F/HO	96-w EEL/HO	91%	15/14
96-w F/SHO	96-w EEL/SHO	90%	30/14
175-w MD	100-w HPS	104%	75/42
400-w MD	200-w HPS	96%	200/50
300-w In	150-w HPS	250%	120/50
750-w In	150-w HPS	104%	570/80
1000-w In	200-w HPS	93%	770/80

[a]Abbreviations:

EEL	energy-efficient fluorescent light (such as Watt-Mizer, Super-Saver, etc.)
ER	elliptical reflector (shape and inside coating of lamp)
F	fluorescent
FL	floodlight
HO	high output: 1000-ma filament
HPS	high-pressure sodium
h	hour (mean life expectancy)
In	incandescent
MD	mercury deluxe: mercury vapor corrected to improve color
PAR	parabolic aluminized reflector (see ER)
RI	reflective coated incandescent
SHO	superhigh output: 1500-ma filament
w	watt

Table 5-10. Lamp Substitutions (including TT)
[ref 7, Table 2]

Standard Lamp	Replacement Lamp	Wattage Savings[2]	Comparative Light Output of Replacement Lamp[3]	Value of Energy Savings over Life of Replacement Lamp at $0.08/kWh	Other Benefits[4]
60W Incandescent	55W Reduced-Wattage Incandescent	5	=	$0.40	
	13W TT Compact Fluorescent with Ballast Adapter	44.5	+	$35.60	x
75W Incandescent	70W Reduced-Wattage Incandescent	5	=	$0.40	
	22W Circline Fluorescent	45	=	$43.20	x
	18W Compact Fluorescent	57	=	$34.20	
100W Incandescent	95W Reduced-Wattage Incandescent	5	=	$0.40	
	44W Circline Fluorescent	56	=	$33.60	x
75W PAR-38 Spot or Flood Incandescent	65W PAR-38 Spot or Flood Incandescent	10	=	$1.60	
	45W Incandescent (Halogen)	30	=	$4.80	
150W R-40 Flood Incandescent[5]	75W ER-30 Incandescent[5]	75	=	$12.00	x
	120W ER-40 Incandescent[5]	30	++	$4.80	
150W PAR-38 Spot or Flood Incandescent	90W PAR-38 Spot or Flood Incandescent (Halogen)	60	=	$9.60	
	120W PAR-38 Incandescent	30	=	$4.80	
300W R-40 Flood Incandescent[5]	120W ER-40 Incandescent[5]	180	=	$28.80	

500W Incandescent	450W Self-Ballasted Mercury Vapor[6]	50	•	$64.00	
1,000W Incandescent	750W Self-Ballasted Mercury Vapor[6]	250	−	$320.00	
F-40 Fluorescent	F-40 Reduced-Wattage, High-Efficiency Fluorescent	7	=	$11.20	
	F-40 Reduced-Wattage, High-Efficiency Cathode-Disconnect Fluorescent	9.5	=	$15.20	
	F-40 Reduced-Wattage, High-Efficiency, Color-improved Fluorescent	7	•	$11.20	x
	F-40 Reduced-Wattage, High-Efficiency, Color-improved Cathode-Disconnect Fluorescent	9.5	•	$15.20	x
	F-40 High-Brightness Fluorescent	0	+	$0.00	x
F-40 Fluorescent (U-Shape)	F-40 Reduced-Wattage, High-Efficiency Fluorescent (U-Shape)	7	=	$11.20	
F-96 Fluorescent	F-96 Reduced-Wattage, High-Efficiency Fluorescent	17.5	•	$16.80	
F-96 HO Fluorescent	F-96 HO Reduced-Wattage, High-Efficiency Fluorescent	21	•	$20.20	
F-96 1,500 MA Fluorescent	F-96 1,500 MA Reduced-Wattage, High-Efficiency Fluorescent	25	•	$20.00	
175W Mercury Vapor	150W Retrofit High-Pressure Sodium	40	+ +	$ 38.40	x
250W Mercury Vapor	215W Retrofit High-Pressure Sodium	65	+ +	$62.40	x

(*Continued*)

Table 5-10. Lamp Substitutions (including TT) (*Conclusion*)

Standard Lamp	Replacement Lamp	Wattage Savings[2]	Comparative Light Output of Replacement Lamp[3]	Value of Energy Savings over Life of Replacement Lamp at 0.08/kWh	Other Benefits
400W Mercury Vapor	325 Retrofit Metal Halide	70	+ +	$112.00	x
	400W Retrofit Metal Halide	0	+ +	$ 0.00	x
	360W Retrofit High-Pressure Sodium	60	+ +	$76.80	x
1,000W Mercury Vapor	880W Retrofit High-Pressure Sodium	160	+ +	$204.80	x
	950W Retrofit Metal Halide	50	+ +	$48.00	x

NOTES

1. This table does not indicate all possible lamp replacement options and, in some cases, replacing the ballast and lamp, or relying on a new fixture, ballast and lamp will provide better overall performance and energy management than the replacement shown. All numbers reported in the table are approximations, and in certain cases assumptions are made about the types of fixtures and other conditions involved. Consult manufacturers for accurate data relative to direct replacements possible for a given installation as well as any ballast operating temperature or other restrictions which may apply.

2. Wattage savings include ballast losses, where applicable, assuming use of a standard ballast. Actual ballast losses to be experienced depend on the specific type of ballast involved and operating conditions which affect its performance. In those cases where wattage savings exceed the difference in lamp wattage (if any), operation of the replacement lamp also has the effect of reducing ballast losses.

3. Symbols used indicate the following: + + (substantially more) + (more) = (about the same) • (less) – (substantially less). Consult manufacturers for accurate information relative to conditions unique to the lamps and installations involved.

4. Other benefits typically provided by retrofit lamps include: maintenance costs due to longer lamp life; improved productivity, safety/ security, quality control, etc., due to higher light output; ability to reduce the number of lamps installed systemwide due to higher output of retrofit lamps, and improved color rendition.

5. When installed in a stack-baffled downlight.

6. For high voltages only.

Example 5-3: Calculate the annual savings from replacing 40-watt F40T12/Workshop lamps with 34-watt energy-saving lamps in two hundred (200) 4-lamp fixtures which are operated continuously. Assume the following:

The F40T12/Workshop lamps cost $1.00 each and last for 12,000 hours. The 34-Watt F40T12 lamps cost $1.50 each and last for 20,000 hours. Electric energy costs $0.05 per kWh. The demand charge is $5.50 per kW. The facility is not air conditioned.

Solution:
Annual Energy Savings (ES):

ES   =   (# of fixtures) × (# of lamps/fixture) × (wattage of low-efficiency lamps–wattage of high-efficiency lamps) × (annual operating hours)

=   (200 fixtures) × (4 lamps/fixture) × (40 - 34) watts/lamp × (8760 hours/yr)

=   800 lamps × 6 watts/lamp × 1 kW/1000 watts × 8760 hr/yr

=   42,048 kWh/yr

Energy Cost Savings (ECS):

ECS =   ES × (cost of electricity)

=   42,048 kWh/yr × $0.05/kWh

= $2102/yr

Demand Reduction (DR):

DR = (# of lamps) × (wattage reduction)

=   (800 lamps) × (6 watts) × 1 kilowatt/1000 watts

=   4.8 kW

Annual Demand Cost Savings (ADCS):

ADCS   =   DR × Demand Charge

=   4.8 kW × $5.50/kW/month × 12 months/yr

= $317/yr

Annual Lamp Cost Savings (ALCS):

The Workshop lights only have a lifetime of 12,000 hours, where the replacement lamps have a lifetime of 20,000 hours. Each of these costs must be annualized to determine the actual cost savings. The total number of lamp hours used in one year is found from multiplying the number of lamps times the hours of use in one year.

Total annual use = 800 × 8760 hr = 7,008,000 lamp hours

1.   Workshop light cost:

To compute the number of Workshop lights needed for one year, divide the total annual lamp hours needed by the life of one Workshop light.

Number of lights needed = 7,008,000/12,000 = 584 lamps

The annual cost is:

584 lamps/yr × $1.00/lamp = $584

2.   34 watt light cost:

Number of lights needed = 7,008,000/20,000 = 350 lamps

The annual cost is:

350 lamps/yr × $1.50/lamp = $525

ALCS = $584 - $525 = $59 per year

The total annual savings from this relamping EMO is the sum of all the savings calculated above.

Total annual cost savings = $2102 + $317 + $59 = $2478

Note that for air-conditioned facilities, the operating costs associated with lower wattage or more efficient lamps will be reduced because they produce less heat. Less air conditioning will be needed with this lower amount of heat produced. This will be discussed further in the air conditioning chapter in Section 6.4.1.5.

Lighting options should be compared to the existing system in a recently relamped and cleaned condition. A reasonable estimate of the lighting levels in the relamped system can be made from measuring lighting levels and noting the age/operation hours of the lamps. Some lamp sales representatives demonstrate how their lamp is superior to the lamps in use by relamping a fixture for comparison with adjacent fixtures. The results can be misleading due to light losses caused by lamp lumen depreciation (i.e., old lamps) and lumen dirt depreciation (i.e., the relamped fixture is usually wiped clean during lamp installation). Make sure both fixtures are clean and contain new lamps before comparing lamp alternatives.

5.6.4 Ballasts

Ballasts are an important part of a lighting system, and each ballast uses from five to twenty percent of the power of the lamp it is associated with. Furthermore, the ballast draws some power even if the lamp has been removed. Therefore, when a lamp is removed from a fixture, the ballast should usually be removed too. The ballast can be stored for future use, saving additional replacement costs. Ballasts should also be replaced if they overheat or smoke.

When older coil and core ballasts in a lighting fixture fail, replacement with an electronic ballast should be considered. New, electronic ballasts are much more efficient than the older magnetic ballasts, and offer desirable features such as dimming capabilities. When T8 fluorescent lamps are used, an electronic ballast is usually specified, too.

5.6.5 Control Technologies

Areas which are seldom occupied do not need constant light, yet lights are frequently left on in such places. Lights should not be left on in warehouses and storage areas unless the lights serve some function—illuminating storage areas for assistance in finding a product or reading labels, security, or other identified functions. Occupancy sensors have benefits to offer in these cases.

Fluorescent lamps should be turned off if they will not be used for five minutes or more. HID lamps should be turned off if they will not be used for about thirty minutes and quick restart time is not critical. HID lamps take up to fifteen minutes to regain full light output after restarting.

There are a number of cost-effective control EMOs that can be used to turn off lights that are not needed, or to utilize daylighting to supplement artificial lighting. These control technologies are discussed below, and additional discussion of controls is covered in chapter nine. Example 9-3 illustrates the savings from an application of occupancy sensors.

Switches: Many types of switches are available for controlling lighting. The simplest is the standard wall-mounted snap switch. Switches should be installed in the areas in which the fixtures are controlled. Rewiring to reduce the number of fixtures controlled by a single switch increases the ability of occupants to control the amount of lighting that is used. Installing switches next to one another frequently results in all the available lighting operating at once because people tend to turn on all the switches at once. If switches are installed next to each other, installing the switch upside-down that controls the least-needed lights will reduce the chance of that switch being turned on accidentally.

Other types of switches control lighting fixture operation on the basis of lighting levels, time, motion, or infrared radiation. Exterior lighting should be controlled by a light-sensitive switch. Photocells operate the lighting between dusk and dawn. They are available in various sensitivities. It is best to use photocells which turn the fixture on when they fail; this provides a good signal that replacement is necessary. The fail-off type can remain undetected and leave a facility without security lights.

Photocell input can also be used as a basis for controlling interior lighting. Some energy management control systems (EMCS) can use photocell input data for automatically adjusting indoor lighting levels to maintain a constant value when dimmable ballasts are used.

Timers: Timers can be used to control outdoor lighting but some are subject to inaccuracy due to seasonal changes in day-length, daylight savings time changes, clock slippage, power outages, and manual override. Adjustments should be made to simple timers about four times per year to prevent unnecessary operation of equipment. Timers can be used in conjunction with photosensors to reduce lighting costs if the lighting can be turned off before dawn.

Occupancy Sensors: Occupancy sensors can also be used to reduce unnecessary lighting use. Infrared sensors are directional and useful for active areas; ultrasonic sensors are fairly non-directional. The sensor's coverage of the area must be complete or nuisance cutoffs will occur and the occupants will remove the sensors.

Example 5-4: The prisoner holding cells in a courthouse utilize two fixtures with two F40T12 lamps each to illuminate an 8'×12' area. The cells are usually occupied a maximum of 45 minutes per day but they are illuminated about 12 hours per day. How much can you save by installing occupancy sensors in these cells? Calculate the simple payback period and return on investment if the sensors cost $70 installed and have a ten year life? Electricity costs 8 cents per kWh.

Solution: Assume that the occupancy sensor has a delay built into its operation, and that the lights will operate 1 hour a day, five days a week, 50 weeks a year. Further assume that each light fixture with its ballast consumes 80 watts for the two lamps, and an additional 15%—or 12 watts—for the ballast. Thus, the total savings from the occupancy sensor is found from:

Cost Savings = (2 fixtures/cell) × (92 watts/fixture) × (1 kW/1000 wh)
× (11 hours/day) ×
(5 days/week) × (50 weeks/year) × ($.08/kWh)

= $40.48/year per cell

SPP   = Installed cost/annual savings

= $70/$40.48/year

= 1.73 years

ROI   = [Annual savings – (cost/life)]/installed cost

= [$40.48 – ($70/10)]/$70

= $33.48/$70 = 47.8%

Dimmers: Dimmers are good for areas which require low ambient lighting levels most of the time with an occasional need for bright lighting. Solid-state dimmers operate by reducing the voltage supplied to the lamps. This reduces energy use and extends lamp life. However, fluorescent and HID lamps cannot be dimmed without dimmable ballasts. Rheostat dimmers are not recommended for any application because they produce considerable heat and do not save energy.

5.6.6 Other Lighting EMOs

5.6.6.1 Exterior Lighting

Exterior lighting is another area in which lighting energy is often wasted. In motels, for example, peripheral lighting is often left on both day and night. Such waste can be easily corrected with a timer or with a light switch turned on and off by a photocell. Each of the perimeter and outside lights should be carefully considered to see when it should be on, how much light is needed for the intended function, and whether more efficient lighting sources would work as well as those now being used.

5.6.6.2 Daylighting

Windows and skylights are often used to add light in a given area. One problem is that windows admit radiant heat as well as light, and it may be more expensive to remove the heat than to supply the light. In that case, the windows should be treated with exterior-mounted solar screens, louvers, or a reflective film with a low shading coefficient and a high percentage transmission of visible light. Daylighting is discussed in more detail in Chapter 13.

5.6.6.3 Environmental Factors

An area can appear to be dark if the walls, floors, or ceilings are painted (or otherwise covered) in dark colors or are greasy or dirty. Using light colors for paint and flooring, or cleaning these surfaces more often can make the existing light more effective and thereby save money.

5.6.7 Selecting Lights for a New Facility

Any time a new facility is built, or an existing facility is expanded, there is a significant opportunity to save on energy costs by selecting and installing cost-effective, energy-efficient lighting systems at the time of construction. It is almost always cheaper to install correct equipment the first time than to retrofit existing equipment. Greater first costs may produce significantly lower operating costs, and provide cost-effective savings for the facility. Unfortunately, many design decisions are made on the basis of first cost rather than life-cycle costs which include operation and maintenance costs. Utilizing life-cycle costing can assure the lowest lighting costs throughout the life of the lighting system.

Example 5-5. Gator Plastics Company is experiencing such a growth in demand for their products that they are planning on adding a new production room. Their Industrial Engineer is responsible for selecting the lighting system to be installed. The IE has identified two alternative lighting systems. Alternative One uses 40 fluorescent light fixtures with four 34-watt lamps in each fixture, together with a four-lamp ballast that consumes a total of 20 watts. This system costs $2000 to purchase and $2000 to install. Alternative Two uses 40 fixtures with three 40-watt T10 lamps, together with one three-lamp electronic ballast that consumes a total of 15 watts. This system costs $2400 to purchase and $2000 to install. If each lighting system lasts six years, the lights are used 2000 hours per year, electricity costs $0.08 per kWh, and the company's investment rate is 8%, which alternative should the IE select? Calculate the three standard economic evaluation measures for each alternative—SPP, ROI and B/C ratio.

Solution:
Alternative One: The operating cost is found as follows:

Annual cost = (40 fixtures) × (156 watts/fixture) × (2000 hours)
$\qquad$ × (1 kWh/1000 wh) × ($0.08/kWh)
$\qquad$ = $998.40

Alternative Two:

Annual cost = (40 fixtures) × (135 watts/fixture) × (2000 hours)
$\qquad$ × (1 kWh/1000 wh) × ($0.08/kWh)
$\qquad$ = $864.00

Economic evaluation: Alternative Two costs $400 more than Alternative One, but it saves ($998.40 - $864.00) = $134.40 per year in operating costs. To determine if this additional cost is a good investment, we need to calculate the three standard economic performance measures.

$\qquad$ SPP = Initial cost/Annual cost savings

$\qquad\qquad$ = $400/$134.40 = 3.0 years

$\qquad$ ROI = (Annual savings - Cost/Life)/Initial cost

$\qquad\qquad$ = ($134.40 - $400/6)/$400 = 16.9%

Using Equation 4-5 from Chapter 4 to find the present worth of the annual savings gives:

$$B/C = \{(Savings/yr) \times [(1 + I)^n - 1]/[I \times (1+I)^n]\}/\$400$$
Since $I = .08$ and $n = 6$,

$$B/C = \$134.40 \times \{[(1 +.08)6 - 1]/[.08 \times (1+.08)6]\}/\$400$$

$$= \$134.40 \times 4.623/\$400 = \underline{1.55}$$

Thus, for many companies the added cost of alternative two would be considered a good investment.

5.7 LIGHTING CHECKLIST

Lighting and maintenance costs can be reduced with a concurrent improvement in worker productivity, safety, and comfort. Figure 5-24 presents a checklist of energy-saving guidelines for lighting developed by the Illuminating Engineering Society.

5.8 EPA GREEN LIGHTS PROGRAM

In 1991, the U.S. Environmental Protection Agency (USEPA) initiated an innovative pollution prevention strategy with its Green Lights Program. Working in partnership with corporations and governmental entities, this program was designed to achieve voluntary reductions in energy use through the adoption of revolutionary lighting technologies.

The EPA has estimated that energy-efficient lighting can reduce lighting electricity demand by over 50 percent which translates into a cut of more than 10 percent in the nation's demand for electricity. This means that the power plants will burn less fuel. For every kilowatt-hour of electricity that is avoided, the EPA estimates that the country avoids the emission of 1.5 pounds of carbon dioxide, 5.8 grams of sulfur dioxide, and 2.5 grams of nitrogen oxides. Other pollution is reduced from mining and transporting power plant fuels and disposing of power plant wastes. As of January 31, 1993, the program had reduced lighting electricity consumption by over 100 million kilowatt-hours per year. By the year 2000, the Green Lights program is expected to save 226.4 billion kWh annually, resulting in total electricity demand savings of 39.8 million kilowatts.

To become a Green Lights Partner, an organization must sign a Memorandum of Understanding (MOU) with the EPA. In the MOU,

Figure 5-24. Checklist of Energy Saving Guidelines for Lighting Ideas
[ref 1, Figure 4-6] [ref 6, figure 2]

Lighting Needs

- **Visual tasks: specification** — Identify specific visual tasks and locations to determine recommended illuminances for tasks and for surrounding areas.

- **Safety and esthetics** — Review lighting requirements for given applications to satisfy safety and esthetic criteria.

- **Overlighted application** — In existing spaces, identify applications where maintained illumination is greater than recommended. Reduce energy by adjusting illuminance to meet recommended levels.

- **Groupings: similar visual tasks** — Group visual tasks having the same illuminance requirements, and avoid widely separated workstations.

- **Task lighting** — Illuminate work surfaces with luminaires properly located in or on furniture; provide lower ambient levels.

- **Luminance ratios** — Use wall washing and lighting of decorative objects to balance brightnesses.

Space Design and Utilization

- **Space plan** — When possible, arrange for occupants working after hours to work in close proximity to one another.

- **Room surfaces** — Use light colors for walls, floors, ceilings and furniture to increase utilization of light, and reduce connected lighting power to achieve required illuminances. Avoid glossy finishes on room and work surfaces to limit reflected glare.

- **Space utilization: branch circuit wiring** — Use modular branch circuit wiring to allow for flexibility in moving, relocating or adding luminaires to suit changing space configurations.

- **Space utilization: occupancy** — Light building for occupied periods only, and when required for security or cleaning purposes (see chapter 31, Lighting Controls).

(Continued)

Figure 5-24. Checklist of Energy Saving Lighting Ideas (*Continued*)

Daylighting
- Daylight compensation   If daylighting can be used to replace some electric lighting near fenestration during substantial periods of the day, lighting in those areas should be circuited so that it may be controlled manually or automatically by switching or dimming.

- Daylight sensing   Daylight sensors and dimming systems can reduce electric lighting energy.

- Daylight control   Maximize the effectiveness of existing fenestration-shading controls (interior and exterior) or replace with proper devices or shielding media.

- Space utilization   Use daylighting in transition zones, in lounge and recreational areas, and for functions where the variation in color, intensity and direction may be desirable. Consider applications where daylight can be utilized as ambient lighting, supplemented by local task lights.

Lighting Sources: Lamps and Ballasts
- Source efficacy   Install lamps with the highest efficacies to provide the desired light source color and distribution requirements.

- Fluorescent lamps   Use T8 fluorescent and high-wattage compact fluorescent systems for improved source efficacy and color quality.

- Ballasts   Use electronic or energy efficient ballasts with fluorescent lamps.

- HID   Use high-efficacy metal halide and high-pressure sodium light sources for exterior floodlighting.

- Incandescent   Where incandescent sources are necessary, use reflector halogen lamps for increased efficacy.

- Compact fluorescent   Use compact fluorescent lamps, where possible, to replace incandescent sources.

- Lamp wattage reduced-wattage lamps   In existing spaces, use reduced-wattage lamps where illuminance is too high but luminaire locations must be maintained for uniformity. *Caution:* These lamps are not recommended where the ambient space temperature may fall below 16°C (60°F).

- Control compatibility   If a control system is used, check compatibility of lamps and ballasts with the control device.

- System change   Substitute metal halide and high-pressure sodium systems for existing mercury vapor lighting systems.

Luminaires

- **Maintained efficiency**

  Select luminaires which do not collect dirt rapidly and which can be easily cleaned.

- **Improved maintenance**

  Improved maintenance procedures may enable a lighting system with reduced wattage to provide adequate illumination throughout system or component life.

- **Luminaire efficiency: replacement or relocation**

  Check luminaire effectiveness for task lighting and for overall efficiency; if ineffective or inefficient, consider replacement or relocation.

- **Heat removal**

  When luminaire temperatures exceed optimal system operating temperatures, consider using heat removal luminaires to improve lamp performance and reduce heat gain to the space. The decrease in lamp temperature may, however, actually increase power consumption.

- **Maintained efficiency**

  Select a lamp replacement schedule for all light sources, to more accurately predict light loss factors and possibly decrease the number of luminaires required.

Lighting Controls

- **Switching: local control**

  Install switches for local and convenient control of lighting by occupants. This should be in combination with a building-wide system to turn lights off when the building is unoccupied.

- **Selective switching**

  Install selective switching of luminaires according to groupings of working tasks and different working hours.

- **Low-voltage switching systems**

  Use low-voltage switching systems to obtain maximum switching capability.

- **Master control system**

  Use a programmable low-voltage master switching system for the entire building to turn lights on and off automatically as needed, with overrides at individual areas.

- **Multipurpose spaces**

  Install multicircuit switching or preset dimming controls to provide flexibility when spaces are used for multiple purposes and require different ranges of illuminance for various activities. Clearly label the control cover plates.

- **"Tuning" illuminance**

  Use switching and dimming systems as a means of adjusting illuminance for variable lighting requirements.

- **Scheduling**

  Operate lighting according to a predetermined schedule.

(Continued)

Figure 5-24. Checklist of Energy Saving Lighting Ideas (*Conclusion*)

- Occupant/motion sensors   Use occupant/motion sensors for unpredictable patterns of occupancy.

- Lumen maintenance   Fluorescent dimming systems may be utilized to maintain illuminance throughout lamp life, thereby saving energy by compensating for lamp-lumen depreciation and other light loss factors.

- Ballast switching   Use multilevel ballasts and local inboard-outboard lamp switching where a reduction in illuminances is sometimes desired.

Operation and Maintenance

- Education   Analyze lighting used during working and building cleaning periods, and institute an education program to have personnel turn off incandescent lamps promptly when the space is not in use, fluorescent lamps if the space will not be used for 5 min or longer, and HID lamps (mercury, metal halide, high-pressure sodium) if the space will not be used for 30 min or longer.

- Parking   Restrict parking after hours to specific lots so lighting can be reduced to minimum security requirements in unused parking areas.

- Custodial service   Schedule routine building cleaning during occupied hours.

- Reduced illuminance   Reduce illuminance during building cleaning periods if building is not otherwise occupied.

- Cleaning schedules   Adjust cleaning schedules to minimize time of operation, by concentrating cleaning activities in fewer spaces at the same time and by turning off lights in unoccupied areas.

- Program evaluation   Evaluate the present lighting maintenance program, and revise it as necessary to provide the most efficient use of the lighting system.

- Cleaning and maintenance   Clean luminaires and replace lamps on a regular maintenance schedule to ensure proper illuminance levels are maintained.

- Regular system checks   Check to see if all components are in good working condition. Transmitting or diffusing media should be examined, and badly discolored or deteriorated media replaced to improve efficiency.

- Renovation of luminaires   Replace outdated or damaged luminaires with modern ones which have good cleaning capabilities and which use lamps with higher efficacy and good lumen maintenance characteristics.

- Area maintenance   Trim trees and bushes that may be obstructing outdoor luminaire distribution and creating unwanted shadows.

Green Lights participants agree to survey their facilities and, within 5 years of signing the MOU, to upgrade lighting in 90 percent of their square footage, where it is profitable and where lighting quality is maintained or enhanced. Participants also agree to appoint an implementation manager who oversees participation in the program. As of March 1993, over 800 organizations had joined the Green Lights program.

The EPA provides the Green Lights partners with a number of products, information, and services:

- state-of-the-art computer software package that enables Partners to survey lighting systems in facilities, assess lighting options, and select the best energy-efficient upgrade.

- Green Lights bulletin board system which provides participants with software files to download to their own computer systems.

- user-friendly data bases of every third party financing program available.

- an Ally Program which induces lighting manufacturers, lighting management companies, and electric utilities that have agreed to educate customers about energy-efficient lighting.

- technical support including a technical services hotline, workshops, and a comprehensive Lighting Upgrade Manual.

- a National Lighting Product Information program which makes information about lighting available to members. This program has published information of electronic ballasts, reflectors, current limiters, occupancy sensors, compact fluorescent packages, and parking lot luminaires. It also publishes a Guide to Performance Evaluation of Efficient Lighting Products.

5.9 SUMMARY

The lighting system in a facility is an important area to examine and to improve in terms of energy efficiency and quality of light. This chapter has discussed the lighting system, described the components of the system, and provided suggestions for ways to improve the system. Lighting technology is changing at a rapid pace, and new lamps and ballasts are being developed and marketed almost daily. Major energy savings opportunities exist in most older lighting systems, and additional cost-effective savings is often possible in relatively new systems since technology is continually improving in this area.

REFERENCES

1. *IES Lighting Handbook, Application and Reference Volume*. Illuminating Engineering Society of North America, New York, 1993.

2. Harrold, Rita M., Lighting, Chapter 13, in *Energy Management Handbook, Second Edition*, Senior Editor W.C. Turner, Fairmont Press/Prentice Hall, Atlanta, GA, 1993.

3. Pansky, Stanley H., *Lighting Standards, Tracing the development of the lighting standard from 1939 to the Present*, Lighting Design & Application, Illuminating Engineering Society, February, 1985.

4. *Lighting Maintenance*, General Electric Technical Pamphlets TP-119 & TP-105, General Electric Co., Cleveland OH, Feb. 1969.

5. *General Electric Company Technical Pamphlet TP-105*, General Electric Company, Cleveland, OH.

6. *Getting the Most from Your Lighting Dollar, 2nd Edition* Pamphlet, National Lighting Bureau.

7. *Profiting from Lighting Modernization* Pamphlet, National Lighting Bureau, 1987.

8. *Electric Utility Guide To Marketing Efficient Lighting*, Western Area Power Administration & American Public Power Administration, 1990.

9. Griffin, C.W., *Energy Conservation in Building: Techniques for Economical Design*, The Construction Specifications Institute, 1974.

10. *Energy Policy Act of 1992*, P.L. 102-486 (H.R. 776), U.S. Congress, October 24, 1992.

BIBLIOGRAPHY
Advanced Lighting Guidelines: 1993, Electric Power Research Institute, EPRI TR-1010225, May, 1993.
Compact Fluorescent Lamps, Electric Power Research Institute, Palo Alto, CA, 1991.

Helms, Ronald N. and M. Clay Belcher, *Lighting for Energy Efficient Luminous Environments*, Prentice-Hall, Englewood Cliffs, NJ, 1991.

Hollister, D.D., Lawrence Berkeley Laboratory, Applied Sciences Division, "Overview of Advances in Light Sources." University of California, 1986.

Lighting Equipment and Accessories Directory, Illuminating Engineering Society of North America, March, 1993.

Lighting Handbook for Utilities, Electric Power Research Institute, Palo Alto, CA, 1993.

Lindsey, Jack L., *Applied Illumination Engineering*, Fairmont Press, Atlanta, GA, 1991.

Retrofit Lighting Technologies, Electric Power Research Institute, Palo Alto, CA, 1991.

Sorcar, P.C., *Rapid Lighting Design and Cost Estimating*. McGraw-Hill, Inc., 1979.

Thumann, Albert, *Lighting Efficiency Applications, Second Edition*, Fairmont Press, Atlanta, GA, 1992.

Verderber, R.R. and O. Morse, Lawrence Berkeley Laboratory, Applied Science Division. "Performance of Electronic Ballasts and Other New Lighting Equipment." University of California, 1988.

Verderber, R.R. and O. Morse, Lawrence Berkeley Laboratory, Applied Science Division. "Cost Effective Lighting," University of California, 1987.

Chapter 6

Heating, Ventilating, and Air Conditioning

6.0 INTRODUCTION

The heating, ventilating, and air conditioning (HVAC) system for a facility is the system of motors, ducts, fans, controls, and heat exchange units which delivers heated or cooled air to various parts of the facility. The purpose of the HVAC system is to add or remove heat and moisture and remove undesirable air components from the facility in order to maintain the desired environmental conditions for people, products or equipment. Providing acceptable indoor air quality is a critical function of the HVAC system, and air movement to remove odors, dust, pollen, etc. is necessary for comfort and health. It may also be necessary to air-condition an area to protect products or to meet unusual requirements such as those in a laboratory or a clean room.

The HVAC system is responsible for a significant portion of the energy use and energy cost in most residential and commercial buildings. Because many industrial facilities do not have heated or cooled production areas, HVAC energy use does not account for as great a portion of the total energy use for these facilities. However, a number of manufacturing plants are fully heated and air conditioned, and almost all industrial facilities have office areas that are heated and cooled. Thus, looking for ways to save on the energy costs of operating a facility's HVAC system is an important part of any energy management program.

Many facilities have HVAC systems that were designed and installed during periods of low energy costs; these are often relatively expensive to operate because energy efficiency was not a consideration in the initial selection of the system. In addition, many HVAC systems are designed to meet extreme load conditions of very hot or very cold weather; they are then poorly matched to the average conditions that are experienced most of the time. Thus, improving the operation of the HVAC system provides many opportunities to save energy and reduce costs. In this chapter we describe how an HVAC system works, discuss the major components of HVAC systems, analyze heating and cooling loads and ventilation requirements, and give methods for improving the energy efficiency of existing HVAC systems.

6.1 HOW AN HVAC SYSTEM WORKS

Air in a facility absorbs heat from lights, people, industrial processes, and the sun, and air conditioning removes the excess heat in order to provide a comfortable working environment. The air conditioning system also removes excess humidity. In periods of cold weather, the heating system adds heat if the working environment is too cold for worker comfort. During the heating season, moisture may be added to increase the humidity. The HVAC system also provides ventilation and air movement even when no heating or cooling load is present.

HVAC systems vary depending on the fluid that is used as a heat exchange medium (usually water or air), on the particular requirements for the system, and on the type of system that was in style when the building was originally built. All heating systems have certain components in common: a source of heat, some means for transferring the heat from the point of generation to the point of use, and a control system. The source of heat is usually a boiler, a furnace, or the sun. For cooling systems, the source of cold temperature is usually a chiller, although cold air can be supplied either as the exhaust air from a cold area or as cool air brought into the facility during periods when the outside air is at a lower temperature than the inside air. The heat or cold is usually transferred from a furnace, a boiler, or a chiller to the air, and this air is distributed to the points of use.

An alternative system may distribute heated or chilled water to the points of use where the water heats or cools the air to be blown into the room. The control system may be as simple as a thermostat that turns on a furnace when it senses room temperature below a preset level, or it may be very elaborate, controlling air volume, humidity, and temperature through monitoring inputs from many sensors and actuating valves, motors, and dampers.

6.1.1 Dual-Duct System Operation
One way to understand how HVAC systems work is to first learn how one system works and then to see other systems as variations of that system. One of the more widely found HVAC systems is the dual-duct

system, which is illustrated in Figure 6-1. Outside air is introduced through dampers (see Figure 6-1) which filter and control the amount of incoming air. This outside air is mixed with return air; the amount which

Figure 6-1 Dual-duct HVAC system.

is mixed is controlled by a return air damper. The air is then blown by a supply fan into both a hot duct and a cold duct. The air in the hot duct passes through heating coils and is sent at a preset temperature to the rooms where it is used. The air in the cool duct goes through conditioning coils and thence to the rooms; its temperature is set when it leaves the cooling coils. At each room, the cold and hot air go into a mixing box controlled by a thermostat; from the mixing box, the air with the desired temperature and moisture level enters the room. Later the air, with the heat, the cold, and the contaminants from the room, is removed through a return air grille and exhausted to the outside and/or returned as part of the intake air.

The dual-duct system has the advantage that it can accommodate widely differing demands for heating and cooling in different zones of a building by changing the ratio of hot to cold air in each zone. Its control system is also easily understood and relatively maintenance-free. The dual duct system has three main disadvantages. First, it requires much ductwork with attendant cost and space usage. Second, when the hot and cold ducts are next to each other, unproductive heat transfer takes place, and energy use and costs increase. Third, the use of energy to heat and cool simultaneously makes the dual-duct system a relatively inefficient system with respect to the energy required to perform the HVAC function.

Most of the original dual-duct systems in facilities have been modified so that the energy efficiency of the system is much better. Often this means that one of the ducts has been shut down so that heating and cooling are not provided at the same time for the entire system. This is an acceptable solution for some facilities with fairly uniform loads in all areas, and where moisture control is not a problem. Some of the systems have been modified to become variable air volume systems which are discussed in Section 6.1.4.1. Other systems use one duct for supplying heat or cooling to the core area of the facility, and perimeter fan-coil units are used to supply heat or cooling to the areas of the facility that are likely to need additional conditioned air.

6.1.2 Single-Duct, Terminal Reheat System Operation

A second type of system is a single-duct, terminal reheat system, and is illustrated in Figure 6-2. In this system, outside air enters through dampers, is mixed with return air in a mixing box or plenum, and is forced by a supply fan through a cooling unit. The air that has been cooled passes through a single supply duct to mixing boxes which contain a heating unit of some type—typically a hot water coil, and the air is then sent into a room. The return air system is similar to that described for the dual-duct system. When the source of heat for the reheat coil is a boiler, a common design fault is to have pumps continuously running water from the boiler

Figure 6-2 Terminal reheat HVAC system.

through the heating coil system. This uses electricity constantly for the pumps but avoids the thermal shock that might occur if cold water were injected into a warm boiler. A better alternative is to install a mixing valve at the boiler so that the pumps can be shut down when heating is not needed.

Many original terminal reheat systems used electric resistance heaters for the source of reheat. When electric energy costs greatly increased, most of these systems were modified to eliminate many of the terminal reheat units, and to use hot water coils and a gas or oil-fired boiler to supply the heat actually needed. Some of these terminal reheat systems were also modified to become variable air volume systems (see Section 6.1.4.1). Reheat systems are being used more now since ventilation standards have changed, and more outside air must now be brought into the facility.

6.1.3 System components

The typical components found in HVAC systems include dampers, grilles, filters, coils, fans, ductwork, and a control system. Each of these contributes to the operation of the HVAC system as follows:

Dampers. A damper controls a flow of air. If the damper is open, the air can flow unimpeded; if it is closed, the flow is reduced to 5-10% of open-damper flow, with the percentage dependent on the construction and maintenance of the damper. Dampers are usually used to regulate the flow of outside air into a system or to control the flow from one part of the system to another part, as in the case of a return air damper (A in Figure 6-2). In Figure 6-2, if the return air damper is closed and the outside air damper is open, all the heat (or cooling) in the return air is lost to the surrounding atmosphere. If the return air damper is open and the outside damper closed, then all the air is recirculated. Most HVAC systems operate somewhere between these two extremes, since some outside air must be supplied to buildings to meet health and safety code requirements.

Grilles. A screen—or grille—is usually placed upstream from a damper to catch bugs, lint, and debris before they go into the air distribution system. When a grille gets plugged or is blocked by objects such as furniture or stored material, this part of the air distribution system does not work. The authors saw a school where the temperature was consistently too hot for comfort although the air conditioning control system and other components seemed to be functioning properly. The grilles, however, were completely plugged with lint. When the grilles were cleaned, the problem disappeared.

Fans. Fans provide the power to move air through the air distribution system. A typical fan has three main parts: a motor, belts or a chain to transmit power from the motor to the fan blades (many small fans are direct drive), and the blades with their housing. If any one of these parts fails or is not connected properly, the fan will not move air, and this part of the air conditioning system will not work.

Heat Exchange Surfaces. The air that circulates through the HVAC system must usually be heated or cooled in order to be useful. This heating or cooling takes place when air is forced around a coil or finned surface containing hot or cold fluid. If these heat exchange surfaces are fouled with dirt, grease, or other materials with poor heat conduction properties, heat exchange will be inefficient, and more heat or cooling energy will have to be used in order to heat or cool air to the desired temperature.

Ductwork. Ductwork directs and conducts air from the heat exchange surfaces to the rooms where the hot or cold air is desired, and it conducts the exhaust air from these rooms back to the mixing plenum and to the outside. This function is impaired if the ducts leak or if loose insulation or other obstructions slow the airflow within the ducts.

Control System. An HVAC control system transforms the operating instructions for desired environmental conditions into the air temperatures and ventilation volumes desired in the working environment. The control system has the task of regulating the HVAC system so that these instructions are met as nearly as possible.

The control system accomplishes its function through a system of sensors, actuators, and communication links. The sensors send appropriate electrical or pneumatic signals when some temperature, pressure, or humidity threshold has been crossed. These signals are sent through a communication network, generally using either an electrical or pneumatic system. Signals can also be multiplexed through electrical supply lines, or radio transmission can be used. Upon reaching their destinations, the control signals are then translated into additional pneumatic or electrical signals that are used to open or close dampers, to regulate fans, and to initiate or stop the source of heating or cooling.

Dampers can be opened or closed to regulate the amount of incoming air, the amount of exhaust air that is mixed with fresh air, and the amount of air that is introduced into an area. Fans can be turned on or off to increase air coming into a room or being exhausted from it, or their speed can be regulated so that the amount of air coming into a room is no more than is needed for proper ventilation and temperature control. Boilers or chillers can be turned on to provide heat or cooling for the heat exchangers. Boiler controls are discussed in more detail in Chapter Eight.

6.1.4 Other HVAC System Types

Section 6.1.1 described the working of a dual-duct system, and Section 6.1.2 described the working of a terminal reheat system. Other common HVAC system types are variable air volume (VAV), fan coil, unit ventilator, induction, steam, and hot water systems. These systems are described briefly below because an energy auditor will invariably run into each of them at some time. For more detail, see References 6 and 7.

6.1.4.1 VAV systems.

In a variable air volume (VAV) system, tempered (heated or cooled) air is forced into a room at a rate dependent on the amount of heating or cooling desired. If the air volume needed is relatively constant, then a simple approach is to use dampers in VAV boxes installed in each duct opening where the air enters the room. If the air volume needed at different times changes significantly, the fan motor must have an adjustable speed drive so that the volume of air moved can be carefully controlled. If less heating or cooling is desired, less hot air is blown into the room. The advantage of this system is that only the amount of air needed is used, and, since power requirements vary as the cube of air volume moved, less air volume means less electrical consumption.

Some of the disadvantages of this system include the complexity and difficulty of maintaining the controls, and the need to use and control high-velocity airstreams. However, new control technologies have improved these problems, and this system is probably the most widely used HVAC system for installation in new buildings. Increased ventilation requirements for health and safety have resulted in more VAV systems needing fan motors with adjustable speed drives. Previously, VAV boxes in rooms could be shut down far enough to provide very low air flow rates; now this cannot be done, so the air flow must be more controllable to meet the required ventilation air flows. In addition, if moisture control is needed in a facility, it is necessary to have reheat to bring the temperature of the overcooled air back up to a comfortable temperature. Thus, even with the use of latest technology VAV systems it is still usually necessary to have reheat available.

6.1.4.2 Fan coil systems.

The fan coil system provides heat or cooling by using a fan to move room air across heating or cooling coils and back into the room. No outside air is introduced for ventilation, and no ductwork for outside air is needed. Air is usually distributed directly from the unit, and no ductwork is needed for supply air or return air. Control of the conditioned air is provided by varying the amount of heating or cooling fluid circulated through the coil and/or the fan speed.

6.1.4.3 Unit ventilators.

In this system, air is brought in directly from the outside and heated or cooled as it enters the room. A window air conditioner falls into this category as do many of the individual packaged room units in motels. The advantage of this system is that each room can be individually and easily controlled; the disadvantages are that installation costs are high and the occupants do not usually control temperatures so as to minimize energy consumption.

6.1.4.4 Induction units.

In these units, high-pressure supply air flows through nozzles and induces additional room air flow into the unit. This secondary air flows over heating or cooling coils and back into the room. This system provides both ventilation and heating or cooling at relatively low capital and energy costs. It also gives good local control of temperature. Its disadvantages are that its controls are complex and that each unit must be maintained regularly to keep it free of lint and dust.

6.1.4.5 Steam units.

In these systems, heat is produced from steam that condenses in radiators and is transferred either by fans or by natural convection. The condensate is then returned to the boiler where the steam was generated. The advantages of such systems include low initial and maintenance expenses for a multiroom installation. Disadvantages may include the need to operate the boiler when only a small part of the boiler design capacity is needed—for example, when only one or two rooms need heat and the boiler was designed to meet the needs of an entire building.

6.1.4.6 Water systems.

Water systems range from complex high-temperature units to the more familiar two-pipe units found in many old apartment buildings. In a typical water system, hot or cold water is pumped through coils and heats or cools air that is drawn around the coil by natural convection or by fans. In a two-pipe system, water enters the radiator through one pipe and leaves by another. In this system, complex valving is necessary to be able to change the system from heating to cooling, and the system operators must be skilled. In a four-pipe system, two pipes take and remove hot water and two take and remove chilled water, with the relative amounts of each depending on the amount of heating or cooling desired. The four-pipe system involves more plumbing than the two-pipe system but avoids the necessity for changing from hot to cold water throughout the system.

The main advantage of water systems is that they move a large amount of heating and cooling energy in return for a small amount of pumping energy; the amount of distribution energy per unit of heating or cooling is significantly less than that of an air system. The main disadvantage is the large amount of plumbing involved. Piping is expensive to buy and to install, and leaks in piping can cause far more expensive consequences than leaks in air-duct systems.

6.1.4.7 Heat pump systems.

A heat pump system is an HVAC system which uses the vapor-compression refrigeration cycle in a reverse mode. A heat pump system can move heat either to the inside or to the outside, so it can provide heating or cooling as the need arises. Single compressor systems up to about 25 to 30 tons are the most common. If a facility is going to be air conditioned, then the heat pump system is often a low cost system to provide the heating needed. In the moderate climate areas of the South and West, air-to-air heat pumps are very effective for heating. Use of water-to-air or ground-source heat pumps greatly expands the area where these systems are cost-effective.

6.2 PRODUCTION OF HOT AND COLD FLUIDS FOR HVAC SYSTEMS

6.2.1 Hot Fluids

Hot air, hot water and steam are produced using furnaces or boilers which are called primary conversion units. These furnaces and boilers can burn a fossil fuel such as natural gas, oil or coal, or use electricity to provide the primary heat which is then transferred into air or water. Direct production of hot air is accomplished by a furnace which takes the heat of combustion of fossil fuels or electric resistance heat, and transfers it to moving air. This hot air is then distributed by ductwork or by direct supply from the furnace to areas where it is needed.

Hot water is produced directly by a boiler which takes the heat from combustion of fossil fuels or electric resistance heat, and transfers it into moving water which is then distributed by pipes to areas where it is needed. A boiler might also be used to add more heat to the water to produce steam which is then distributed to its area of need. The combustion process and the operation of boilers and steam distribution systems are described in detail in Chapters Seven and Eight.

6.2.2 Cold Fluids

Cold air, cold water and other cold fluids such as glycol are produced by refrigeration units or by chillers, which are the primary conversion units. Refrigeration units or chillers commonly use either a vapor-compression cycle or an absorption cycle to provide the primary source of cooling which is then used to cool air, water or other fluids to be distributed to areas in which they are needed.

6.2.2.1 The basic vapor-compression cycle.

Room air conditioners and electrically powered central air conditioners with capacities up to 20 -30 tons (or up to 100 tons with multiple compressors) operate using the basic vapor-compression cycle which is illustrated in Figure 6-3. There are four main components in a refrigeration unit using the vapor-compression cycle: the compressor, the condenser, the expansion valve and the evaporator. There is also a working fluid which provides a material that experiences a phase change from liquid to gas and back in order to move heat from one component of the system to another. The working fluid is typically a chlorofluorocarbon or CFC, but these CFCs are being phased out because of the damage they cause to the ozone layer. Hydrofluorocarbons or HFCs, or Hydrogen-

Figure 6-3. Vapor compression refrigeration cycle.

CFCs or HCFCs are already in use in some systems, and are expected to serve as replacements for CFCs until totally ozone-safe working fluids are developed. The replacement of CFCs is discussed in more detail in Section 6.2.3.

We can start the description of the vapor-compression cycle operation at any point in the diagram, so let's start at the compressor. As the working fluid enters the compressor, it is in the state of a low temperature and low pressure vapor. After compression, the fluid becomes a high pressure, superheated vapor. This vapor then travels to the condenser which is a heat transfer coil that has outside air blowing through it. As the heat from the vapor is transferred to the outside air, the vapor cools and condenses to a liquid. This liquid then travels to the expansion valve where both its pressure and temperature are reduced. Next, the low pressure, low temperature liquid travels to the evaporator, which is another heat transfer coil. Air from inside the conditioned space is blown through the evaporator coil, and heat from this air is absorbed by the working fluid as it continues to expand to a low pressure and low temperature vapor as it passes through the evaporator. The inside air has now been cooled as a result of some of its heat content being absorbed by the evaporator, and this cool air can be distributed to the area where it is needed. Finally, the cycle repeats as this low pressure and low temperature vapor from the evaporator returns to the compressor.

Rapid heat transfer from the inside air to the evaporator coil, and from the condenser to the outside air is critical to the proper operation and energy efficiency of a refrigeration unit using the vapor-compression cycle. These coils must be kept clean to allow rapid heat transfer. If the coils are dirty or air flow is partially blocked because of physical obstructions or shrubbery, the refrigeration unit will not work as effectively or as efficiently as it should. Duct leakage is also a common reason for poor cooling or low air flow. Proper operation also requires the correct amount of working fluid in the system. If leaks have allowed some of the fluid to escape, then the system should be recharged to its rated level.

6.2.2.2 Chillers.

A typical chiller provides cold water or some cold fluid such as glycol which is supplied to areas where secondary units such as fan-coil units are used to provide the cooling that is needed at each location. Chillers have capacities that vary widely, from a few hundred tons to several thousand tons. The majority of chillers use either the vapor-compression cycle or the absorption cycle as the basic cooling mechanism, and have secondary fluid loops that reject the unwanted heat to the outside air or water, and provide the cold fluid to the areas where it is needed. The schematic diagram illustrated in Figure 6-4 is typical of a water chiller that is water cooled.

In Figure 6-4 the condenser cooling water is usually supplied by a closed loop that goes to a cooling tower. The cooling tower is an evaporative cooler that transfers the heat from the water to the outside air through the process of evaporation as the water is sprayed or falls through the air. If lake water or ground water were used in an open loop, the water would simply be supplied from one location and returned to a different location in the lake or in the ground.

The chilled water produced by the evaporator is circulated in another secondary closed loop to the parts of the facility where it will be used to provide air conditioning or process cooling. Individual fan coil units can be used in rooms, or centralized air handling units can be used to take a larger quantity of cooled air and distribute it to various parts of the facility. Part of the chilled water may be used to circulate through production machines such as plastic injection molding machines, welders, or metal treatment baths. Chilled water or other chilled fluids may be used to provide refrigeration or freezing capability for various types of food processing, such as meat packing or orange juice processing.

There are three types of mechanical compressors used in chillers. Small compressors used in chillers with capacities up to about 50 tons are almost always reciprocating compressors; they may also be used in chill-

Figure 6-4. Diagram of a typical chiller.

ers up to around 250 tons. Rotary screw compressors are used in chillers as small as 40-50 tons, and in large units up to and somewhat over 1000 tons. Centrifugal compressors are used in chillers as small as 75 tons, and in very large units up to and over 5000 tons.

6.2.3 Replacement of CFCs with alternative working fluids

Releases of many working fluids or refrigerants commonly used in HVAC systems can cause damage to the ozone layer in the earth's stratosphere [9]. The refrigerants degrade to release chlorine molecules which then proceed to break down ozone. Ozone, a molecule containing three oxygen atoms, is effective in filtering out ultra-violet (UV) radiation. The hole in the ozone layer over the Antarctic has allowed a large increase in the amount of UV radiation reaching the earth's surface. Sheep in southern Argentina received sunburns for the first time in recorded history when the ozone hole drifted northward. More importantly, the increase in UV radiation may damage marine crustaceans which form a vital link near the bottom of the food chain in the world's oceans.

The production of chlorinated fluorocarbons (CFC) used in refrigeration and air-conditioning systems will cease on January 1, 1996. Several of the most common refrigerants in use (i.e., R-11 and R-12) will no longer be manufactured. They will be available from recyclers for a short period of time but probably at high cost. Several alternative refrigerants under development may be suitable substitutes.

There are problems associated with most of the alternatives in many applications. The energy efficiency and capacity of the systems may be less than that of systems using CFC's. This will require more energy use and its associated pollution and production of greenhouse gases. Other problems include: incompatibility with commonly used lubricants and gasket materials, toxicity, and difficulty in retrofitting existing systems (e.g., replacement of sensors, motors, impellers, gears, etc.).

The current frontrunners in the field of substitutes are R-123 and R-245ca for R-11 and R-134a for R-12. Another short-term remedy is to use halogenated chlorofluorocarbons (HCFC) such as the commonly used R-22 or a mixture of R-22 with other compounds (i.e., near azeotropic compounds). HCFC compounds have less ozone depletion potential (ODP) than CFC compounds (i.e., ODP of R-11 is 1.0 by definition). For example, R-22 has an ODP of 0.05. HCFC production is slated to be phased out around 2030, but there is pressure to accelerate the timetable.

Large industrial refrigeration needs can be met with existing compounds such as ammonia or water/lithium bromide. These are well-developed technologies which predate the common use of CFC compounds for refrigeration.

Many facilities will replace their existing equipment with new systems containing refrigerants with low ODP and then recover the existing refrigerants for use in their other equipment.

6.3 ENERGY EFFICIENCY RATINGS FOR HVAC SYSTEM COMPONENTS

6.3.1 Boilers and furnaces

The efficiency of furnaces and boilers is specified in terms of the ratio of the output energy supplied to the input energy provided. This efficiency is shown in equation 6-1 below:

$$\text{Efficiency (\%)} = [\text{heat output}/\text{heat input}] \times 100 \qquad (6\text{-}1)$$

Efficiency specifications differ depending on where the heat output is measured. Combustion efficiency will be the highest efficiency number since it measures heat output at the furnace or boiler. Combustion efficiency can be measured with a stack gas analyzer, or it can be measured by determining the temperature and flow rate of the air or water from the furnace or boiler, and then calculating the heat output.

Furnace efficiencies can range from 65% to 85% for most standard furnaces, and up to 98% for pulse-combustion, condensing furnaces. Boiler efficiencies range from about 65% for older, smaller boilers to 85% for newer, larger models. Furnace and boiler efficiencies should be checked periodically, and tuned up to keep the efficiencies at their upper levels and reduce energy costs.

6.3.2 Air conditioners

The efficiencies of air conditioners are usually measured in terms of their Energy Efficiency Ratios or EERs, or their Seasonal Energy Efficiency Ratios or SEERs. In either case, the efficiency is specified as:

$$\text{EER} = \text{Btu of cooling}/(\text{watt-hours of electric energy input}) \qquad (6\text{-}2)$$

The EER value is measured at a single temperature for the outside air, while the SEER involves a weighted average of the EERs over a typical season with a range of outside temperatures. SEERs usually range from 0.5 to 1.0 units higher than the corresponding EERs. Air conditioning units with capacities of five tons or less are rated with SEERs, while units over five tons are rated with EERs. Current levels of SEERs for air conditioners can reach 15 or greater, but most units have SEERs around 10. Federal appliance efficiency standards and ASHRAE standards have increased in the last few years, and will increase further with standards set in the 1992 National Energy Policy Act [10].

Example 6-1: A five ton air conditioner has an average electric load of 8 kW. What is its SEER?

Solution: Using Equation 6-2 gives:

SEER $\quad$ = (5 tons) (12,000 Btu/hr/ton)/(8 kW)(1 hr) (1000 Wh/kWh)

$\qquad$ = 60,000 Btu/8,000 Wh

$\qquad$ = 7.5 Btu/Wh

6.3.3 Chillers

Chiller efficiency is usually measured in terms of a Coefficient of Performance (COP) which is expressed as:

$$\text{COP} = \text{heat absorbed by the evaporator}/[(\text{heat rejected by the condenser}) - (\text{heat absorbed by the evaporator})] \qquad (6\text{-}3)$$

Chiller efficiencies depend on what type of compressor is used in the chiller, and whether air or water cooling is used. COPs may be as low as 2.5 for small chillers, and up to 5.6 for large, water-cooled, centrifugal or screw compressor chillers. Absorption chillers have COPs that range from 0.4 to 1.2.

Chiller efficiencies may also be expressed as EERs, where

$$\text{EER} = \text{COP} \times 3.412 \text{ Btu/Wh} \qquad (6\text{-}4)$$

Example 6-2: A 100-ton chiller has a COP of 3.5. What is its electrical load?

Solution: Using Equation 6-2 and rearranging it gives:
$\qquad$ electrical load = cooling capacity/EER
$\qquad\qquad$ = (100 tons)(12,000 Btu/hr/ton)/(3.5)(3412 Btu/kWh)
$\qquad\qquad$ = 100.5 kW

Chiller efficiencies are greatly affected by the amount of load on the system. Part-load COPs are as much as one-fourth or one-fifth of the full-load COPs. Typically, a chiller must operate at 70% or greater load to be close to its full-load COP.

6.4 HEATING, COOLING, AND VENTILATING LOADS

One of the easier ways to reduce costs in HVAC systems is to reduce the amount of energy that must be added to or extracted from an area to bring the area to the desired temperature range. Two major strategies for accomplishing this are available: (1) reduce the heating or cooling load; (2) change the targeted temperature range. The amount of *cooling* needed in an area can be reduced by reducing the amount of heat brought into the area through the walls, by reducing the number of people present, by reducing the number of heat sources such as lighting, or by modifying the energy consumption characteristics of the industrial processes. The amount of *heating* needed can be reduced by increasing the amount of machinery located indoors, by capturing some of the heat from lights, by reducing infiltration of cold outside air, or by insulating roofs or ceilings so that less heat escapes. In any of these examples, the cooling or heating load is being changed.

The second strategy is to change the temperature range that is considered to be desirable. This means changing the temperature limit above which air cooling occurs—the upper set point, and the temperature limit below which heating occurs—the lower set point. If heating does not start until the temperature is 55°F or lower, less heat will be used than if the threshold temperature for heating is 65°F. Similarly, an upper control limit of 85°F is more economical than a limit of 75°F. Changing these temperatures has the effect of changing the heating or cooling load imposed upon the HVAC system, although none of the heat sources are changed.

6.4.1 Heating and Cooling Load Calculations

The heating and cooling loads in a building occur because of (1) heat given off by people; (2) radiant energy from the sun that enters through windows, is absorbed by furniture, walls, and equipment, within the building, and is later radiated as heat within the building; (3) heat conducted through the building envelope (walls, roofs, floors, and windows) to or from the environment around the building; (4) waste heat given off by processes and machinery within the building; (5) heat given off by lighting; and (6) heat or cooling lost to ventilation or infiltration air. In this section we emphasize managing the energy costs of an existing building by examining those aspects of the heating and cooling load that can be changed by moderate or low expense or by scheduling. Reference 1 can be used for design purposes because it contains a sufficiently detailed methodology for calculating these heat and cooling losses and their interactions.

6.4.1.1 Heating and cooling load: people.

People give off heat, and the amount they give off depends on the type of work they are doing, the temperature of their surroundings, and whether they are men or women. Table 6-1 gives representative values for the heat given off under various conditions. If no cooling or heating takes place during nonworking hours, the figures in Table 6-1 can be used directly. If cooling or heating takes place when the work force is not present, the later reradiation of heat given off by people and absorbed by equipment and surroundings must be taken into account as described in the ASHRAE Fundamentals Handbook [1].

Table 6-1. Rates of Heat Gain from People

Activity	Total heat gain for male adults (Btu/h)
Seated at rest	400
Seated, writing	480
Seated, typing	640
Standing, light work or slow walking	800
Light bench work	880
Normal walking, light machine work	1040
Heavy work, heavy machine work, lifting	1600

Note: Heat gain from adult females is assumed to be 85% of that for adult males.
Source: From *1993 Fundamentals Handbook,* ©1993. Reprinted with permission from the American Society of Heating, Refrigerating, and Air Conditioning Engineers, Atlanta, Ga.

People-generated heat can be managed in several ways. The first management technique is that of scheduling: decreasing the number of people in an air-conditioned area during the time of peak energy consumption will decrease the amount of heat that must be removed and thus will decrease this component of the peak demand. If the work of people can be scheduled when the outside temperature is lower than the inside temperature, it is often possible to remove people-generated heat by introducing colder outside air rather than by using mechanical refrigeration. Another technique is to remove people. This is accomplished by replacing people by automated equipment such as automated storage and retrieval systems. Removing the people decreases the cooling load and may completely eliminate the necessity for cooling (or heating) an area.

6.4.1.2 Heating and cooling load: solar radiation.

The cooling load due to solar radiation through windows can be calculated by

$$q = \Sigma \,(A \times SC \times MSHG \times CLF) \qquad\qquad (6\text{-}5)$$

where
- q = cooling load (Btu/hr)
- A = window area (square feet)
- SC = shading coefficient
- $MSHG$ = maximum solar heat gain (Btu/hr/ft^2)
- CLF = cooling load factor

The total cooling load is found by summing the cooling loads for all surfaces of the building envelope. In Equation 6-5, the window area A must be measured. The shading coefficient SC depends on the kind of glass, the indoor shading (roller blinds or venetian blinds), and the average outside wind velocity. For single-pane glass, typical values range from 0.53 for 1/2-inch heat-absorbing glass to 1.00 for 1/8-inch clear glass; a white roller shade inside a clear 3/8-inch glass pane gives a shading coefficient of 0.25. Additional values of the shading coefficient can be obtained from glass vendors and from Reference 1. Changing glass types and installing shades are energy management measures that can make a significant change in the shading coefficient. Adding sun screens can reduce the shading coefficient to about 0.2, and using reflective film on windows can provide a shading coefficient of around 0.4.

The maximum solar heat gain and the cooling load factors, however, are fixed by the location and design of the structure. The maximum solar heat gain (MSHG) depends on the month, the hour, the latitude, and the direction the surface is facing. For 40° north latitude (Denver and Salt Lake City), these factors are shown in Table 6-2. The cooling load factor (CLF) measures the fraction of energy absorbed at a given time that is radiated as heat at a later time. This factor, like the shading coefficient, depends on the interior shading. Representative values for this factor are given in Table 6-3 for a building with medium-weight construction (approximately 70 lb of building material per square foot of floor area). Additional values for these factors can be found in Reference 1.

Table 6-2. Maximum Solar Heat Gain Factors
(Btu/h • ft² for 40° North Latitude)

			Surface orientation			
	N	NE/NW	E/W	SE/SW	S	Horizontal
Jan.	20	20	154	241	254	133
Feb.	24	50	186	246	241	180
March	29	93	218	236	206	223
April	34	140	224	203	154	252
May	37	165	220	175	113	265
June	48	172	216	161	95	267
July	38	163	216	170	109	262
Aug.	35	135	216	196	149	247
Sept.	30	87	203	226	200	215
Oct.	25	49	180	238	234	177
Nov.	20	20	151	237	250	132
Dec.	18	18	135	232	253	113

Source: From Reference 1. Reprinted with permission from the American Society of Heating, Refrigerating, and Air Conditioning Engineers, Atlanta, Ga.

Equation 6-5 shows what factors can be altered to manage the solar component of the cooling load. For buildings in a hot climate, it is often cost effective to cover all outside windows and thus reduce the area A through which solar radiation enters the building. If, however, heating is more of an expense than cooling, it may be possible to increase the area and use passive solar heating.* The shading coefficient (SC) can be changed by installing shades, heat-absorbing glass, or reflective coatings. The maximum solar heat gain (MSHG) can be used to help decide where to locate people and equipment within a building relative to the location of windows. More heat will come in if the windows are facing south than

---

*For more detail on the possibilities of this approach, see Reference 2.

Table 6-3. Cooling Load Factors for Glass

Solar time	Direction window is facing								
	N	NE	E	SE	S	SW	W	NW	Hor.
Without Interior Shading									
2 a.m.	.20	.06	.06	.08	.11	.13	.13	.12	.14
4	.16	.05	.05	.06	.08	.10	.10	.09	.11
6	.34	.21	.18	.14	.08	.09	.09	.09	.11
8	.46	.44	.44	.38	.14	.12	.10	.11	.24
10	.59	.40	.51	.54	.31	.15	.12	.14	.43
12	.70	.33	.39	.51	.52	.23	.14	.17	.59
2 p.m.	.75	.30	.32	.40	.58	.44	.29	.21	.67
4	.74	.26	.26	.33	.47	.58	.50	.42	.62
6	.79	.21	.21	.25	.36	.53	.55	.53	.47
8	.50	.15	.15	.18	.25	.33	.33	.32	.32
10	.36	.11	.11	.14	.18	.24	.23	.22	.24
12	.27	.08	.08	.10	.14	.18	.17	.16	.18
With Interior Shading									
2 a.m.	.07	.02	.02	.03	.04	.05	.05	.04	.05
4	.06	.02	.02	.02	.03	.04	.04	.03	.04
6	.73	.56	.47	.30	.09	.07	.06	.07	.12
8	.65	.74	.80	.74	.22	.14	.11	.14	.44
10	.80	.37	.62	.79	.58	.19	.15	.19	.72
12	.89	.27	.27	.49	.83	.38	.17	.21	.85
2 p.m	.86	.24	.22	.28	.68	.75	.53	.30	.81
4	.75	.20	.17	.22	.35	.81	.82	.73	.58
6	.91	.12	.11	.13	.19	.45	.61	.69	.25
8	.18	.05	.05	.07	.09	.12	.12	.12	.12
10	.13	.04	.04	.05	.07	.08	.08	.08	.08
12	.09	.03	.03	.04	.05	.06	.06	.06	.06

Note: Solar time = local standard time (LST) + 4 min × (LST meridian − local longitude) + correction for the month (from − 13.9 min in February to + 15.4 min in October).
Source: From Reference 1. Reprinted with permission from the American Society of Heating, Refrigerating, and Air Conditioning Engineers, Atlanta, Ga.

if the windows are facing north. If a close visual contact with the outside is desired but the sun heat is not, both conditions can be met with windows facing the north. The cooling load factor is reduced by shading, since less heat is absorbed by furniture and walls to be reradiated later.

6.4.1.3 Heating and cooling load: conduction

In addition to inside heating caused by radiation of solar energy absorbed by inside materials, the sun heat combines with the outside temperatures to create heating due to conduction through the walls and the roof. Although writing and solving the heat conduction equations are beyond the scope of this text, several observations can be made that are relevant to energy management. First, the amount of heat gain or loss through a wall depends on the thermal conductivity—or U value—of the wall. Adding insulation to walls or roofs can significantly reduce the unwanted heat gain or heat loss through a wall and can be very cost effective if the main component of the cooling load is from conduction rather than from inside sources. Thermal conductivity, insulation, and the calculation of heat transfer through walls and roofs are discussed in detail in Chapter 11.

Second, the construction of the wall and roof is also an important factor, with the material used to build the structure having a potential thermal storage effect that can be utilized. The so-called *flywheel effect* describes this: In the same way that a flywheel at motion tends to remain in motion, a warm wall tends to remain warm and to radiate heat after the sun is down. The amount of heat and the time the wall continues to radiate depend on the construction of the wall in the same sense that the speed and running time of a released flywheel depend on the construction of the wheel. If an existing building has massive walls, it may be possible to schedule work hours so as to take advantage of the flywheel effect and obtain free heating or cooling. These points are discussed in Reference 3.

Heating or cooling loads due to conduction through a wall can be approximated using Heating Degree Day (HDD) or Cooling Degree Day (CDD) data. Definitions of HDDs and CDDs and calculation methods were given in Section 2.1.1.4. The method involves the use of the heat flow equation 11-9 given in Chapter 11. In this equation, A is the area of the wall, U is the thermal conductivity of the wall, Q is the heat flow through the wall, and the temperature difference ΔT is replaced by the number of heating or cooling degree days, HDD or CDD.

$$Q = U \times A \times DD \times 24 \text{ h/day} \tag{6-6}$$

Example 6-3: A wall has an area of 100 ft^2 and has a thermal conductivity of 0.25 Btu/ft$^2 \bullet$ h $\bullet$ °F. If there are 3000 degree days in the heating season, what is the total amount of heat lost through the wall?

Solution: The heat lost through the wall is found using equation 6-6 as:

$$Q = (0.25 \text{ Btu/ft}^2 \bullet \text{h} \bullet °\text{F}) \times 100 \text{ ft}^2 \times 3000 °\text{F days} \times 24 \text{ h/day}$$
$$= \underline{1,800,000 \text{ Btu/year}}$$

The degree day method has many potential problems that limit its value in finding the total heating needs of a building or facility. First, it only provides an approximate value for the heat flow due to conduction. It does not take into account the moisture in the air; internal heat loads that might provide less than or greater than a 5°F inside temperature gain; solar heat gain; or many other factors.

6.4.1.4 Heating and cooling load: equipment.
    The fourth major source of heating is from equipment. The energy consumption in Btu per hour from ovens, industrial processes such as solder pots that use much heat, and many other types of equipment can either be read directly from nameplates or can be approximated from gas or electricity usage by assuming that every kWh of electricity contributes 3412 Btu of heat at the point of use and that each Mcf (thousand cubic feet) of gas has an energy content of 1 million Btu.
    The efficiency for a single-phase motor is usually 50-60%, and the efficiency for a three-phase motor is usually 60-95%. The energy used by other kinds of equipment can be found in Reference 1, from equipment vendors, and from gas and electric utilities. Decreasing or rescheduling the amount of equipment using electricity or gas at a given time has a twofold effect. First, the actual energy used at that time is decreased. Second, the heat introduced into the space is also reduced. If this heat must be removed by cooling, the cooling load is thus reduced by turning off this equipment. If this heat is desirable, Equations 6-9, 6-10 and 6-11 can be used to indicate how much waste heat can be made available by scheduling or reducing the number of electric motors.
    The amount of heat given off by a motor can also be determined from its nameplate rating in kW if there is one, or it can be estimated from the nameplate voltage and current used by the motor. If it is a single-phase motor, the energy consumed per hour by the motor is given by

kWh/h = kW × use factor (fraction of time that
$$\text{the motor is in use)} \qquad (6\text{-}7)$$

or by the calculation involving the voltage, current and power factor

$$\text{kWh/h} = \text{voltage} \times \text{current} \times \text{power factor} \times \text{use factor} \qquad (6\text{-}8)$$

The amount of heat that is given off in the space being analyzed depends on the job the motor is doing. In most cases, all of the energy used by the motor shows up as heat added to the space. This may seem odd at first, but when the physical factors are considered, the result is quite clear. First, the motor is not 100% efficient, so some of the motor energy is lost directly to heat through losses in the armature and field windings, losses in the motor core and friction losses in bearings. This heat makes the motor feel hot to the touch, and directly heats up the space.

The remaining motor energy is then used for some application. However, in most cases, that application is such that the remaining motor energy is also converted to heat somewhere in the space. Consider a typical application where a motor is being used to drive a conveyor belt in a production room. The useful energy from the motor is used to overcome the friction of the rollers and belt, and the inertia of the items being moved on the conveyor, and thus is converted completely to heat. Unless the conveyor belt extends outside the space being considered, all of the motor's useful energy becomes heat inside that space. Thus, all of the energy supplied to the motor eventually becomes heat somewhere in the space being considered.

It is useful to consider the few instances where the energy from a motor is not converted completely to heat in the immediate space. This can only occur if some of the energy of the motor is stored in the product being made and leaves the space as embedded energy in the product. For example, if a motor is being used to compress a spring that is then used in a device which is further assembled, then a small part of the motor's energy is transferred to the spring in the form of potential energy. Some other examples might be energy in compressed gases, or energy in frozen foods. Although there are some cases where this energy storage and removal are involved, it is usually a very small part of the overall energy used by the motor, and for all practical considerations, all of the energy used by a motor ends up as heat in the immediate space.

The heat given off by a single phase motor—in Btu per hour—can be found as follows

$$\text{Btu/h} = \text{kW} \times \text{use factor} \times 3412 \text{ Btu/kWh} \tag{6-9}$$

or

$$\text{Btu/h} = \text{voltage} \times \text{current} \times (\text{power factor}) \times (3412$$
$$\text{Btu/kWh}) \times \text{use factor}/(1000 \text{ wh/kWh}) \tag{6-10}$$

For a three-phase motor, the energy converted to heat per hour is given by equation 6-9 since the calculation assumes the kW rating of the motor is known. If the kW rating is not listed on the nameplate for the motor, then the voltage, current and power factor must be used, along with the factor 1.732 for proper determination of the power in a three phase motor.

$$\text{Btu/h} = \text{voltage} \times \text{current per phase} \times (\text{power factor})$$
$$\times 1.732 \times 3.412 \times \text{use factor} \tag{6-11}$$

<u>6.4.1.5 Heating and cooling load: lighting.</u>

Heat generated from lighting is another example in which all of the energy used is generally converted to heat in the immediate space. A very small part of the energy supplied to lights appears in the form of visible light. Incandescent lamps convert about 2-3% of the energy they use into visible light. The remaining 97% is immediately converted to heat which enters the surrounding space. Fluorescent lights are more efficient, and they convert around 10% of their input energy into visible light. Even so, the remaining 90% is converted directly into heat from the lamps and from the ballasts.

Next consider what happens to the visible light once it is produced. The photons of light strike surfaces such as floors, walls, desks, machines and people where some of the energy is absorbed and becomes heat. Some of the light energy is reflected, but it then strikes the same surfaces and more energy is converted to heat. Unless some of the light escapes outside the area of interest, all of the energy in the light eventually becomes heat. Thus, except in some relatively rare instances, all of the energy supplied to lights in an area quickly becomes heat in that same area.

In some cases, heating systems have been designed and sized to utilize the heat from lights as a significant source of heat. This is not an efficient heating strategy, but it may need to be recognized and dealt with to improve the energy efficiency of this kind of system. In other cases, light fixtures may be ventilated to the outside, or to another unheated area. Here, the heat from the lights does not contribute much of a load to the space when air conditioning is considered. However, during the heating season, this is a significant loss of heat that increases the energy that must be supplied by the regular heating system. If the ventilating ducts in the lighting fixtures can be easily closed during the heating season and reopened during the cooling season, this is an energy efficient operation strategy.

To calculate the amount of heat that is added to a space from lighting, remember that each kWh of electricity is equivalent to 3412 Btu. Thus the heat produced each hour is found by

$$Btu/h = 3412 \; Btu/kWh \times K \; kW \qquad\qquad (6\text{-}12)$$

where K is the number of kW of lighting load. Note that K must include the power consumption for any ballasts that are connected to the lights. To obtain the total heat produced from lighting, Equation 6-12 is calculated for each hour and the total obtained.

Some of the ways of reducing the amount of lighting energy needed to illuminate a space to a prescribed level include replacing lamps with more efficient lamps, cleaning the luminaries, and painting adjacent surfaces. These and other measures were discussed in detail in Chapter 5. The air conditioning savings from reducing lighting energy is illustrated in the example below.

Example 6-4: A common relamping EMO was described in Example 5-3 in Chapter Five. In that example, 200 40-watt lamps were replaced with 34-watt lamps in a facility that was not air conditioned. If the facility had been air conditioned, there would be an additional savings depending on the number of hours that the air conditioning was needed. Calculate this additional savings assuming the air conditioner has a COP of 2.8.

Solution: Example 5-3 calculated that the energy saved from the relamping was 42,048 kWh/yr. As discussed earlier, all of this lighting energy becomes heat that must be removed by the air conditioner. If the facility is one that has a number of heat-generating machines and processes which results in the facility having to be cooled 24 hours per day, all of the lighting energy saved translates to air conditioning savings. The heat reduction from the lighting energy savings can be found using Equation 6-12 re-written as:

$$Btu \quad = 3412 \; Btu/kWh \times K \; kW \times h$$

$$= 3412 \text{ Btu/kWh} \times 42{,}048 \text{ kWh}$$

$$= 143.5 \times 10^6 \text{ Btu}$$

The electric energy savings from the air conditioner can now be found by dividing this Btu quantity by the EER of the air conditioner.

A/C energy savings $= 143.5 \times 10^6 \text{ Btu}/(2.8 \times 3.412 \text{ Btu/wh})$

$$= 15.02 \times 10^6 \text{ wh}$$

$$= 15{,}020 \text{ kWh}$$

A more direct way to get this same result is just to divide the lighting energy savings by the air conditioner COP.

A/C energy savings $= 42{,}048 \text{ kWh}/2.8$

$$= 15{,}020 \text{ kWh}$$

This increased savings from considering air conditioning makes a significant change in the cost-effectiveness of the relamping program. In this case, it increases the savings by over one-third.

6.4.1.6 Heating and cooling load: air.

The sixth major category of heating or cooling load comes from energy used to heat, cool, or humidify air. This part of the heating or cooling load can be reduced by weatherstripping, caulking, and tightening windows; by installing loading dock shelters; by replacing broken windows; and by other measures designed to reduce or eliminate air leakage to or from the outside. Air infiltration can occur through the envelope at many places, and there are infiltration-reducing techniques unique to each place. Such techniques range from caulking cracks and openings to sealing loading docks. It can also prove worthwhile to prevent airflow from conditioned areas of a plant to unconditioned areas (or vice versa) by installing airflow barriers indoors such as a plastic curtain.

Infiltration or exfiltration generally involves air moving through openings. The annual amount of energy lost through a hole or crack can be estimated from

$$Btu/year = V \times 1440 \times .075 \times .24 \times (HDD + CDD) \qquad (6-13)$$

where V   = volume of air entering or leaving, in cubic feet per minute
       1440 = number of minutes per day
       .075 = pounds of dry air per cubic foot
       .24   = specific heat of air, in Btu per pound per degree F
       HDD   = heating degree days per year, in days × degrees F
       CDD   = cooling degree days per year, in days × degrees F

This formula is a modification of that found in Reference 1. The impact of water vapor is ignored in Equation 6-13 because in many climatic regions exhausting it is often a benefit for cooling and a disbenefit for heating, and the total effect is usually negligible. In climatic areas where moisture is a problem for heating or cooling, the effect of water vapor may need to be considered.

6.4.2 Ventilation Requirements for Health

In addition to satisfying the need for a comfortable working environment, the HVAC system must also provide ventilation to remove noxious substances from the air. Many ventilation requirements are specified by state and local health and safety codes. Table 6-4 shows some of the minimum outdoor air requirements specified by ASHRAE Standard 62-1989 [11]. These requirements can often be met without having to heat or cool excessive amounts of outside air by installing special equipment such as self-ventilating hoods, or by isolating contaminant areas from other parts of a plant.

Getting rid of cigarette smoke is another problem. Where people are not smoking, the recommended ventilation standards range from 15-30 cfm/person; in smoking lounges, the recommended standard is 60 cfm/person. In this situation, the amount of ventilation air is reduced by having separate areas for smokers; a separation which is enforced by law in many states.

In Table 6-4, some applications involving air contaminants such as those from dry cleaners, and smoking areas may require additional ventilation, or at least special equipment to remove the smoke or contaminants.

6.4.3 Ventilation Standards for Comfort

One of the main reasons for using air conditioning is to keep people comfortable. The definition of a comfortable condition changes with clothing styles, the number and quantity of local drafts, and the price of energy for heating, cooling, and humidifying. Comfort is generally defined, however, by a temperature range that varies depending on the relative humid-

Table 6-4 Outdoor Air Requirements for Ventilation
(From ASHRAE Standard 62-1989)

Applications	Outdoor air requirements
Offices	
Office space	20
Reception area	15
Conference room	20
Data entry area	20
Dry cleaners, Laundries	
Commercial laundry	25
Commercial dry cleaner	30
Smoking lounge	60
Bars, cocktail lounges	30

ity of the air. Three measures take advantage of the comfort zone: lowering the minimum temperature at which heating is initiated, raising the maximum temperature at which cooling is initiated, and changing the humidity in the air to more nearly conform to that of the outside environment. If outside air can be used without either humidifying or dehumidifying, considerable energy can be saved. The exact amount depends on the HVAC system, but note that evaporating 1 pound of water uses 1040 Btu. The method of cooling air so that condensation takes place and then reheating the dried air to a comfortable temperature is also costly. An alternative approach in some cases is the use of heat pipes which is discussed in Section 6.6. Another method for saving energy is to reduce the amount of ventilation air to that required by the given environment; the cube law for fan horsepower (Equation 6-14) can be used to estimate possible savings.

Note that management bears the burden of selling the employees on energy-saving measures or of introducing the changes so gradually that they are not noticed. Thus, one good method of energy management that relates to comfort is to allow temperatures to change slowly. Several studies have shown that people are relatively insensitive to slow changes in air temperature [4,5]. This insensitivity can be used to advantage by turning the air conditioning down or off 30 minutes to an hour before quitting time and letting the temperature drift.

6.5 IMPROVING THE OPERATION OF THE HVAC SYSTEM

6.5.1 Basic Operating Rules

The objective of learning how an HVAC system operates is to manage that system more efficiently. HVAC system management can be improved by careful attention to the following operating rules.

Operating Rule 1. Heat to the lowest temperature possible, and cool to the highest temperature possible. Set the hot and cold air temperatures on the hot and cold sides of a dual-duct system so that one zone is receiving only hot air and one zone is receiving only cold air. The hot temperature is thus set so that the system meets the heating needs of the coldest room and cooling needs of the warmest zone. It automatically meets the temperature needs of all the other zones.

Operating Rule 2. Avoid heating or cooling when heating or cooling is not needed. For example, heating or cooling for people is not needed when people are not in a building at times such as weekends or at night. At those times, a building temperature can be allowed to drift with the only constraint being the safe temperature of building components or other material contained within the building. Avoid heating or cooling warehouses unless they contain people or materials sensitive to heat or to cold.

Operating Rule 3. Learn how your control system is supposed to work and then maintain it properly. A consistent problem that plagues buildings is a control system that does not work. For example, a control system is not functioning as intended if return air dampers are blocked open, and the HVAC system heats or cools all the air used for ventilation. People seldom understand the way a two-thermostat system is supposed to work, and they turn the wrong thermostat, causing heating at night but none when the heating is actually needed.

Operating Rule 4. To insure that the minimum amount of ventilation air is being used, adjust the ventilation system by altering the control system settings or by changing pulleys on fans or their drive motors. One very useful relationship is

$$hp_A/hp_B = (cfm_A/cfm_B)^3 \qquad (6\text{-}14)$$

This is the cube law for fan horsepower [8]. It is useful in calculating the energy consumption to be saved from reducing the ventilation requirements.

Example 6-5: ACE Industries presently has a 5-HP ventilating fan that draws warm air from a production area. The motor recently failed and they think they can replace it with a smaller motor. They have determined that they can reduce the amount of ventilation air by one third. What size motor is needed now?

Solution: Use Equation 6-14, and note that the ratio of the new to old cfm rate is 2/3. Thus, the new HP needed is:

New HP = $(2/3)^3 \times 5$ HP = 0.3×5 HP = 1.5 HP

Since 1 HP is equivalent to .746 kW at 100% efficiency, the power savings achieved by a reduction of H hp is given by

Power saved (kW) = (H hp ×.746 kW/hp)/EFF   (6-15)

where EFF is the motor efficiency, usually between .70 and .90. This expression can only be used if the old and new motors have the same efficiencies. If this motor is running constantly, the action of reducing the fan horsepower reduces both the demand and the energy part of the electric bill; otherwise it may affect the demand but must be multiplied by the hours of use to determine the amount and cost of energy saved.

Example 6-6: What is the electrical load reduction for the smaller fan motor in Example 6-5 if the 5-HP motor had an efficiency of 84%? The new 1.5-HP motor has an efficiency of 85.2%.

Solution: We cannot use Equation 6-15 since the two motors do not have the same efficiency. We must calculate the electrical load from each motor and determine the difference.

Old load = (5 HP) × (0.746 kW/HP)/(0.84) = 4.44 kW

New load = (1.5 HP) × (0.746 kW/HP)/(0.852) = 1.31 kW

Electric load reduction = <u>3.13 kW</u>

However, since the two motor efficiencies are approximately equal, we can make an approximation by using Equation 6-15:

Motor load reduction = (3.5 HP) × (0.746 kW/HP)/(0.84) = <u>3.11 kW</u>

This is very close to the correct value.

Operating Rule 5. If you do not need heating, cooling or ventilation, turn off the HVAC system. Unnecessary conditioning and ventilation cost money. Find out when the conditioning and ventilation is needed and arrange to have the HVAC system running only at those times.

6.5.2 Inspecting the HVAC system

First, the auditor should determine whether the room is being cooled or heated more than is necessary. Ideally, a person in a jacket should be comfortable in the winter, and a person in shirt sleeves should be comfortable in the summer. Excessive heating or cooling is another unnecessary energy cost.

Next, the energy auditor must inspect the HVAC system thoroughly to determine whether it is operating properly. Every HVAC system has heat transfer surfaces to enable heating or cooling to take place, every HVAC system has some means of transporting its working fluid to the point of use from the point where heating or cooling is supplied, and every HVAC system has a set of controls which govern its operation. Inspecting each one of these component areas is necessary in a complete energy audit. Since the most common systems are those in which air is the working fluid, the rest of this discussion will be confined to air-cooled systems.

6.5.2.1 Heat transfer surfaces

A heat transfer surface is the surface where a hot or cold fluid gives up or receives heat from air that is passing around it. Typically, heat transfer surfaces are designed so that a hot or cold liquid flows through pipes surrounded by fins. The fins are used to increase the heat transfer rate. These heat transfer surfaces must be examined periodically and be maintained to continue to work properly. The points to be inspected in this part of the HVAC system are the fluid flow lines and the heat transfer surfaces. Do relevant gauges show that fluid is flowing? Are "hot water" pipes actually hot? Can you hear the sound of fluid flowing? When you examine the heat transfer surfaces, are the fins and coils clean, or are they fouled with dirt or grease or dust? Fins must be clean to function effectively and efficiently.

6.5.2.2 Air transportation system

The air transportation system moves air from the outside, mixes outside air with return air, and removes used air either to the outside or to the supply fans for use as return air. The main components of this system are the ductwork, dampers, and filters and the fans, blowers, and associated motors. The ductwork can have insulation hanging loose. The ducts can leak air through untaped seams, or they can be crushed by adjacent piping. Loose insulation can be detected by removing one duct panel and examining the inside of the duct with a flashlight; untaped seams and crushed ducts can be detected with a quick visual inspection. Dampers and filters can also be inspected visually. The dampers should be clean and should close, and their mechanical linkages should be connected to the control actuators. Filters should be installed and reasonably clean; return air grilles should also be inspected to see whether they need to be cleaned. Filters should be cleaned or changed at periodic intervals.

The HVAC fans and blowers should be examined carefully. Fans and blowers should be operational, and the belts should be aligned correctly so that the fan pulley and the motor pulley are in a straight line. Be particularly thorough in examining fan belts and motor connections; the authors have observed instances where motors had been installed and were running but were not connected to the fans they were supposed to drive. Motors should also be inspected to see that they are properly connected with balanced voltages to all three legs of a three-phase system. Motors should also be free from excess bearing noise. Fans or blowers should be reasonably clean, since accumulated dust detracts from their efficiency. The fan or blower should rotate in the correct direction, and the fan shaft should not bind.

6.5.2.3 The control system

The HVAC control system detects pressures or temperatures and compares these with preset values. Depending on the result, the control system sends electrical or pneumatic signals to open or close dampers, open or close valves, and turn furnaces, chillers, and blower motors on or off. Clearly it is important that the control system function properly; otherwise, the HVAC system does not work as intended, and will not be energy efficient.

The first step in inspecting the control system is to examine the thermostats. A reliable industrial thermometer can be used to calibrate each thermostat thermometer; if the temperature difference between the two temperature readings is significant, the thermostat should be checked by a vendor. Next the thermostat set temperature should be raised and lowered to see if the heat or cold comes on or shuts off; this procedure tests the entire HVAC control system.

Gauges should be checked to see whether they are connected and whether they are reading within the correct operating ranges. The compressor that supplies compressed air to the control system should be inspected to see that it is working properly and not leaking oil or water into the control system; if either oil or water gets into the controls, a complete replacement of the control system is often necessary. An air dryer is almost always a necessity on the air supply system to keep the controls working properly.

6.6 HEAT PIPES

In many areas of the country, removing humidity is the chief energy cost of air conditioning. This is because moisture is condensed out of the air by the cooling coil and the colder the air when it passes over the cooling coil, the more moisture is removed. Therefore, in areas where the humidity is high, the air must be cooled much lower than the desired temperature in order to remove the moisture and then heated back to the desired temperature. This means that energy is required to overcool the air; additional energy is required to reheat the air; and the equipment must be oversized in order to overcool the air which increases the power demand of the air conditioning system.

The older, energy-inefficient air-conditioners were designed with very cold cooling coils; newer, high-efficiency models are often designed with warmer coils. These coils are larger and require less energy to operate, and one of the reasons why they save energy is because they do not remove as much moisture. Although this is not a problem in some parts of the country, in areas with high humidity the energy-efficient air-conditioners often leave the conditioned air uncomfortably humid. This problem is generally solved by lowering the temperature setting of the thermostat which uses additional cooling energy and may negate the savings from the energy efficient model.

One energy savings solution is the heat pipe [12]. Heat pipes are relatively new on the commercial air conditioning scene. Although they were first developed near the turn of the century, their commercial use only recently became feasible as a result of research on a NASA contract. A heat pipe is a metal tube that is filled with an evaporative fluid and then sealed at both ends. When heat is applied to one end of the tube, the fluid inside evaporates and the vapor moves to the other, cooler end. The vapor then condenses in the cooler end and returns to the warmer end by gravity or by some sort of capillary action through an inside wick. The heat pipe is activated by the temperature difference between the two ends and does not use energy to operate. See Figure 6-5.

When used in an air conditioning system, one end of the heat pipe is placed in the return air system and is heated by the warm return air. That

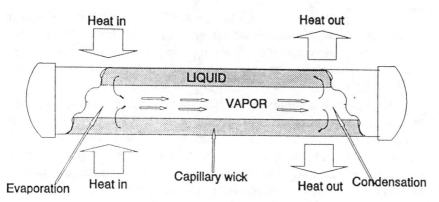

Figure 6-5. Diagram of heat pipe.

heat is "free" because no additional energy is expended to generate the heat. Because the heat pipe absorbs heat from the return air, that air is cooler when it goes to the cooling coil so the cooling coil can work at a lower temperature. That means that the cooling coil can remove more moisture from the air; it also means that the cooling load on the compressor is lower. The other end of the heat pipe is placed in the supply, or conditioned, air system. The heat from the first section in the return air stream is transferred to this section which is used to reheat the chilled air and lower the humidity before the air is distributed throughout the system. The process of reheating the air causes that end of the heat pipe to become chilled. See Figure 6-6.

Figure 6-6.
A high-efficiency air conditioner/dehumidifier using heat pipes.
Courtesy of Florida Solar Energy Center.

Heat pipes are also useful for areas which need large amounts of outside air for ventilation. Fresh air brings in both heat and humidity which means that larger air conditioning systems are usually needed. The addition of heat pipes to the air conditioning system allows the addition of up to 20% fresh air without increasing the size of the system.

Heat pipes are suitable for a number of air conditioning situations. Industries which require a low humidity level or humidity control include: electronic component production, assembly and storage; film drying, processing and storage; drug, chemical and paper manufacturing and storage; printing; candy and chocolate processing and storage. Other examples of industries or areas which would benefit from heat pipes are hospital operating rooms, libraries, grocery stores, telephone exchanges and relay stations, clean rooms, underground silos, and places with indoor swimming pool or spa facilities.

6.7 THERMAL STORAGE

Thermal energy storage offers one of the most promising technologies for effective electrical peak load management in buildings and other facilities [13]. Heating and cooling energy needs provided by electrically powered equipment usually correspond to the time of a facility's peak demand, and contribute to increased electric costs that occur from the demand charge. A thermal energy storage system may produce chilled water or ice off-peak for later use in cooling a facility, or produce heated air or water off-peak for later use in heating a facility.

Cool storage systems operate by producing and storing chilled water or ice during the evening and night when electric rates are low, and drawing on that stored water or ice during the day when the cooling load is greatest. The high-demand electrically powered chillers are shifted to operation during off-peak periods, and are especially cost-effective for facilities that have time-of-day rates which offer low-cost energy during the night. The storage of chilled water or ice also allows the facility to operate with smaller-sized chillers since the peak cooling load is handled with the combination of the small chiller and the cool storage. For new buildings or facilities, this reduction in the size of the chillers often pays most of the cost for the cool storage system. Other cooling system operating cost savings also result from this approach because of the downsizing of fans, pumps and ducts due to the lower temperatures of distributed air because of the very low temperatures of the stored water or ice.

Thermal energy storage systems typically increase the overall system energy consumption because of the storage losses, but they also significantly decrease the costs associated with peak load or peak demand charges. The increase in energy for cooling with storage is often moderated due to the higher efficiency of the chiller system operating at lower night time temperatures.

Many electric utilities offer rebates or incentives for facilities to install thermal storage systems. The incentive may be in the form of low rates for off-peak energy use, or in the form of a direct rebate based on the number of kW moved off-peak.

6.8 SUMMARY

In this chapter, we have explained the functions and the components of HVAC systems. Once the functions and components are understood, the energy manager can find ways to reduce the heating and cooling loads and thus reduce this element of energy costs. When the manager also understands how each of the HVAC system components works, he or she is then prepared to improve the operation of the physical system. By using the operating rules presented, and by adding additional rules unique to your own system, this understanding can be translated into improved operating policies for both HVAC equipment and for the people affected by the HVAC system.

REFERENCES

1. *ASHRAE Handbook of Fundamentals*, American Society of Heating, Refrigerating, and Air Conditioning Engineers, Atlanta, Ga., 1993.
2. Donald Rapp, *Solar Energy*, Prentice-Hall, Englewood Cliffs, N.J., 1981.
3. W.J. Kennedy, Jr., and Richard E. Turley, "Scheduling and the Energy Economics of Wall Design," in *Proceedings of the 1981 Spring Annual Conference & World Productivity Conference*, Detroit, Michigan, Institute of Industrial Engineers, Atlanta, Ga., 1981.
4. Larry G. Berglund and Richard R. Gonzalez, "Human Response to Temperature Drift," *ASHRAE Journal*, Aug. 1978, pp. 38-41.
5. L.G. Berglund, and R.R. Gonzalez, "Application of Acceptable Temperature Drifts to Built Environments as a Mode of Energy Conservation," *ASHRAE Transactions*, Vol. 84, Part 1, 1978.
6. National Electric Contractors Association and National Electrical Manufacturers Association, *Total Energy Management: A Practical Handbook on Energy Conservation and Management*, Third Edition,

National Electrical Contractors Association, Washington, D.C., 1986.

7. *ASHRAE Handbook for Systems and Equipment*, American Society of Heating, Refrigerating,, and Air Conditioning Engineers, Atlanta, Ga., 1992.
8. *ASHRAE Handbook for HVAC Applications*, American Society of Heating, Refrigerating, and Air Conditioning Engineers, Atlanta, Ga., 1991.
9. *CFC's: Time of Transition*, American Society of Heating, Refrigerating and Air Conditioning Engineers, Atlanta, GA, 1989.
10. Energy Policy Act of 1992, P.L. 102-486 (HR 776), U.S. Congress, October 24, 1992.
11. ANSI/ASHRAE Standard 62-1989, Ventilation for Acceptable Indoor Air Quality, American Society of Heating, Refrigerating, and Air Conditioning Engineers, Atlanta, Ga., 1989.
12. *Commercial High Efficiency Dehumidification Systems Using Heat Pipes, A Technical Manual*, Heat Pipe Technology Company, Alachua, FL, 1993.
13. "Thermal Storage Forum," *ASHRAE Journal*, April, 1990, pp 26-66, American Society of Heating, Refrigerating and Air Conditioning Engineers, Atlanta, GA.

BIBLIOGRAPHY

Frank E. Beaty, *Sourcebook of HVAC Details*, McGraw-Hill Book Company, New York, NY, 1987.

Robert Chatenver, *Air Conditioning and Refrigeration for the Professional*, Wiley, New York, NY, 1988.

Harold R. Colen, *HVAC System Evaluation*, R.S. Means Company, Kingston, MA, 1990.

Roger W. Haines, *HVAC System Design Handbook*, TAB Professional and Reference Books, Blue Ridge Summit, PA, 1988.

HVAC Systems—Applications, Sheet Metal and Air Conditioning Contractors National Association, Vienna, VA, 1987.

Industrial Ventilation, Committee on Industrial Ventilation, American Conference of Governmental Industrial Hygienists, Lansing MI, 1979.

John I. Levenhagen, *HVAC Controls and Systems*, McGraw-Hill Book Company, New York, NY, 1993.

Raymond K. Schneider, *HVAC Control Systems, Second Edition*, Wiley, New York, NY, 1988.

Variable Air Volume Systems Manual, 1980 ed., Honeywell, Inc., Minneapolis, 1980.

Chapter 7

Combustion Processes and The Use of Industrial Wastes

7.0 INTRODUCTION

Many commercial and industrial facilities use fossil fuel-fired boilers to produce steam for space heating or for process heating. These boilers are usually the major consumers of the fuel used in the facility. Therefore, any person involved in energy management needs to know how the performance of a boiler can be improved, and in particular, what operating parameters of a boiler are most important. One critical parameter is the amount of combustion air that is mixed with the fuel. The proper amount of air for optimum boiler performance depends on the fuel type, the amount of energy demanded from the boiler per hour, the boiler type, and other considerations, many of which are related to the chemistry of the combustion process. Combustion products, some of which are subject to governmental regulation, occur in differing amounts and types depending on the chemical reactions of combustion. A major function of this chapter is to describe the basic chemistry of combustion in a boiler. Flue gas analysis is discussed as well.

This chapter also looks at the chemistry and combustion problems associated with alternative fuels, including industrial wastes. Until recently, industrial wastes were not often used for fuel because of technical problems associated with burning those wastes. However, the increasing costs of waste disposal and of energy have caused a reexamination of alternatives with sometimes surprising results. This chapter gives the reader a procedure for comparing alternative fuel proposals.

7.1 COMBUSTION CHEMISTRY

7.1.1 Basic Reactions and Quantities

When fuels burn, oxygen combines with chemical components of the fuel and releases energy in the form of heat. Since there are many components in fuel, there are many chemical reactions associated with combustion. Some of them are shown here:

$$C + O_2 \rightarrow CO_2 + 14,093 \text{ Btu/lb of carbon} \qquad (7\text{-}1)$$

$$2C + O_2 \rightarrow 2CO + 3960 \text{ Btu/lb of carbon} \qquad (7\text{-}2)$$

$$2H_2 + O_2 \rightarrow 2H_2O + 61,100^* \text{ Btu/lb of hydrogen} \qquad (7\text{-}3)$$

$$CH_4 + 2O_2 \rightarrow 2H_2O + CO_2 + 23,879 \text{ Btu/lb of methane} \qquad (7\text{-}4)$$
$$\text{(methane reaction)}$$

$$S + O_2 \rightarrow SO_2 + 3980 \text{ Btu/lb of sulfur} \qquad (7\text{-}5)$$

The detailed chemistry of boiler combustion is complicated and includes many other reactions such as those that result in the formation of nitrogen and sulfur oxides. The reactions presented here, however, are adequate for energy management purposes and give good quantitative results. For more detail, see Reference 1.

To understand the implications of these reactions for boiler operation, it is necessary to know the constituents of any boiler fuel by weight percentage and to understand the weight balances implied by the reactions. It is also necessary to know which of the reactions are favored at a given temperature and to know what temperatures are to be used in the computations. With this information, one can calculate the optimum quantity of air needed for combustion of each ton of fuel consumed. One can also determine the cost of operating at air quantities that exceed this optimum.

Start this calculation by considering Reaction 7-4. This represents a balance in molecules; i.e., one molecule of methane combines with two molecules of oxygen to yield two molecules of water and one molecule of carbon dioxide. Avogadro's law for ideal gases states that equal numbers of molecules of dissimilar gases occupy the same volume. Therefore,

---

*These figures assume that all of the water leaves as water vapor. The actual value is up to 10% less than this, depending on the amount of water vapor that condenses. Since this amount is usually not known, the authors have chosen to use the higher figure.

Reaction 7-4 is also a statement about volumes; i.e., 1 ft^3 of methane combines with 2 ft^3 of oxygen to yield 2 ft^3 of water vapor and 1 ft^3 of carbon dioxide. Converting Reaction 7-4 into a statement about weights requires a knowledge of the molecular weights of each component. The atomic weights of carbon, hydrogen, and oxygen are, respectively, 12.005, 1.008, and 16.000.

The molecular weights of each component in Equation 7-4 can be calculated as follows:

methane = atomic weight of carbon + (4 atoms of hydrogen per molecule of methane) × atomic weight of hydrogen
= 12.005 lb + 4 × 1.008 lb
= 16.037 lb/mole

oxygen = 2 (the number of oxygen molecules) × 2 (atoms of atomic oxygen per atom of molecular oxygen) × 16.000 lb
= 64 lb/mole

H_2O = 2 × (2 × 1.008 + 16.000)
= 36.032 lb/mole

CO_2 = 12.005 + 2 × 16.000
= 44.005 lb/mole

We must also assume some volume. One mole of gas (the number of pounds of a gas equal to its molecular weight) occupies 359 ft^3 (the *molal volume*) at standard temperature and pressure (32°F and 14.7 psi absolute). For a molal volume, Equation 7-4 becomes the following mass balance:

$$16.037 \text{ lb methane} + 64 \text{ lb oxygen} \rightarrow 36.032 \text{ lb water vapor}$$
$$+ 44.005 \text{ lb } CO_2 \quad (7\text{-}6)$$

From Reaction 7-4, the heat generated in the combustion of methane is calculated as

$$16.037 \text{ lb} \times 23{,}879 \text{ Btu/lb} = 382{,}947 \text{ Btu}$$

and the combustion equation is

$$16.037 \text{ lb methane} + 64 \text{ lb oxygen} \rightarrow 36.032 \text{ lb water vapor}$$
$$+ 44.005 \text{ lb } CO_2 + 382{,}947 \text{ Btu}$$

These relationships assume complete combustion of all carbon in the fuel, such that the fuel is completely converted into water and CO_2. Such complete combustion requires perfect mixing of exact amounts of oxygen and fuel. Perfect mixing would ensure that every molecule of fuel is surrounded by the right amount of oxygen for complete combustion to take place, but this situation is rarely, if ever, attained. Complete combustion can also be achieved by adding extra oxygen (or excess air) to the combustion process. However, when excess air is added, some of the heat generated by the combustion process is lost to the excess air. Since the products of incomplete combustion are less desirable than the heat loss associated with excess air, excess air is preferred, in the amounts shown in Table 7-1.

Table 7-1. Recommended Excess Air for Various Fuels

Fuels		% excess air
Solid	Coal (pulverized)	15-30
	Coke	20-40
	Wood	25-50
	Bagasse	25-45
Liquid	Oil	3-15
Gaseous	Natural gas	5-10
	Refinery gas	8-15
	Blast-furnace gas	15-25
	Coke-oven gas	5-10

Source: Courtesy of Combustion Engineering, Inc.

Usually, the objective of combustion air control is to introduce enough air to keep the concentration of CO below 400 ppm (for gas fuels) or the smoke color clear or light brown (for liquid fuels or coal). A safety factor is added to this air requirement to keep the combustion in an oxygen-rich environment. In designing a combustion system, the components must be sized larger than minimum capacity to assure that sufficient capacity is always available to deliver the required amount of combustion air. If the combustion air supplies 100% excess oxygen the equation for combustion becomes

16.037 lb methane + 128 lb oxygen
→ 36.032 lb water vapor + 44.005 lb CO_2 +
64 lb excess oxygen + 382,947 Btu   (7-7)

To translate this into an equation related to combustion air, it is necessary to know the number of pounds of oxygen per pound of air. Since oxygen makes up approximately 23.15% of the weight of air, Equation 7-8 gives

16.037 lb methane + 552.916 lb air
$$\rightarrow 36.032 \text{ lb water vapor} + 44.005 \text{ lb } CO_2 +$$
$$276.458 \text{ lb excess air} + 382{,}947 \text{ Btu} \qquad (7\text{-}8)$$

Another useful relationship is developed by using the density of air, .07655 lb/ft^3 for dry air at sea level, to get

16.037 lb CH_4 + 7222.9 ft^3 dry air
$$\rightarrow 36.032 \text{ lb water vapor} + 44.005 \text{ lb } CO_2 +$$
$$3611.5 \text{ ft}^3 \text{ dry air} + 382{,}947 \text{ Btu} \qquad (7\text{-}9)$$

The preceding analysis gives a method for determining the amount of dry air and the amount and type of chemical products of combustion for a given fuel constituent, in this case methane. Table 7-2 provides this information for many substances and is often used to estimate the heat liberated and the excess air needed when a particular gas of known composition is considered as a fuel.

Example 7-1: Consider a gas (typical of a natural gas being supplied to users) with the chemical composition shown in Figure 7-1. Calculate the Btu output for the gas, and the theoretical amount of air needed for combustion of the gas.

Figure 7-1. Composition of a typical fuel gas, by volume.

Constituent	Percent by volume	Btu/ft^3 constituent	Btu/ft^3 fuel	Ft^3 air/ft^3 constituent	Ft^3 air/ft^3 fuel
CH_4	86.4	1013.2	875.4	9.528	8.23
C_2H_6	8.4	1792	150.5	16.675	1.40
C_3H_8	1.5	2590	38.9	23.821	0.36
C_4H_{10}	1.1	3370	37.1	30.967	0.34
N_2	0.5	—		—	
CO_2	2.1	—		—	
			1101.9		10.33

Note: High heat values (i.e. before the heat of vaporization of water vapor) were used in these computations so as to be able to use water vapor energy in a later computation.

Table 7-2. Combustion Constants

No.	Substance	Formula	Molecular Weight[a]	Lb per Cu Ft[b]	Cu Ft per Lb[b]	Sp Gr Air = 1.000[b]	Heat of Combustion[c] Btu per Cu Ft Gross	Net[d]	Btu per Lb Gross	Net[d]
1	Carbon	C	12.01	–	–	–	–	–	14,093[e]	14,093[e]
2	Hydrogen	H_2	2.016	0.005327	187.723	0.06959	325.0	275.0	61,100	51,623
3	Oxygen	O_2	32.000	0.08461	11.819	1.1053	–	–	–	–
4	Nitrogen (atm)	N_2	28.016	0.07439[e]	13.443[e]	0.9718[e]	–	–	–	–
5	Carbon monoxide	CO	28.01	0.07404	13.506	0.9672	321.8	321.8	4,347	4,347
6	Carbon dioxide	CO_2	44.01	0.1170	8.548	1.5282	–	–	–	–
	Paraffin series C_nH_{2n+2}									
7	Methane	CH_4	16.041	0.04243	23.565	0.5543	1013.2	913.1	23,879	21,520
8	Ethane	C_2H_6	30.067	0.08029[f]	12.455[e]	1.04882[e]	1792	1641	22,320	20,432
9	Propane	C_3H_8	44.092	0.1196[e]	8.365[e]	1.5617[e]	2590	2385	21,661	19,944
10	n-Butane	C_4H_{10}	58.118	0.1582[e]	6.321[e]	2.06654[e]	3370	3113	21,308	19,680
11	Isobutane	C_4H_{10}	58.118	0.1582[e]	6.321[e]	2.06654[e]	3363	3105	21,257	19,629
12	n-Pentane	C_5H_{12}	72.144	0.1904[e]	5.252[e]	2.4872[e]	4016	3709	21,091	19,517
13	Isopentane	C_5H_{12}	72.144	0.1904[e]	5.252[e]	2.4872[e]	4008	3716	21,052	19,478
14	Neopentane	C_5H_{12}	72.144	0.1904[e]	5.252[e]	2.4872[e]	3993	3693	20,970	19,396
15	n-Hexane	C_6H_{14}	86.169	0.2274[n]	4.398[e]	2.9704[e]	4762	4412	20,940	19,403
	Olefin series C_nH_{2n}									
16	Ethylene	C_2H_4	28.051	0.07456	13.412	0.9740	1613.8	1513.2	21,644	20,295
17	Propylene	C_3H_6	42.077	0.1110[e]	9.007[e]	1.4504[e]	2336	2186	21,041	19,691
18	n-Butene (Butylene)	C_4H_8	56.102	0.1480[e]	6.756[e]	1.9336[e]	3084	2885	20,840	19,496
19	Isobutene	C_4H_8	56.102	0.1480[e]	6.756[e]	1.9336[e]	3068	2869	20,730	19,382
20	n-Pentene	C_5H_{10}	70.128	0.1852[e]	5.400[e]	2.4190[e]	3836	3586	20,712	19,363
	Aromatic series C_nH_{2n-6}									
21	Benzene	C_6H_6	78.107	0.2060[e]	4.852[e]	2.6920[e]	3751	3601	18,210	17,480
22	Toluene	C_7H_8	92.132	0.2431[e]	4.113[e]	3.1760[e]	4484	4284	18,440	17,620
23	Xylene	C_8H_{10}	106.158	0.2803[e]	3.567[e]	3.6618[e]	5230	4980	18,650	17,760
	Miscellaneous gases									
24	Acetylene	C_2H_2	26.036	0.06971	14.344	0.9107	1499	1448	21,500	20,776
25	Naphthalene	$C_{10}H_8$	128.162	0.3384[e]	2.955[e]	4.4208[e]	5854[l]	5654[l]	17,298[l]	16,708[l]
26	Methyl alcohol	CH_3OH	32.041	0.0846[e]	11.820[e]	1.1052[e]	867.9	768.0	10,259	9,078
27	Ethyl alcohol	C_2H_5OH	46.067	0.1216[e]	8.221[e]	1.5890[e]	1600.3	1450.5	13,161	11,929
28	Ammonia	NH_3	17.031	0.0456[e]	21.914[e]	0.5961[e]	441.1	365.1	9,668	8,001
29	Sulfur	S	32.06	–	–	–	–	–	3,983	3,983
30	Hydrogen sulfide	H_2S	34.076	0.09109[e]	10.979[e]	1.1898[e]	647	596	7,100	6,545
31	Sulfur dioxide	SO_2	64.06	0.1733	5.770	2.264	–	–	–	–
32	Water vapor	H_2O	18.016	0.04758[e]	21.017[e]	0.6215[e]	–	–	–	–
33	Air	–	28.9	0.07655	13.063	1.0000	–	–	–	–

All gas volumes corrected to 60F and 30 in. Hg dry. For gases saturated with water at 60F, 1.73% of the Btu value must be deducted.

[a] Calculated from atomic weights given in "Journal of the American Chemical Society", February 1937.

[b] Densities calculated from values given in grams per liter at 0C and 760 mm in the International Critical Tables allowing for the known deviations from the gas laws. Where the coefficient of expansion was not available, the assumed value was taken as 0.0037 per C. Compare this with 0.003662 which is the coefficient for a perfect gas. Where no densities were available the volume of the mol was taken as 22.4115 liters.

[c] Converted to mean Btu per lb (1/180 of the heat per lb of water from 32F to 212F) from data by Frederick D. Rossini, National Bureau of Standards, letter of April 10, 1937, except as noted.

Table 7-2. Combustion Constants (*Continued*)

| Cu Ft per Cu Ft of Combustible | | | | | | Lb per Lb of Combustible | | | | | | Experimental Error in Heat of Combustion Percent + or − |
| Required for Combustion | | | Flue Products | | | Required for Combustion | | | Flue Products | | | |
O₂	N₂	Air	CO₂	H₂O	N₂	O₂	N₂	Air	CO₂	H₂O	N₂	
—	—	—	—	—	—	2.664	8.863	11.527	3.664	—	8.863	0.012
0.5	1.882	2.382	—	1.0	1.882	7.937	26.407	34.344	—	8.937	26.407	0.015
—	—	—	—	—	—	—	—	—	—	—	—	—
0.5	1.882	2.382	1.0	—	1.882	0.571	1.900	2.471	1.571	—	1.900	0.045
—	—	—	—	—	—	—	—	—	—	—	—	—
2.0	7.528	9.528	1.0	2.0	7.528	3.990	13.275	17.265	2.744	2.246	13.275	0.033
3.5	13.175	16.675	2.0	3.0	13.175	3.725	12.394	16.119	2.927	1.798	12.394	0.030
5.0	18.821	23.821	3.0	4.0	18.821	3.629	12.074	15.703	2.994	1.634	12.074	0.023
6.5	24.467	30.967	4.0	5.0	24.467	3.579	11.908	15.487	3.029	1.550	11.908	0.022
6.5	24.467	30.967	4.0	5.0	24.467	3.579	11.908	15.487	3.029	1.550	11.908	0.019
8.0	30.114	38.114	5.0	6.0	30.114	3.548	11.805	15.353	3.050	1.498	11.805	0.025
8.0	30.114	38.114	5.0	6.0	30.114	3.548	11.805	15.353	3.050	1.498	11.805	0.071
8.0	30.114	38.114	5.0	6.0	30.114	3.548	11.805	15.353	3.050	1.498	11.805	0.11
9.5	35.760	45.260	6.0	7.0	35.760	3.528	11.738	15.266	3.064	1.464	11.738	0.05
3.0	11.293	14.293	2.0	2.0	11.293	3.422	11.385	14.807	3.138	1.285	11.385	0.021
4.5	16.939	21.439	3.0	3.0	16.939	3.422	11.385	14.807	3.138	1.285	11.385	0.031
6.0	22.585	28.585	4.0	4.0	22.585	3.422	11.385	14.807	3.138	1.285	11.385	0.031
6.0	22.585	28.585	4.0	4.0	22.585	3.422	11.385	14.807	3.138	1.285	11.385	0.031
7.5	28.232	35.732	5.0	5.0	28.232	3.422	11.385	14.807	3.138	1.285	11.385	0.037
7.5	28.232	35.732	6.0	3.0	28.232	3.073	10.224	13.297	3.381	0.692	10.224	0.12
9.0	33.878	42.878	7.0	4.0	33.878	3.126	10.401	13.527	3.344	0.782	10.401	0.21
10.5	39.524	50.024	8.0	5.0	39.524	3.165	10.530	13.695	3.317	0.849	10.530	0.36
2.5	9.411	11.911	2.0	1.0	9.411	3.073	10.224	13.297	3.381	0.692	10.224	0.16
12.0	45.170	57.170	10.0	4.0	45.170	2.996	9.968	12.964	3.434	0.562	9.968	−[f]
1.5	5.646	7.146	1.0	2.0	5.646	1.498	4.984	6.482	1.374	1.125	4.984	0.027
3.0	11.293	14.293	2.0	3.0	11.293	2.084	6.934	9.018	1.922	1.170	6.934	0.030
0.75	2.823	3.573	—	1.5	3.323	1.409	4.688	6.097	— (SO₂)	1.587	5.511	0.088
—	—	—	— (SO₂)	—	—	0.998	3.287	4.285	1.998 (SO₂)	—	3.287	0.071
1.5	5.646	7.146	1.0 (SO₂)	1.0	5.646	1.409	4.688	6.097	1.880 (SO₂)	0.529	4.688	0.30
—	—	—	—	—	—	—	—	—	—	—	—	—
—	—	—	—	—	—	—	—	—	—	—	—	—

[d] Deduction from gross to net heating value determined by deducting 18,919 Btu per pound mol of water in the products of combustion. Osborne, Stimson, and Ginnings, "Mechanical Engineering", p. 163, March 1935, and Osborne, Stimson, and Fiock, National Bureau of Standards Research Paper 209.

[e] Denotes that either the density or the coefficient of expansion has been assumed. Some of the materials cannot exist as gases at 60F and 30 in. Hg pressure, in which case the values are theoretical ones given for ease of calculation of gas problems. Under the actual concentrations in which these materials are present their partial pressure is low enough to keep them as gases.

[f] From Third Edition of "Combustion."

[g] National Bureau of Standards, RP 1141.

Reprinted from "Fuel Flue Gases", 1941 Edition, courtesy of American Gas Association.

Solution: From the Heat of Combustion column of Table 7-2, the Btu output for the gas is found by summing the Btu outputs of the components as a percent of the volume of the gas.

$$\text{Btu output} = (.864 \times 1013.2) + (.084 \times 1792) +$$
$$(.015 \times 2590) + (.011 \times 3370)$$

$$= 1101.9 \text{ Btu/ft}^3 \text{ gas}$$

The amount of air needed for combustion per cubic foot of gas, assuming perfect mixing, is found in a similar manner using the column of Table 7-2 for Air Required for Combustion.

$$\text{Combustion air} = (.864 \times 9.528) + (.084 \times 16.675) +$$
$$(.015 \times 23.821) + (.011 \times 30.967)$$

$$= 10.33 \text{ ft}^3 \text{ of air/ft}^3 \text{ of gas}$$

These components and final results are also shown in Figure 7-1.

We can also compute the actual amount of air needed for combustion assuming that some excess air is required to assure complete combustion. A typical value for excess air is about 12%. This result can be used to estimate the number of cubic feet per minute of combustion air that must be supplied for a given Btu output.

Example 7-2: If a boiler using the gas from Figure 7-1 is to deliver 2 million Btu/h and the boiler efficiency is .80, find the actual amount of combustion air needed per minute.

Solution: If we assume that 12% excess air is needed to assure complete combustion, then:

$$\text{Excess air} = (0.12) \times (10.33 \text{ ft}^3 \text{ air}) = 1.24 \text{ ft}^3 \text{ air}$$

The total air required is then:

$$\text{Total air} = 10.33 + 1.24 = 11.57 \text{ ft}^3 \text{ of air/ft}^3 \text{ of gas.}$$

The cubic feet per minute (CFM) of combustion air needed for the boiler Btu output is found by:

$$\frac{2,000,000 \, \text{Btu/h}}{.8} \times \frac{1 \, \text{h}/60 \, \text{min}}{1015 \, \text{Btu/ft}^3 \, \text{gas}} \times 11.6 \, \text{ft}^3 \, \text{air/ft}^3 \, \text{gas}$$

$$= 476 \, \text{cfm} \, (\text{ft}^3/\text{min})$$

(7-10)

Table 7-2 can also be used to help calculate the composition of the flue gas from the combustion process. The calculations are shown in Example 7-3 (Figure 7-2) under the assumed conditions of 12 percent excess air. Air is 20.99% oxygen (O_2) and 78.03% nitrogen (N_2) by volume, with about .98% inert content [2].

Example 7-3: Calculate the flue gas composition of the combustion products for the gas in Figure 7-1 with excess air of 12%.

Solution: The volumes for the combustion products are found by using Table 7-2, and finding the volume of product for each cubic foot of constituent of the fuel. For example, the products resulting from combusting methane (CH_4) are found by multiplying 0.864 ft^3 of methane by the numbers of cubic feet of combustion products shown in Table 7-1. The table value for H_2O is 2.0, so:

H_2O from methane = $0.864 \times 2.0 = 1.728$ ft^3 of H_2O

CO_2 from methane = $0.864 \times 1.0 = 0.864$ ft^3 of CO_2

N_2 from methane = $0.864 \times 7.528 = 6.504$ ft^3 of N_2

These results are summarized in Figure 7-2 for each of the components of the gas. Finally, air is approximately 76.85% N_2 and 23.15% O_2 by weight. Therefore, the excess air of .12 lb/lb air needed for combustion translates into .0922 lb N_2 and .0278 lb O_2.

Figure 7-2. Volumetric analysis of flue gas for a gas fuel.

Constituent	(1)	(2)	CO_2	(3)	N_2	(4)	H_2O	O_2 (in excess air)
Excess air (from preceding example)	1.24		—		0.980		—	0.260
CH_4	0.864	1.0	0.864	7.528	6.504	2.0	1.728	
C_2H_6	0.084	2.0	0.168	13.175	1.107	3.0	0.252	
C_3H_8	0.015	3.0	0.045	18.821	0.282	4.0	0.060	
C_4H_{10}	0.011	4.0	0.044	24.467	0.269	5.0	0.055	
N_2	0.005				0.005			
CO_2	0.021		0.021					
Totals			1.142		9.147		2.095	0.260
Fraction of total volume			0.090	0.000	0.723	0.000	0.166	0.021

Total 12.644

(1) Cu ft of constituent per cu ft of fuel gas
(2) Cu ft of CO_2 in flue gas per cu ft of combustible (Table 7.2)
(3) Cu ft of N_2 in flue gas per cu ft of combustible
(4) Cu ft of H_2O in flue gas per cu ft of combustible

If an analysis of the flue gas by weight is desired, two methods are available. The first method repeats the preceding calculations from Figure 7-2 using weight as the basis instead of volume. To calculate the weight of combustion air needed for each constituent, determine the weight of excess air, calculate the number of pounds of N_2 and O_2 represented by this weight, and then tabulate the results. The second method converts the total volume fractions from the preceding calculations directly into weight calculations. The latter method is shown in Figure 7-3.

Figure 7-3. Weight analysis of flue gas for a gas fuel.

Constituent	ft^3/ft^3 flue gas	Density (lb/ft^3)	lb/ft^3 flue gas	Weight %
CO_2	.090	.1170	.01053	14.2
N_2	.723	.07439	.05378	72.7
O_2	.021	.08461	.00178	2.4
H_2O	.166	.04758	.00790	10.7
Total weight, lb/ft^3:			.07399	100.0

The methods used to analyze coal and other nongas fuels differ from the gas analysis already shown. For these fuels, a *proximate* analysis gives the percentage by weight of moisture, volatile matter, fixed carbon, and ash. A *proximate* analysis is usually provided at the time coal is purchased and may be part of the basis for the price paid for the coal. An *ultimate* analysis is the determination, from a dried sample, of carbon, hydrogen, sulfur, and nitrogen and an estimate of the amount of oxygen as the difference between the starting weight and the weight of all the other constituents. Typical, though hypothetical, values are shown in Figure 7-4, together with a typical heating value.

The data from Figure 7-4 can be used to calculate the amount of combustion air as shown in Figure 7-5.

The dry air equivalent of the amount of oxygen shown in Figure 7-5 weighs 2.471 lb $O_2 \times$ 100 lb dry air/23.15 lb O_2 = 2.471 × 4.32 = 10.675 lb. At 125% total air (25% excess air), 13.343 lb of dry air would be needed per pound of fuel, and the flue gas would contain 13.343 − 10.675 = 2.668 lb of excess air with .25 × 2.471 =.618 lb excess O_2 per pound of fuel. These calculations are useful in analyzing the flue gas.

Figure 7-4. Proximate and ultimate analyses for a typical coal (heating value 12,700 Btu/lb)

Proximate analysis		Ultimate analysis	
Component	Weight %	Component	Weight %
Moisture	3.5	Moisture	3.5
Volatile matter	42.4	Carbon	76.2
Fixed carbon	51.3	Hydrogen	6.2
Ash	2.8	Sulfur	1.8
	Total: 100.0	Nitrogen	2.6
		Oxygen	6.9
		Ash	2.8
			Total: 100.0

Figure 7-5. Calculation of combustion air quantities.

	Ultimate analysis: lb/lb fuel as fired	lb O_2/lb combustible element (from Table 7-2)	Required for combustion, lb O_2/lb fuel (perfect air mixing)
C	.762	2.664	2.030
H_2	.062	7.937	.492
S	.018	.998	.018
N_2	.026		—
O_2	.069		—
H_2O	.035		—
Ash	.028		—
	Total: 1.000		Total: 2.540
		Less O_2 in fuel:	−.069
			2.471

For the 2 million Btu/h boiler discussed earlier, the combustion air requirement, assuming 25% excess air by weight, is

$$\frac{2,000,000\,\text{Btu/h}}{.80} \times \frac{1\,\text{h/60 min}}{12,700\,\text{Btu/lb}} \times 13.34\,\text{lb air/lb fuel}$$

$$= 43.76\,\text{lb dry air/min)} \qquad\qquad (7\text{-}11)$$

$$= 571.6\,\text{cfm}$$

The amount of each combustion product in the flue gas can also be calculated from the ultimate analysis using the "flue products" columns of Table 7-2 and assuming that *standard air* (air with .013 lb H_2O/lb dry air at 80°F dry bulb, relative humidity 60% and pressure 14.7 psia) is used. These calculations proceed as shown in Figure 7-6 for the coal described. Dividing the component weights by the total wet weight gives the components by percentage: CO_2, 19.3%; H_2O, 5.3%; SO_2, .25%; N_2, 70.9%; O_2 4.3%. (Dry weight could also be used as the basis for the analysis method).

7.1.2 Energy Loss Calculations

In Section 7.1.1 we described some of the basic calculations associated with chemical reactions of combustion. A person involved in energy management, however, needs to know more than combustion chemistry. He or she is interested in the aspects of combustion that can be controlled to make the boiler more efficient so that it uses less fuel for a given output of usable heat energy. This efficiency is governed by the amount of heat that is lost in flue gas, which is dependent on the amount of combustion air provided. The heat loss can be computed from the flue gas composition and from the stack gas temperature.

As mentioned in Section 7.1.1, using excess air leads to complete combustion but causes heat losses to the environment. Combustion raises the temperature of all combustion products sent up the flue, and the heat loss in these products is a significant source of boiler inefficiency. Some of this loss cannot be avoided—combustion products must be removed, and higher temperatures help the flue to exhaust these gases efficiently. Excess combustion air is also exhausted through the flue after it has absorbed enough heat to raise its temperature from ambient to stack temperature, so excess air also causes some heat losses. Without some excess air, however, some of the carbon in the fuel produces CO rather than CO_2, and since the reaction giving CO yields 3960 Btu/lb rather than the 14,093

Figure 7-6. Calculation of flue gas composition by weight.

Analysis:

Constituent	lb/lb fuel as fired	Product	lb product/lb constituent		lb product/lb fuel (assuming 25% excess air)	Product weight %
C	.762	CO_2	3.664 × .762 = 2.792		2.792	19.3
H_2	.062	H_2O	8.94 × .062 = .554			
H_2O (in fuel)	.035		.035			
Air (H_2O)	13.343		.013 × 13.343 = .173			
component)			Total H_2O .762		.762	5.3
S	.018	SO_2	.018 × 1.998 = .036		.036	0.25
N_2 (in fuel)	.026	N_2	.026			
Air (N_2)	13.343		.7685 × 13.343 = 10.254			
component)			Total N_2 10.280		10.280	70.9
O_2 (excess)					.618	4.3
					14.488	100.0
			Weight (wet)			
			Weight (dry) = 14.488-.762 =		13.726	

(Note that .173 lb extra H_2O is included. The weight balance is 13.343 lb dry air +.173 lb water in air + 1 lb fuel –.028 lb ash = 14.488 lb. This checks with the wet weight of total products.)

Btu/lb associated with CO_2, lack of excess air causes a loss of heat.

To calculate the heat lost in the flue gas (the combustion products) for the coal example given above, you must first obtain the percent by volume of each flue gas component, either by an Orsat analysis (a standard wet chemical analysis method) or by converting the weight analysis shown in Figure 7-6 into a volume analysis after dividing the weight of each component by its molecular weight. (Remember that the weight of a given volume of gas divided by its molecular weight gives the number of *lb-mol* of the gas, a value proportional to the volume the gas occupies.) You must also know the exit temperature of the flue gas, the specific heat of each component at the average of the ambient and exit temperatures, and the heating value of the fuel. Figure 7-8 provides the necessary specific heat values.

Two other terms associated with heat calculations need to be defined. One is *sensible heat*, which is that heat which when added to or removed from a substance results in a change of temperature. If the temperature of liquid water increases, it has gained sensible heat. The other term is *latent heat*, which is the heat that changes the physical state of a substance without changing its temperature. If liquid water at 212°F changes to vapor at 212°F, it has gained latent heat.

Example 7-4: Using the coal described in Figure 7-4, calculate the heat loss from the flue gas. Assume that the exit temperature for the flue gas is 500°F, the ambient temperature is 80°F, and the heating value is 12,700 Btu/lb.

Solution: Step 1. Compute the relative volumes of the flue gas components. These volumes are calculated as shown in Figure 7-7. Weights of flue gas components were calculated just prior to Section 7.1..2. Specific heats for the components are given in Figure 7-8.

Step 2. Determine the energy needed to produce 1 lb of flue gas. As was shown in Figure 7-6, 1 lb of this coal combines with the combustion air to create 14.488 lb of wet flue gas or 13.726 lb of dry gas. 1 lb of wet gas represents the exhaust equivalent of 12,700/14.488 = 876.6 Btu.

Step 3. Determine the energy (heat) loss in the flue gas. The calculations for determining the sensible heat loss in the flue gas are shown in Figure 7-7.

Figure 7-7. Energy losses in flue gas.

Component	lb/lb flue gas /	Molecular weight	lb mol/lb flue gas ×	Specific heat Btu/lb mol /deg F. ×	500-80 =	Btu
CO_2	0.193	44	0.00439	10.2	420	18.8
H_2O	0.053	18	0.00294	8.3	420	10.3
(sensible heat)						
SO_2	0.0025	64	0.00004	10.7	420	0.2
N_2	0.709	28	0.02532	7.1	420	75.5
O_2	0.043	32	0.00134	7.2	420	4.1
					Total:	108.8

Figure 7-8. Mean molal specific heat of gases between final temperature (t_2) and 80°F at standard atmospheric pressure.

The total heat loss includes the latent heat necessary to convert water into water vapor. There are three sources for this water: water created as a combustion product, moisture in the fuel, and moisture in combustion air. From Figure 7-6, these total 0.762 lb per lb of fuel. Since one lb of fuel becomes 14.488 lb of wet flue gas, the amount of water in one lb of wet flue gas is 0.762/14.488, or .0526 lb. When this water is converted to water vapor, 1040 Btu/lb is used, or a total latent heat of .0526 × 1040 = 54.7 Btu per lb of flue gas. Together with the heat losses from Figure 7-7, the total heat lost per lb of flue gas is:

Total heat loss = 54.7 + 108.8 = <u>163.5 Btu/lb of flue gas</u>

Since each lb of flue gas represents an input of 876.6 Btu, the efficiency loss is:

Efficiency loss = 163.5/876.6 = <u>18.7%</u>.

The preceding analysis can be used as a basis for evaluating many proposed energy management improvement measures. For example, if waste heat recovery can be used in the flue to reduce the exhaust gas temperature to 350°F, the heat lost will be 54.7 + [(350 − 80)/(500 − 80)] × 108.8, or 124.6 Btu/lb of flue gas, for a loss of 124.6/876.6, or 14.2%. This represents an efficiency improvement of 4.5%—a substantial sum when multiplied by the amount paid each year for boiler fuel. This approach can also be used to determine the energy loss effects of two changes in operational parameters.

Example 7-5: Suppose that two measures were performed simultaneously on the coal-fired boiler described earlier (results shown in Figures 7-4, 7-5 and 7-6): (1) the amount of excess air was decreased until the first traces of CO appeared, and this was at 10% excess air; and, (2) a waste heat recovery device was installed in the flue and reduced the exhaust temperature from 500°F to 430°F.

Solution: From the calculations in the previous section just prior to Figure 7-5, the amount of air needed for complete combustion was 10.675 lb per lb of fuel gas. With 10% excess air, the total amount of air is then 1.10 × 10.675 = 11.74 lb air per lb of fuel. The other computations are shown in Figures 7-9 and 7-10 (pages 269 and 270).

Since the initial heating value of the coal is 12,700 Btu/lb, and since 1 lb of fuel gas gives 12.885 lb of flue gas products, the heat per lb of flue gas products is:

Heat lost = 12,700 Btu/12.885 lb gas = 985.6 Btu/lb gas

and the loss of 150.9 Btu/lb of flue gas represents a loss of efficiency of:

Loss of efficiency = 150.9/985.6 = 15.3%.

Before leaving combustion chemistry, it is important to consider the case of incomplete combustion. For any fuel, combustion is incomplete when CO is present in the flue gas. Oxidation of carbon into CO yields 3960 Btu/lb C; oxidation into CO_2 yields 14,093 Btu/lb C. Thus the percentage loss in efficiency due to combustion to CO is given by:

Figure 7-9. Calculation of flue gas composition by weight for Example 7-5. The weight balance is calculated as in Figure 7-6.

Constituent	lb/constituent /lb fuel as fired	Product	Lb Product /lb constituent	lb product /lb fuel (10% excess air)	Total lb product /lb fuel	Lb product /lb flue gas
C	0.762	CO_2	3.664	2.792	2.792	0.217 (=2.792/12.885)
H_2	0.062	H_2O	8.940	0.554		
H_2O (in fuel)	0.053			0.053		
Air (additional) H_2O	11.743 (=10.675 × 1.10)		0.013	0.153 / 0.760	0.760	0.059
S	0.018	SO_2	1.998	0.036	0.036	0.003
N_2 (in fuel)	0.026			0.026		
Air (N_2 component)	11.743	N_2	0.769	9.024 / 9.050	9.050	0.702
O_2 (excess)					0.247 (=0.10 × 2.471)	0.019

Weight (wet) 12.885
Weight (dry) 12.125 (= 12.885 − 0.760)

Figure 7-10. Combined heat loss calculations for exhaust gas.

Flue gas Component	lb component /lb flue gas	/	Molecular weight	×	Specific heat, Btu/lb mol-deg	×	430-80	=	Btu
CO_2	0.217		44		9.9		350		17.1
H_2O (sensible)	0.059		18		8.2		350		9.4
SO_2	0.003		64		10.1		350		0.2
N_2	0.702		28		7.0		350		61.4
O_2	0.019		32		7.2		350		1.5

Total sensible heat loss (Btu/lb flue gas): 89.6

Latent heat loss = 0.059 lb × 1040 Btu/lb: 61.36

Total heat loss/lb flue gas 150.9

$$\text{percent loss} = \frac{14{,}093 - 3{,}960}{14{,}093} \times \frac{\text{ppm CO}}{\text{ppm CO} + \text{ppm CO}_2} \times 100 \qquad (7\text{-}12)$$

where the concentrations of CO_2 and CO come from the flue gas analysis. If, for example, the concentrations of CO_2 and CO were, respectively, 90,000 and 800 ppm (parts per million, by volume), then the percent loss in efficiency would be

$$\text{percent loss} = \frac{.719 \times 800 \times 100}{90800}$$

$$= .633\%$$

This loss is small, but there are other reasons for avoiding incomplete combustion. Incomplete combustion is associated with CO or smoke or both. CO is poisonous even in small quantities, and standards generally keep it below 400 ppm in residences. Smoke fouls the interior of boilers; it can also make the facility an undesirable member of a community. Furthermore, the presence of products of incomplete combustion can cause inherently unstable situations chemically; boiler explosions, though infrequent, can result. It is possible to have *both* incomplete combustion *and* excess combustion air in older boiler units if the burner forces combustion products onto a cold surface.

7.2 TYPES OF BOILERS

Boilers can be categorized into several types: packaged boilers, field-erected boilers, and fluidized-bed boilers. Packaged boilers are delivered to a customer with all the controls attached, ready for installation upon arrival. They are manufactured, low-capacity (350,000-35,000,000 Btu/h at 300 psig maximum) units and are designed to be operated with a minimum of skill and attention. They can be either a fire tube type (with water surrounding the tubes), as shown in Figure 7-11, or a water tube type.

In contrast to the packaged boiler, the field-erected boiler is a large-capacity, custom-built unit with specialized controls; it usually requires specially trained operators. The water in these boilers is in the tubes that make up the wall of the combustion chamber and in the economizers and other boiler components designed to contribute directly to steam distribu-

Figure 7-11. Packaged Scotch boiler.
(Courtesy of American Boiler Manufacturer's Association.)

tion. Such boilers are complicated and are discussed in detail in References 2 and 3.

Fluidized-bed boilers are becoming more important in energy management because they can burn many kinds of industrial wastes as fuel with very little air pollution. In most boilers, fuel is either burned in front of burners within the combustion chamber or burned on a traveling grate. In a fluidized-bed boiler, however, air is forced upward through a series of holes into the bottom of a bed of broken limestone, and fuel is inserted on top of the limestone. The air holds the limestone in suspension and provides combustion air for the fuel; the limestone captures sulfur compounds in the fuel and combines with them to form sulfates, solid products that can be removed and used for fertilizer. In addition to solving the problem of sulfur, fluidized-bed boilers also burn at a temperature lower than ash-softening temperatures, resulting in much simpler ash handling than with conventional boilers. Also, soot blowers are not required.

7.3 BOILER FUELS

The cost and time needed to understand combustion chemistry are justified by the cost penalties associated with inefficient operation. In the search for more efficient ways to operate a facility, the energy manager should examine the boiler fuel presently used to see whether it is cost-effective to replace the fuel with a waste product or with a less expensive or more available fuel. In Section 7.3.1, conventional fuels are discussed together with their general advantages and disadvantages; this discussion forms a base for Section 7.3.2 on less conventional fuels, including various types of industrial waste.

7.3.1 Characteristics of Common Industrial Fuels

A comparison of industrial fuels must examine the following characteristics of each fuel: (1) cost per Btu as a raw material; (2) *availability* in any kind of weather and any international political climate; (3) complexity of the on-site *equipment* needed to transport and burn; (4) problems associated with *storage* of the fuel; (5) type of *emissions* caused by combustion; (6) and, historical success of the *technology* for boilers using this fuel. Consider coal, fuel oil, and natural gas in the light of these characteristics.

7.3.1.1 Coal.

The cost of coal as delivered ranges from \$35 to \$70/ton, about \$1.40-\$2.80/million Btu. It is readily available, but it is subject to labor strikes and occasionally to weather-caused transportation problems. At the boiler plant, the coal must be removed from its trucks, railroad cars, or slurry pipeline and transported to the boiler, often with intermediate processing. Burning coal usually involves complicated mechanical equipment such as spreader-stokers or atomizing burners. The technology of this equipment is, however, well developed, and the procedures for its maintenance are well-known. Coal storage does not present a major problem, although it (1) requires space, (2) requires material handling, (3) can be subject to spontaneous combustion, and (4) can be subject to freezing. Again, these problems are not new, and their solutions are well-known.

Since the Clean Air Act of 1977, the amount of NO_x and SO_x (nitrogen and sulfur oxides) in flue gas emissions have been regulated by law, and hence the amount of these oxides for a given fuel is a matter of some concern. Most coal-fired boilers operate more efficiently at high (2100-2500°F) temperatures than at lower temperatures. Unfortunately, more NO_x forms at these temperatures, and the residence time in the boiler is not sufficient for it to dissociate into O_2 and N_2. SO_x is also a problem, depending on the sulfur content of the coal being burned. The reduction of NO_x and SO_x to legal levels requires expensive flue gas treatment and has made low-sulfur fuels and low temperature combustion processes more competitive economically. The technology for mining and burning coal has evolved over many centuries; both the problems and their solutions are well known. Figure 7-4 gave a chemical analysis for a typical coal.

The Clean Air Act Amendments of 1990 (P.L 101-549) is considered to be the most complex and wide-reaching environmental law ever passed by the U.S. Congress [4]. Title IV of the CAAA-90 relates to Acid Deposition Control. Acid deposition is more commonly called acid rain, and occurs as a result of SO_2 and NO_x emissions from coal-fired power plants, automobiles and industrial boilers reacting with water vapor in the atmosphere, and returning to ground as destructive acids. Coal-fired power plants are the major source of SO_2 emissions. CAAA-90 mandates a 10-million ton reduction in annual SO_2 emissions from the 1980 baseline level—a 50% reduction.

Title IV is designed to encourage energy conservation and renewable and clean alternative fuel technologies as ways to reduce air pollution generated by energy production processes. The CAAA-90, Title IV, created an innovative trading system with a fixed number of fully marketable "allowances." An "allowance" authorizes the emission of one ton of SO_2, and utilities must not emit SO_2 in amounts that exceed the number of allowances they hold [3]. Existing utility power plants are authorized to emit certain amounts of SO_2 per year, based on historical operation levels and specific emission rates. CAAA-90 structured a two-phased approach to meet the ultimate goal of a fixed cap of emissions of 8.95 million tons—or 8.95 million allowances.

Title IV provides utilities with market-based incentives for meeting—and exceeding—the required reductions in SO_2 emissions. Utilities who choose to use energy conservation and Demand Side Management programs as ways to meet these requirements can develop a compliance flexibility that may help them meet CAAA-90 goals in a less costly manner to their customers. The ability to use energy conservation as a CAAA-90 compliance strategy will result in an overall increase in energy efficiency and DSM programs among utilities, and will improve the cost-effectiveness of many of these programs [5].

7.3.1.2 Fuel oil.

This classification includes everything from No. 1, a distillate, such as the kerosene used for home heating, to No. 6, a heavy residual oil. Prices for No. 6 fluctuate around $20-$50/42-gal barrel. The availability of this fuel is not affected by weather, but the world political climate can change the supply dramatically as was proved by the Arab Oil Embargo of 1973 and the Persian Gulf Conflict of 1991. The offloading and storage of fuel oil is usually uncomplicated, but its viscosity increases in cold weather, and, depending on the grade, it may be necessary to provide auxiliary heating before it will flow (see Table 7-3). Methods for solving the viscosity problem are well known. As with coal, NO_x and SO_x emissions can present a problem.

7.3.1.4 Natural gas.

Gas has many advantages: it is clean; it has no particulate emissions and its exhaust can therefore be used in gas turbines; it is easy to transport with pipelines; and it is easy to burn. Because of its convenience and because its price was kept low by government regulation for a long time, it has been the preferred fuel for many applications. The decontrol of natural gas has, however, allowed the price of gas to increase and to fluctuate, and has made other sources of energy more attractive than before. The Clean Air Act Amendments of 1990 are greatly restricting the emissions of SO_2 and NO_x, and this is providing a renewed growth rate for gas use. NO_x and SO_x do not seem to cause the same problems when gas is the fuel as they do for coal and fuel oil, and gas remains the fuel of choice for many users. The technology is well-known and less complicated than that for coal or fuel oil. The components of a gas fuel include various hydrocarbons; a typical example is the natural gas that was shown in Figure 7-1.

7.3.2 Unconventional Fuels

Recently companies have begun to search for fuels less expensive than fuel oil, coal, and gas. A natural source for an inexpensive fuel is manufacturing waste, and this can be almost anything. Waste materials being used as fuels include pulp mill liquor, sawdust, food processing waste, municipal garbage [6], coal wash water, coffee grounds, cardboard, bark, and bagasse (sugar cane after the liquid has been extracted). Some sample analyses are given in Table 7-4, with natural gas included for comparison purposes. Using industrial waste as fuel can simplify the refuse disposal problem for a company as well as provide it with an inexpensive source of heat.

Still, there are some problems associated with burning any new fuel. The technology for dealing with coal, gas, or fuel oil is well-known. Using new fuels has, however, raised questions such as: (1) How high in the combustion chamber should the new fuel be injected into the boiler? (This is critical in burning municipal waste.) (2) What kind of problems will the ash or residue create? (3) What modifications are needed to burners and other boiler components? (4) What storage problems can be expected? (5) How regular will the supply be? It often pays to examine the operation of a successful facility before making a significant new investment, and it also pays to have the best possible engineering advice.

Table 7-5 shows costs and savings for a typical boiler and refuse-handling system which burns industrial waste.

Table 7-3. Characteristics of Fuel Oils[a]

Grade of fuel oil	Flash point, °C (°F), min.	Pour point, °C (°F), max.	Water and Sediment (vol. %), max.	Carbon residue on 10% bottoms (%), max.	Ash weight (%), max.	Distillation temp. °C (°F) 10% Point, max.	90% point min.	90% point max.
No. 1: Distillate oil intended for vaporizing pot-type burners and other burners requiring this grade	38 or legal (100)	-18^c (0)	.05	.15	—	215 (420)	—	288 (550)
No. 2: Distillate oil for general purpose heating for use in burners not requiring No. I	38 or legal (100)	-6^c (20)	.05	35	—	—	282^c (540)	338 (640)
No. 4: Preheating not usually required for handling or burning	55 or legal (130)	-6^c (20)	.50	—	.10	—	—	—
No. 5 (light): Preheating may be required depending on climate and equipment	55 or legal (130)	—	1.00	—	.10	—	—	—

Table 7-3. Characteristics of Fuel Oils[a] (Continued)

Grade of fuel oil	Flash point, °C (°F), min.	Pour point, °C (°F), max.	Water and Sediment (vol. %), max.	Carbon residue on 10% bottoms (%), max.	Ash weight (%), max.	Distillation temp. °C (°F) 10% Point, max.	90% point min.	90% point max.
No. 5 (heavy): Preheating may be required for burning and, in cold climates, may be required for handling	55 or legal (130)	—	1.00	—	10	—	—	—
No. 6: Preheating required for burning and handling	60 (140)	d	2.00[e]	—	—	—	—	—

[a] It is the intent of these classifications that failure to meet any requirement of a given grade does not automatically place an oil in the next lower grade unless in fact it meets all requirements of the lower grade.

[b] Viscosity values in parentheses are for information only and not necessarily limiting.

[c] Lower or higher pour points may be specified whenever required by conditions of storage or use. When pour point less than −18°C (0°F) is specified, the minimum viscosity for Grade No. 2 shall be 1.8 cST(32.0 SUS) and the minimum 90% point shall be waived.

[d] Where low-sulfur fuel oil is required, Grade 6 fuel oil will be classified as low pour +15°C (60°F) max. or high pour (no max.). Low pour fuel oil should be used unless all tanks and lines are heated.

[e] The amount of water by distillation plus the sediment by extraction shall not exceed 2.00%. The amount of sediment by extraction shall not exceed .50%. A deduction in quantity shall be made for all water and sediment in excess of 1.0%.

Table 7-3. (Continued)

Saybolt viscosity, S^b				Kinematic viscosity, cST^b				Specific gravity 60/60°F (deg API), max.	Copper strip corrosion, max.	Sulfur (%), max.
Universal at 38°C (100°F)		Furol at 50°C (122°F)		At 38°C (100°F)		At 50°C (122°F)				
min.	max.	min.	max.	min.	max.	min.	max.			
—	—	—	—	1.4	2.2	—	—	.8499 (35 min.)	No. 3	.5 or legal
(32.6)	(37.9)	—	—	2.0^c	3.6	—	—	.8762 (30 min.)	No. 3	.5g or legal
(45)	(125)	—	—	5.8	26.4^f	—	—	—	—	Legal
(>125)	(300)	—	—	>26.4	65^f	—	—	—	—	Legal
(>300)	(900)	(23)	(40)	>65	194^f	(42)	(81)	—	—	Legal
(>900)	(9000)	(>45)	(300)	>92	638^f	50°C	(122°F)	—	—	Legal

fWhere low-sulfur fuel is required, fuel oil falling in the viscosity range of a lower-numbered grade down to and including No. 4 may be supplied by agreement between purchaser and supplier. The viscosity range of the initial shipment shall be identified and advance notice shall be required when changing from one viscosity range to another. This notice shall be in sufficient time to permit the user to make the necessary adjustments.

gIn countries outside the United States other sulfur limits may apply.

Source: Reference I. From *Combustion Fossil Power Systems*, Combustion Engineering, Inc. Used with permission. Excerpted from *ASTM Standards D 396*, Specification for Fuel Oils. Copyright, American Society for Testing and Materials, Philadelphia. Reprinted with permission.

Table 7-4. Composition of Nontraditional Fuels

Fuel	Sulfur (S)	Hydrogen (H₂)	Carbon (C)	Component oxygen (O₂)	Moisture (H₂O)	Ash	Heating value (Btu/lb)
Pine bark (dry basis)	.1%	5.6%	53.4%	37.9%	(50%)	2.9%	9,030
Natural gas	—	23.30	74.72	1.22	—	(.76% N₂)	22,904
Fuel oil No. 6	12.0	10.5	85.7	.92	2.0	.08	18,270
Coke breeze	.6	.3	80.0	.5	7.3	11.0	11,670
Bagasse	—	2.8	23.4	20.0	52.0	1.7	4,000
Municipal garbage (metals removed)	.1-.4	3.4-6.3	23.4-42.8	15.4-31.9	19.7-31.3	9.4-26.8	3100-6500

Source: References 1 and 2.

Table 7-5. Waste-burning boiler economics, sample case.

Savings (1980 dollars)	Costs (1980 dollars)
Coal and natural gas: $250,000/year	Site preparation: $235,000
	Building to house system: $465,000
Trash hauling and landfill: $362,000/year	Equipment support structures: $125,000
	Boiler and trash-handling equipment: $1,250,000
	Piping: $200,000
	Instrumentation: $160,000
	Crew locker room: $140,000
	Miscellaneous mechanical equipment: $85,000
	Spare parts: $40,000

Before leaving this section, it should be noted that other factors must be taken into account in the decision to use an alternate fuel. First is the need to have some kind of backup boiler if the waste fuel is not available. This problem is particularly severe when one company uses the waste that is produced by another company; it was also responsible for some of the early difficulties experienced in the attempt to use municipal refuse as a fuel. A second major factor is the political climate. It is necessary to determine what governmental agencies must give their approval before a particular plan can be put into effect. In the case of municipal refuse, political problems probably delayed more projects than did technological difficulties. Where to store the refuse on a interim basis can also be a problem.

7.3.3 Cost Comparison Example

To illustrate some of the factors involved in a typical choice of boiler fuels, consider the following example. A company is using process steam at rates that sometimes reach 300,000 lbs/hr. The company presently has a gas-fired boiler capable of meeting its needs. Someone has noticed, however, that this and similar local companies in the same industry are sending significant amounts of non-toxic combustible waste to a local landfill. This person has suggested that these wastes might be profitably used as a replacement source of fuel. A preliminary study has indicated that this fuel usage will be acceptable to EPA and other local and federal authorities, that the other companies will buy into this solution, and that there seem to be no negative environmental aspects to such usage. The study showed three alternatives to be viable: (1) Continue buying gas and sending the waste to a local landfill; (2) Construct two boilers, one for waste and capable of efficient operation from 90,000 to 200,000 lb/hr, and one burning coal with an efficient operating range of 30,000 to 100,000 lb/hr; (3) Construct a single waste-fired boiler with an efficient capacity of 210,000 to 300,000 lb/hr, and charge $15.00/T for burning acceptable industrial wastes, estimated at 30,000 T/yr from nearby companies.

This company uses a 15% after-tax return on investment as a minimum attractive rate of return, based on the first 10 years of projected cash flows. Costs shown are in addition to a projected 4% annual inflation rate. Boilers are assumed to be non-residential real property subject to straight-line depreciation over 31.5 years. It is assumed that the company is in a 34% tax bracket.

The first step is to determine the details and the costs of each alternative. Alternative 1, the present system, uses purchased gas and has costs of trash hauling and landfill fees added to the usual operation and maintenance of the boiler. Gas presently costs $10.00/Mcf (at 1 million Btu/Mcf, taking boiler efficiency into account). Present gas costs are $5,000,000/yr. This represents about 50×10^{10} usable Btu/yr. The company presently produces 40,000 tons of combustible waste per year. This waste has been analyzed and found to contain 16% ash by weight and to have a heating value of 6390 Btu/lb as fired. It is estimated that a waste-fired boiler of the type contemplated would have an efficiency of 75%. The usable heat content of the company waste is therefore 6390 Btu/lb $\times$ 40,000 T $\times$ 2,000 lb/T $\times .75 = 38.3 \times 10^{10}$ Btu/yr, an amount which would need to be supplemented by some other energy source to meet the needs of the plant. This waste is presently transported to a landfill at a cost of $1.25/ton and then landfilled at $2.50 per ton (tipping fee). Waste hauling costs for this company are not expected to increase, but landfill costs are expected to increase 30% per year for the next five years and 10% per year thereafter. These same rates will hold for any ash that is landfilled.

Alternative 2, the two boiler combination, avoids the gas cost and all of the cost of waste haulage and waste landfilling. This alternative, however, causes the company to incur the initial capital cost of the boilers and higher operating and maintenance costs than under the present system. In addition, there is the cost of hauling and landfilling the ash. The coal to be used has a heating value of 12,780 Btu/lb and an ash content of 9.6%. The coal boiler efficiency is estimated as 82%, giving the effective heating value of the coal as 21 million Btu/T. The amount of coal needed per year is calculated from $(50 - 38.3) \times 10^{10}$ Btu/(12,780 Btu/lb $\times$ 2000 lb/T $\times .82$

efficiency) = 5580 T/yr. Coal costs are projected to be $55.00/T for the near
future. Ash comes from the waste and from the coal; the amount from the
waste is 40,000 T ×.16 = 6400 T/yr; the amount from the coal is 5580 T
×.096 = 540 T/yr.

Alternative 3, the large waste-fired boiler, avoids the gas cost and all
of the waste haulage and landfilling expense but incurs a larger capital
cost. This alternative will help pay for itself with the revenue generated
from industrial customers, in addition to the company costs it avoids. This
revenue is estimated as 30,000 T/yr × $15.00/T, or $450,000/yr. This
alternative has ash haulage and landfilling cost. Since the total amount of
waste burned per year is 70,000 T, the ash to be disposed of is 70,000 T ×.16
= 11,200 T/yr.

These costs are shown in Table 7.6, and the analysis in Table 7.7. The
depreciation rates were taken from Reference 7. The analysis shows a
distinct economic advantage in favor of Alternative 3, the large single
boiler. To complete this analysis it would be necessary to do a number of
sensitivity analyses, testing the effect of different foreseeable circumstances on the economics shown.

Table 7-6. Costs of waste-burning boiler alternatives.

	Present System	Two Boilers	One Big Boiler
FIRST COST	None	$12,500,000	$14,000,000
ANNUAL COSTS			
Gas	$5,000,000	$0	$0
Coal	$0	$306,900	$0
Boiler Maintenance	$50,000	$300,000	$250,000
Waste transportation	$50,000 (40,000 T × $1.25)	$0	$0
Waste landfilling (First year)	$20,000 (40,000 T × $.50)	$0	$0
Ash transportation	$0	$8,700 (6940 T × $1.25)	$14,000 (11,200 T × $1.25)
Ash landfilling (First year)	$0	$17,350 (6940 T × $2.50)	$28,000 (11,200 T × $2.50)
ANNUAL REVENUES			
Waste from other companies			$450,000 (30,000 T × $15.00/T)

T = ton

Table 7-7. Present worth calculations for example problem (costs/1000)

Present System

End of Year	First Cost	Fuel, transportation, and maintenance	Waste Landfilling	Revenues (waste)	Total	Depreciation	Net Before-tax Costs	Taxes	Net Costs
0	$0				$0	$0	$0	$0	$0
1		$5,100	$20	$0	$5,120	$0	$5,120	($1,741)	$3,379
2		$5,100	$26	$0	$5,126	$0	$5,126	($1,743)	$3,383
3		$5,100	$34	$0	$5,134	$0	$5,134	($1,745)	$3,388
4		$5,100	$44	$0	$5,144	$0	$5,144	($1,749)	$3,395
5		$5,100	$57	$0	$5,157	$0	$5,157	($1,753)	$3,404
6		$5,100	$63	$0	$5,163	$0	$5,163	($1,755)	$3,407
7		$5,100	$69	$0	$5,169	$0	$5,169	($1,757)	$3,412
8		$5,100	$76	$0	$5,176	$0	$5,176	($1,760)	$3,416
9		$5,100	$84	$0	$5,184	$0	$5,184	($1,762)	$3,421
10		$5,100	$92	$0	$5,192	$0	$5,192	($1,765)	$3,427
11		$5,100	$101	$0	$5,201	$0	$5,201	($1,769)	$3,433

Net present worth of costs at 15%: $17,788

Table 7-7. Present worth calculations for example problem (costs/1000) (*Continued*)

Two Boilers

End of Year	First Cost	Fuel, transportation, and maintenance	Waste Landfilling	Revenues (waste)	Total	Depreciation (Note)	Before-tax Costs	Taxes	Net Costs
0	$12,500				$12,500		$12,500	$0	12,500
1		$616	$17	$0	$633	$198	$435	($148)	$287
2		$616	$23	$0	$638	$397	$241	($82)	$159
3		$616	$29	$0	$645	$397	$248	($84)	$164
4		$616	$38	$0	$654	$397	$257	($87)	$169
5		$616	$50	$0	$665	$397	$268	($91)	$177
6		$616	$55	$0	$670	$397	$273	($93)	$180
7		$616	$60	$0	$676	$397	$279	($95)	$184
8		$616	$66	$0	$682	$397	$285	($97)	$188
9		$616	$73	$0	$688	$397	$291	($99)	$192
10		$616	$80	$0	$695	$397	$298	($101)	$197
11		$616	$88	$0	$703	$397	$306	($104)	$202

Net present worth of costs at 15%: $13,518

Note: First year depreciation is $1/2 \times 12,500/31.5$.
Next and following years, depreciation is 12,500/31.5.

Table 7-7. Present worth calculations for example problem (costs/1000) (*Continued*)

One Big Boiler

End of Year	First Cost	Fuel, transportation, and maintenance	Waste Landfilling	Revenues (waste)	Total	Depreciation	Before-tax Costs	Taxes	Net Costs
0	$14,000				$14,000	$0	$14,000	$0	14,000
1		$264	$28	($450)	($158)	$222	($380)	$129	($251)
2		$264	$36	($450)	($150)	$444	($594)	$202	($392)
3		$264	$47	($450)	($139)	$444	($583)	$198	($385)
4		$264	$62	($450)	($124)	$444	($568)	$193	($375)
5		$264	$80	($450)	($106)	$444	($550)	$187	($363)
6		$264	$88	($450)	($98)	$444	($542)	$184	($358)
7		$264	$97	($450)	($89)	$444	($533)	$181	($352)
8		$264	$106	($450)	($80)	$444	($524)	$178	($346)
9		$264	$117	($450)	($69)	$444	($513)	$174	($339)
10		$264	$129	($450)	($57)	$444	($501)	$170	($331)
11		$264	$142	($450)	($44)	$444	($488)	$166	($322)

Net present worth of costs at 15%: $12,191

7.4 SUMMARY

In this chapter we have presented the reactions and the standard analysis techniques needed for a fundamental understanding of combustion. Specific techniques discussed were weight balances, heat balances, and flue gas analysis. These techniques were used to demonstrate the importance of proper control of air quantities for combustion of common fuels and as a background for the discussion of unconventional fuels. An example was presented that showed the analysis steps and some of the problems to be considered when evaluating industrial waste as an alternative fuel source.

REFERENCES
1. Irvin Glassman, *Combustion*, Second Edition, Academic Press, Orlando, FL, 1989.

2. *Steam/Its Generation and Use*, Babcock and Wilcox, New York, 1978.

3. *Heat Engineering*, Foster Wheeler Corporation, Livingston, NJ, 1982.

4. U.S. Department of Energy, Clean Air Act Amendments of 1990: A Reference Handbook, Office of Conservation and Renewable Energy, Washington, DC, March 12, 1992, p. 1.

5. Barney L. Capehart, Improved Energy Efficiencies: Response to the Clean Air Act Amendments of 1990, Presented at the 1992 World Energy Engineering Congress, Atlanta, GA, October, 1992.

6. P.J. Schindler, Municipal Waste Combustion Assessment, U.S. Environmental Protection Agency, EPA/600/S8-89/58, Research Triangle Park, NC, 1990.

7. G.J. Theusen and W.J. Fabrycky, *Engineering Economy*, Eighth Edition, Prentice-Hall, Englewood Cliffs, NJ, 1993.

BIBLIOGRAPHY
Cleaver-Brooks Training Center Manual for Boilers, Cleaver-Brooks Company, Lebanon, PA, 1979.

Combustion and Fuels, Chapter 15, *ASHRAE Handbook: 1993 Fundamentals*, American Society for Heating, Refrigerating, and Air Conditioning Engineers, Atlanta, Ga., 1993.

Dukelow, Sam G., *The Control of Boilers*, Instrument Society of America Press, Research Triangle Park, NC, 1986.

Galvin, Cindy, "Trash-Fired Boiler Nets Paper Firm $600,000 Savings," *Energy User News*, Vol. 7, No. 4, p. 8, Jan. 25, 1982.

Payne, William F., *Efficient Boiler Operations Sourcebook, Third Edition*, Fairmont Press, Atlanta, GA, 1991.

Schroeder, George, *Combustion and Energy Loss in Boilers*, Mountain Fuel Supply Company, Salt Lake City, 1977.

Scollon, R.B., and R.D. Smith, "Boilers and Fired Systems," in Chap. 5, *Energy Management Handbook, Second Edition*, Wayne C. Turner, Senior Editor, Fairmont Press, Atlanta, GA, 1993.

Singer, Joseph G., *Combustion/Fossil Power Systems*, Combustion Engineering, Inc., Windsor, Conn., 1981.

Taplin, Harry, *Combustion Efficiency Tables*, Fairmont Press, Atlanta, GA, 1991.

Chapter 8
Steam Generation and Distribution

8.0 INTRODUCTION

One major use of fuel in many facilities is to generate steam which is then used to provide space heat, process heat and mechanical power. Many significant opportunities for energy cost reduction can be found in a good technical examination of the steam generation and distribution systems in buildings and industries. Such an examination first estimates the amount of energy coming into a steam generation system and then determines where the energy goes. These estimates provide a guide to possible waste heat utilization and other ways to improve the efficiency of a boiler, and can be used to evaluate the insulation possibilities for steam distribution lines. Maintenance of the steam system is discussed in some detail, as are waste heat recovery systems and applications. Cogeneration is also discussed, and conditions for feasibility are presented.

8.1 THE HEAT BALANCE

An underlying principle of energy analysis is that the most effort should be placed where the largest opportunity exists. For boiler operations, there are two areas with significant energy savings potential— where the boiler energy comes from—the energy sources—and where it goes—the energy sinks. The fundamental tool for analyzing boilers and steam distribution systems is the heat balance which is used to determine the energy sources and energy sinks within the system.

A heat balance equates the energy entering a system to the energy leaving the system. The energy sources are ranked in order of the amount of energy they supply, and the energy sinks are ranked by the amount of heat they exhaust to the environment. Then the source components are examined to see whether each of their functions can be performed with less energy, and the sink components are examined to see if they can feed their energy back in some useful way to one or more of the source components. Because this process shows where energy is utilized or wasted in the boiler and steam distribution system, the benefits to be gained from insulation and from waste heat utilization can be determined.

A heat balance is intended to account for all the heat that goes into a system and to find where all of it leaves the system. This same accounting principle was used in the mass balance of chemical reactions in Chapter 7. Using the heat balance requires understanding the heat content of steam—a concept embodied in the term *enthalpy*, and understanding the basic principles of heat transfer.

8.1.1. Enthalpy.

At atmospheric pressure it takes about 970 Btu to convert 1 lb of water to steam at 212°F. This energy, called *latent* heat, is given up whenever the steam condenses. Steam also carries *sensible* heat, heat proportional to the temperature difference through which the steam was heated. Steam under pressure contains additional heat due to the mechanical work done on the vapor. (The latent heat is generally lower at higher pressures.) The sum of the latent heat, the sensible heat, and the mechanical work is called *enthalpy* and, when expressed in Btu/lb, *specific enthalpy*. (This definition of enthalpy does not take into account the internal kinetic energy of the steam, but this can usually be neglected in energy management [1].) When water is present with the steam, as in most steam distribution systems, the steam is said to be saturated, and the pressure increases as the temperature increases. Furthermore the specific volume, ft^3/lb, decreases as the temperature increases. The enthalpy and specific volume are found in Table 8-1 for various saturation pressure conditions and in Table 8-2 for various saturation temperature conditions.

These tables have many uses. For example, suppose that a steam leak is estimated to be losing 150 lb of steam per hour from a 1200-psi line to an ambient temperature of 70°F. From Table 8-1, the enthalpy at 1200 psi is 1183.4 Btu/lb and, from Table 8-2, the enthalpy at 70°F is 38 Btu/lb. (From 70 to 212°F, the water is all liquid, and the only heat is sensible. Raising the temperature of 1 lb of *water* 1°F takes 1 Btu.) Thus the total change in enthalpy is 1183 − 38, or 1145 Btu/lb. A loss of 150 lb/h represents a loss of 150 lb/h × 1145 Btu/lb, or 171,750 Btu/h. If fuel costs $8.00/million Btu, this represents a cost of $1.37/h or almost $33/day on days when the facility is running constantly. The amount of *flash steam* released to the atmosphere by this leak is the volume of steam at 212°F multiplied by the estimated pounds of steam lost. Using Table 8-2 yields

26.80 ft^3/lb × 150 lb/h, or 4020 ft^3/h.

In some applications, steam is present without water. Such steam is said to be superheated, and its enthalpy and specific volume at various temperatures and pressures are given in Table 8-3.

Example 8-1: If a throttling valve reduces the temperature and pressure of superheated steam from 1300°F and 1200 psi to 700°F and 600 psi, how much energy is being lost?

Solution: Using Table 8-3, the enthalpy of superheated steam at 1200 psi and 1300°F is 1671.6 Btu/lb; the enthalpy at 600 psi and 700°F is 1351.8 Btu/lb. The difference in enthalpy is 1671.6 – 1351.8, or 319.8 Btu/lb. This represents heat energy that could possibly be used in some other part of the facility.

8.1.2 Heat gains.

To estimate the heat energy entering a steam generation and distribution system, the first step is to determine the amount and form of all heat entering the boiler and the second, to express heat inputs in the same units as outputs, namely Btu/h. The greatest source of heat input to a boiler (except for some waste heat boilers) is the boiler fuel itself. The heat energy per pound of fuel is generally known because it is part of the design criteria for the boiler.

The boiler system also contains heat in the combustion air, the returned condensate, and the boiler makeup water. The enthalpy in each of these sources is used as the measure of the heat energy the source contributes to the system. The enthalpy of combustion air depends on its moisture content and its temperature. For air containing .013 lb water vapor/lb air (the moisture content at 80°F and 60% relative humidity), the enthalpy at near atmospheric pressure is given in Table 8-4. (For a different humidity, the enthalpy of the moisture can be calculated using the formula for the specific heat of water vapor given in Reference 3.)

The enthalpy of the returned condensate can be determined from the temperature or pressure in the condensate line using Table 8-1 or Table 8-2. The enthalpy of the boiler makeup water, per pound, is the difference in °F between its temperature and 32°F.

Table 8-1. Properties of saturated steam and saturated water (pressure)

Press. psia	Temp F	Volume, ft³/lb			Enthalpy, Btu/lb			Entropy, Btu/lb, F			Energy, Btu/lb		Press. psia
		Water v_f	Evap v_{fg}	Steam v_g	Water h_f	Evap h_{fg}	Steam h_g	Water s_f	Evap s_{fg}	Steam s_g	Water u_f	Steam u_g	
0.0886	32.018	0.01602	3302.4	3302.4	0.00	1075.5	1075.5	0	2.1872	2.1872	0	1021.3	0.0886
0.10	35.023	0.01602	2945.5	2945.5	3.03	1073.8	1076.8	0.0061	2.1705	2.1766	3.03	1022.3	0.10
0.15	45.453	0.01602	2004.7	2004.7	13.50	1067.9	1081.4	0.0271	2.1140	2.1411	13.50	1025.7	0.15
0.20	53.160	0.01603	1526.3	1526.3	21.22	1063.5	1084.7	0.0422	2.0738	2.1160	21.22	1028.3	0.20
0.30	64.484	0.01604	1039.7	1039.7	32.54	1057.1	1089.7	0.0641	2.0168	2.0809	32.54	1032.0	0.30
0.40	72.869	0.01606	792.0	792.1	40.92	1052.4	1093.3	0.0799	1.9762	2.0562	40.92	1034.7	0.40
0.5	79.586	0.01607	641.5	641.5	47.62	1048.6	1096.3	0.0925	1.9446	2.0370	47.62	1036.9	0.5
0.6	85.218	0.01609	540.0	540.0	53.25	1045.5	1098.7	0.1028	1.9186	2.0215	53.24	1038.7	0.6
0.7	90.09	0.01610	466.93	466.94	58.10	1042.7	1100.8	0.3	1.8966	2.0083	58.10	1040.3	0.7
0.8	94.38	0.01611	411.67	411.69	62.39	1040.3	1102.6	0.1117	1.8775	1.9970	62.39	1041.7	0.8
0.9	98.24	0.01612	368.41	368.43	66.24	1038.1	1104.3	0.1264	1.8606	1.9870	66.24	1042.9	0.9
1.0	101.74	0.01614	333.59	333.60	69.73	1036.1	1105.8	0.1326	1.8455	1.9781	69.73	1044.1	1.0
2.0	126.07	0.01623	173.74	173.76	94.03	1022.1	1116.2	0.1750	1.7450	1.9200	94.03	1051.8	2.0
3.0	141.47	0.01630	118.71	118.73	109.42	1013.2	1122.6	0.2009	1.6854	1.8864	109.41	1056.7	3.0
4.0	152.96	0.01636	90.63	90.64	120.92	1006.4	1127.3	0.2199	1.6428	1.86-26	120.90	1060.2	4.0
5.0	162.24	0.01641	73.515	73.53	130.20	1000.9	1131.1	0.2349	1.6094	1.8443	130.18	1063.1	5.0
6.0	170.05	0.01645	61.967	61.98	138.03	996.2	1134.2	0.2474	1.5820	1.8294	138.01	1065.4	6.0
7.0	176.84	0.01649	53.634	53.65	144.83	992.1	1136.9	0.2581	1.5587	1.8168	144.81	1067.4	7.0
8.0	182.86	0.01653	47.328	47.35	150.87	988.5	1139.3	0.2676	1.5384	1.8060	150.84	1069.2	8.0
9.0	188.27	0.01656	42.385	42.40	156.30	985.1	1141.4	0.2760	1.5204	1.7964	156.28	1070.8	9.0
10	193.21	0.01659	38.404	38.42	161.26	982.1	1143.3	0.2836	1.5043	1.7879	161.23	1072.3	10
14.696	212.00	0.01672	26.782	26.80	180.17	970.3	1150.5	0.3121	1.4447	1.7568	180.12	1077.6	14.696
15	213.03	0.01673	26.274	26.29	181.21	969.7	1150.9	0.3137	1.4415	1.7552	181.16	1077.9	15
20	227.96	0.01683	20.070	20.087	196.27	960.1	1156.3	0.3358	1.3962	1.7320	196.21	1082.0	20
30	250.34	0.01701	13.7266	13.744	218.9	945.2	1164.1	0.3682	1.3313	1.6995	218.8	1087.9	30
40	267.25	0.01715	10.4794	10.497	236.1	933.6	1169.8	0.3921	1.2844	1.6765	236.0	1092.1	40

50	281.02	0.01727	8.4967	8.514	250.2	923.9	1174.1	0.4112	1.2474	1.6586	250.1	1095.3	50
60	292.71	0.01738	7.1562	7.174	262.2	915.4	1177.6	0.4273	1.2167	1.6440	262.0	1098.0	60
70	302.93	0.01748	6.1875	6.205	272.7	907.8	1180.6	0.4411	1.1905	1.6316	272.5	1100.2	70
80	312.04	0.01757	5.4536	5.471	282.1	900.9	1183.1	0.4534	1.1675	1.6208	281.9	1102.1	80
90	320.28	0.01766	4.8777	4.895	290.7	894.6	1185.3	0.4643	1.1470	1.6113	290.4	1103.7	90
100	327.82	0.01774	4.4133	4.431	298.5	888.6	1187.2	0.4743	1.1284	1.6027	298.2	1105.2	100
120	341.27	0.01789	3.7097	3.728	312.6	877.8	1190.4	0.4919	1.0960	1.5879	312.2	1107.6	120
140	353.04	0.01803	3.2190	3.219	325.0	868.0	1193.0	0.5071	1.0681	1.5752	324.5	1109.6	140
160	363.55	0.01815	2.8155	2.834	336.1	859.0	1195.1	0.5206	1.0435	1.5641	335.5	1111.2	160
180	373.08	0.01827	2.5129	2.531	346.2	850.7	1196.9	0.5328	1.0215	1.5543	345.6	1112.5	180
200	381.80	0.01839	2.2689	2.287	355.5	842.8	1198.3	0.5438	1.0016	1.5454	354.8	1113.7	200
250	400.97	0.01865	1.8245	1.8432	376.1	825.0	1201.1	0.5679	0.9585	1.5264	375.3	1115.8	250
300	417.35	0.01889	1.5238	1.5427	394.0	808.9	1202.9	0.5882	0.9223	1.5105	392.3	1117.2	300
350	431.73	0.01913	1.3064	1.3255	409.8	794.2	1204.0	0.6059	0.8909	1.4968	408.6	1118.1	350
400	444.60	0.0193	1.14162	1.1610	424.2	780.4	1204.6	0.6217	0.8630	1.4847	422.7	1118.7	400
450	456.28	0.0195	1.01224	1.0318	437.3	767.5	1204.8	0.6360	0.8378	1.4738	435.7	1118.9	450
500	467.01	0.0198	0.90787	0.9276	449.5	755.1	1204.7	0.6490	0.8148	1.4639	447.7	1118.8	500
550	476.94	0.0199	0.82183	0.8418	460.9	743.3	1204.3	0.6611	0.7936	1.4547	458.9	1118.6	550
600	486.20	0.0201	0.74962	0.7698	471.7	732.0	1203.7	0.6723	0.7738	1.4461	469.5	1118.2	600
700	503.08	0.0205	0.63505	0.6556	491.6	710.2	1201.8	0.6928	0.7377	1.4304	488.9	1116.9	700
800	518.21	0.0209	0.54809	0.5690	509.8	689.6	1199.4	0.7111	0.7051	1.4163	506.7	1115.2	800
900	531.95	0.0212	0.47968	0.5009	526.7	669.7	1196.4	0.7279	0.6753	1.4032	523.2	1113.0	900
1000	544.58	0.0216	0.42436	0.4460	542.6	650.4	1192.9	0.7434	0.6476	1.3910	538.6	1110.4	1000
1100	556.28	0.0220	0.37863	0.4006	557.5	631.5	1189.1	0.7578	0.6216	1.3794	553.1	1107.5	1100
1200	567.19	0.0223	0.34013	0.3625	571.9	613.0	1184.8	0.7714	0.5969	1.3683	566.9	1104.3	1200
1300	577.42	0.0227	0.30722	0.3299	585.6	594.6	1180.2	0.7843	0.5733	1.3577	580.1	1100.9	1300
1400	587.07	0.0231	0.27871	0.3018	598.8	576.5	1175.3	0.7966	0.5507	1.3474	592.9	1097.1	1400
1500	596.20	0.0235	0.25372	0.2772	611.7	558.4	1170.1	0.8085	0.5288	1.3373	605.2	1093.1	1500
2000	635.80	0.0257	0.16266	0.1883	672.1	466.2	1138.3	0.8625	0.4256	1.2881	662.6	1068.6	2000
2500	668.11	0.0286	0.10209	0.1307	731.7	361.6	1093.3	0.9139	0.3206	1.2345	718.5	1032.9	2500
3000	695.33	0.0343	0.05073	0.0850	801.8	218.4	1020.3	0.9728	0.1891	1.1619	782.8	973.1	3000
3208.2	705.47	0.0508	0	0.0508	906.0	0	906.0	1.0612	0	1.0612	875.9	875.9	3208.2

Table 8-2. Properties of saturated steam and saturated water (temperature)

Temp F	Press psia	Volume, ft3/lb			Enthalpy, Btu/lb			Entropy, Btu/lb, F			Temp F
		Water v_f	Evap v_{fg}	Steam v_g	Water h_f	Evap h_{fg}	Steam h_g	Water s_f	Evap s_{fg}	Steam s_g	
32	0.08859	0.01602	3305	3305	-0.02	1075.5	1075.5	0.0000	2.1873	2.1873	32
35	0.09991	0.01602	2948	2948	3.00	1073.8	1076.8	0.0061	2.1706	2.1767	35
40	0.12163	0.01602	2446	2446	8.03	1071.0	1079.0	0.0162	2.1432	2.1594	40
45	0.14744	0.01602	2037.7	2037.8	13.04	1068.1	1081.2	0.0262	2.1164	2.1426	45
50	0.17796	0.01602	1704.8	1704.8	18.05	1065.3	1083.4	0.0361	2.0901	2.1262	50
60	0.2561	0.01603	1207.6	1207.6	28.06	1059.7	1087.7	0.0555	2.0391	2.0946	60
70	0.3629	0.01605	868.3	868.4	38.05	1054.0	1092.1	0.0745	1.9900	2.0645	70
80	0.5068	0.01607	633.3	633.3	48.04	1048.4	1096.4	0.0932	1.9426	2.0359	80
90	0.6981	0.01610	468.1	468.1	58.02	1042.7	1100.8	0.1115	1.8970	2.0086	90
100	0.9492	0.01613	350.4	350.4	68.00	1037.1	1105.1	0.1295	1.8530	1.9825	100
110	1.2750	0.01617	265.4	265.4	77.98	1031.4	1109.3	0.1472	1.8105	1.9577	110
120	1.6927	0.01620	203.25	203.26	87.97	1025.6	1113.6	0.1646	1.7693	1.9339	120
130	2.2230	0.01625	157.32	157.33	97.96	1019.8	1117.8	0.1817	1.7295	1.9112	130
140	2.8892	0.01629	122.98	123.00	107.95	1014.0	1122.0	0.1985	1.6910	1.8895	140
150	3.718	0.01634	97.05	97.07	117.95	1008.2	1126.1	0.2150	1.6536	1.8686	150
160	4.741	0.01640	77.27	77.29	127.96	1002.2	1130.2	0.2313	1.6174	1.8487	160
170	5.993	0.01645	62.04	62.06	137.97	996.2	1134.2	0.2473	1.5822	1.8295	170
180	7.511	0.01651	50.21	50.22	148.00	990.2	1138.2	0.2631	1.5480	1.8111	180
190	9.340	0.01657	40.94	40.96	158.04	984.1	1142.1	0.2787	1.5148	1.7934	190
200	11.526	0.01664	33.62	33.64	168.09	977.9	1146.0	0.2940	1.4824	1.7764	200
210	14.123	0.01671	27.80	27.82	178.15	971.6	1149.7	0.3091	1.4509	1.7600	210
212	14.696	0.01672	26.78	26.80	180.17	970.3	1150.5	0.3121	1.4447	1.7568	212
220	17.186	0.01678	23.13	23.15	188.23	965.2	1153.4	0.3241	1.4201	1.7442	220
230	20.779	0.01685	19.364	19.381	198.33	958.7	1157.1	0.3388	1.3902	1.7290	230
240	24.968	0.01693	16.304	16.321	208.45	952.1	1160.6	0.3533	1.3609	1.7142	240
250	29.825	0.01701	13.802	13.819	218.59	945.4	1164.0	0.3677	1.3323	1.7000	250

Temp	P	v_f	v_{fg}	v_g	h_f	h_{fg}	h_g	s_f	s_{fg}	s_g	Temp
260	35.427	0.01709	11.745	11.762	228.76	938.6	1167.4	0.3819	1.3043	1.6862	260
270	41.856	0.01718	10.042	10.060	238.95	931.7	1170.6	0.3960	1.2769	1.6729	270
280	49.200	0.01726	8.627	8.644	249.17	924.6	1173.8	0.4098	1.2501	1.6599	280
290	57.550	0.01736	7.443	7.460	259.4	917.4	1176.8	0.4236	1.2238	1.6473	290
300	67.005	0.01745	6.448	6.466	269.7	910.0	1179.7	0.4372	1.1979	1.6351	300
310	77.67	0.01755	5.609	5.626	280.0	902.5	1182.5	0.4506	1.1726	1.6232	310
320	89.64	0.01766	4.896	4.914	290.4	894.8	1185.2	0.4640	1.1477	1.6116	320
340	117.99	0.01787	3.770	3.788	311.3	878.8	1190.1	0.4902	1.0990	1.5892	340
360	153.01	0.01811	2.939	2.557	332.3	862.1	1194.4	0.5161	1.0517	1.5578	360
380	195.73	0.01836	2.317	2.335	353.6	844.5	1198.0	0.5416	1.0057	1.5473	380
400	247.26	0.01864	1.8444	1.8630	375.1	825.9	1201.0	0.5667	0.9607	1.5274	400
420	308.78	0.01894	1.4808	1.4997	396.9	806.2	1203.1	0.5915	0.9165	1.5080	420
440	381.54	0.01926	1.1976	1.2169	419.0	785.4	1204.4	0.6161	0.8729	1.4890	440
460	466.9	0.0196	0.9746	0.9942	441.5	763.2	1204.8	0.6405	0.8299	1.4704	460
480	566.2	0.0200	0.7972	0.8172	464.5	739.6	1204.1	0.6648	0.7871	1.4518	480
500	680.9	0.0204	0.6545	0.6749	487.9	714.3	1202.2	0.6890	0.7443	1.4333	500
520	812.5	0.0209	0.5386	0.5596	512.0	687.0	1199.0	0.7133	0.7013	1.4146	520
540	962.8	0.0215	0.4437	0.4651	536.8	657.5	1194.3	0.7378	0.6577	1.3954	540
560	1133.4	0.0221	0.3651	0.3871	562.4	625.3	1187.7	0.7625	0.6132	1.3757	560
580	1326.2	0.0228	0.2994	0.3222	589.1	589.9	1179.0	0.7876	0.5673	1.3550	580
600	1543.2	0.0236	0.2438	0.2675	617.1	550.6	1167.7	0.8134	0.5196	1.3330	600
620	1786.9	0.0247	0.1962	0.2208	646.9	506.3	1153.2	0.8403	0.4689	1.3092	620
640	2059.9	0.0260	0.1543	0.1802	679.1	454.6	1133.7	0.8686	0.4134	1.2821	640
660	2365.7	0.0277	0.1166	0.1443	714.9	392.1	1107.0	0.8995	0.3502	1.2498	660
680	2708.6	0.0304	0.0808	0.1112	758.5	310.1	1068.5	0.9365	0.2720	1.2086	680
700	3094.3	0.0366	0.0386	0.0752	822.4	172.7	995.2	0.9901	0.1490	1.1390	700
705.5	3208.2	0.0508	0	0.0508	906.0	0	906.0	1.0612	0	1.0612	705.5

Table 8-3. Properties of superheated steam and compressed water.

Abs. press. lb/sq.in. (sat.temp)		\[Temperature, F\] 100	200	300	400	500	600	700	800	900	1000	1100	1200	1300	1400	1500
1 (101.74)	v	0.0161	392.5	452.3	511.9	571.5	631.1	690.7								
	h	68.00	1150.2	1195.7	1241.8	1288.6	1336.1	1384.5								
	s	0.1295	2.0509	2.1152	2.1722	2.2237	2.2708	2.3144								
5 (162.24)	v	0.0161	78.14	90.24	102.24	114.21	126.15	138.08	150.01							
	h	68.01	1148.6	1194.8	1241.3	1288.2	1335.9	1384.3	1433.6							
	s	0.1295	1.8716	1.9369	1.9943	2.0460	2.0932	2.1369	2.1776							
10 (193.21)	v	0.0161	38.84	44.98	51.03	57.04	63.03	69.00	74.98	80.94	86.91	92.87	98.84	104.80	110.76	116.72
	h	68.02	1146.6	1193.7	1240.6	1287.8	1335.5	1384.0	1433.4	1483.5	1534.6	1586.6	1639.5	1693.3	1747.9	1803.4
	s	0.1295	1.7928	1.8593	1.9173	1.9692	2.0166	2.0603	2.1011	2.1394	2.1757	2.2101	2.2430	2.2744	2.3046	2.3337
15 (213.03)	v	0.0161	0.0166	29.899	33.963	37.985	41.986	45.978	49.964	53.946	57.926	61.905	65.882	69.858	73.833	77.807
	h	68.04	168.09	1192.5	1239.9	1287.3	1335.2	1383.8	1433.2	1483.4	1534.5	1586.5	1639.4	1693.2	1747.8	1803.4
	s	0.1295	0.2940	1.8134	1.8720	1.9242	1.9717	2.0155	2.0563	2.0946	2.1309	2.1653	2.1982	2.2297	2.2599	2.2890
20 (227.96)	v	0.0161	0.0166	22.356	25.428	28.457	31.466	34.465	37.458	40.447	43.435	46.420	49.405	52.388	55.370	58.352
	h	68.05	168.11	1191.4	1239.2	1286.9	1334.9	1383.5	1432.9	1483.2	1534.3	1586.3	1639.3	1693.1	1747.8	1803.3
	s	0.1295	0.2940	1.7805	1.8397	1.8921	1.9397	1.9836	2.0244	2.0628	2.0991	2.1336	2.1665	2.1979	2.2282	2.2572
40 (267.25)	v	0.0161	0.0166	11.036	12.624	14.165	15.685	17.195	18.699	20.199	21.697	23.194	24.689	26.183	27.676	29.168
	h	68.10	168.15	1186.6	1236.4	1285.0	1333.6	1382.5	1432.1	1482.5	1533.7	1585.8	1638.8	1992.7	1747.5	1803.0
	s	0.1295	0.2940	1.6992	1.7608	1.8143	1.8624	1.9065	1.9476	1.9860	2.0224	2.0569	2.0899	2.1224	2.1516	2.1807
60 (292.71)	v	0.0161	0.0166	7.257	8.354	9.400	10.425	11.438	12.446	13.450	14.452	15.452	16.450	17.448	18.445	19.441
	h	68.15	168.20	1181.6	1233.5	1283.2	1332.3	1381.5	1431.3	1481.8	1533.2	1585.3	1638.4	1692.4	1747.1	1802.8
	s	0.1295	0.2939	1.6492	1.7134	1.768	1.8168	1.8612	1.9024	1.9410	1.9774	2.0120	2.0450	2.0765	2.1068	2.1359
80 (312.04)	v	0.0161	0.0166	0.0175	6.218	7.018	7.794	8560	9319	10075	10829	11.581	12.331	13081	13829	14.577
	h	68.21	168.24	269.74	1230.5	1281.3	1330.9	1380.5	1430.5	1481.1	1532.6	1584.9	1638.0	1692.0	1746.8	1802.5
	s	0.1295	0.2939	0.4371	1.6790	1.7349	1.7842	1.8289	1.8702	1.9089	1.9454	1.9800	2.0131	2.0446	2.0750	2.1041

Press. (Sat. Temp.)																
100 (327.82)	v	0.0161	0.0166	0.0175	4.935	5.588	6.216	6.833	7.443	8.050	8.655	9.258	9.860	10.460	11.060	11.659
	h	68.26	168.29	269.77	1227.4	1279.3	1329.6	1379.5	1429.7	1480.4	1532.0	1584.4	1637.6	1691.6	1746.5	1802.2
	s	0.1295	0.2939	0.4371	1.6516	1.7088	1.7586	1.8036	1.8451	1.8839	1.9205	1.9552	1.9883	2.0199	2.0502	2.0794
120 (341.27)	v	0.0161	0.0166	0.0175	4.0786	4.6341	5.1637	5.6831	6.1928	6.7006	7.2060	7.7096	8.2119	8.7130	9.2134	9.7130
	h	68.31	168.33	269.81	1224.1	1277.4	1328.1	1378.4	1428.8	1479.8	1531.4	1583.9	1637.1	1691.3	1746.2	1802.0
	s	0.1295	0.2939	0.4371	1.6286	1.6872	1.7376	1.7829	1.8246	1.8635	1.9001	1.9349	1.9680	1.9996	2.0300	2.0592
140 (353.04)	v	0.0161	0.0166	0.0175	3.4661	3.9526	4.4119	4.8585	5.2995	5.7364	6.1709	6.6036	7.0349	7.4652	7.8946	8.3233
	h	68.37	168.38	269.85	1220.8	1275.3	1326.8	1377.4	1428.0	1479.1	1530.8	1583.4	1636.7	1690.9	1745.9	1801.7
	s	0.1295	0.2939	0.4370	1.6085	1.6686	1.7196	1.7652	1.8071	1.8461	1.8828	1.9176	1.9508	1.9825	2.0129	2.0421
160 (363.55)	v	0.0161	0.0166	0.0175	3.0060	3.4413	3.8480	4.2420	4.6295	5.0132	5.3945	5.7741	6.1522	6.5293	6.9055	7.2811
	h	68.42	168.42	269.89	1217.4	1273.3	1325.4	1376.4	1427.2	1478.4	1530.3	1583.0	1636.3	1690.5	1745.6	1801.4
	s	0.1294	0.2938	0.4370	1.5906	1.6522	1.7039	1.7499	1.7919	1.8310	1.8678	1.9027	1.9359	1.9676	1.9980	2.0273
180 (373.08)	v	0.0161	0.0166	0.0174	2.6474	3.0433	3.4093	3.7621	4.1084	4.4505	4.7907	5.1289	5.4657	5.8014	6.1363	6.4704
	h	68.47	168.47	269.92	1213.8	1271.2	1324.0	1375.3	1426.3	1477.7	1529.7	1582.4	1635.9	1690.2	1745.3	1801.2
	s	0.1294	0.2938	0.4370	1.5743	1.6376	1.6900	1.7362	1.7784	1.8176	1.8545	1.8894	1.9227	1.9545	1.9849	2.0142
200 (381.80)	v	0.0161	0.0166	0.0174	2.3598	2.7247	3.0583	3.3783	3.6915	4.0008	4.3077	4.6128	4.9165	5.2191	5.5209	5.8219
	h	68.52	168.51	269.96	1210.1	1269.0	1322.6	1374.3	1425.5	1477.0	1529.1	1581.9	1635.4	1689.8	1745.0	1800.9
	s	0.1294	0.2938	0.4369	1.5593	1.6242	1.6776	1.7239	1.7663	1.8057	1.8426	1.8776	1.9109	1.9427	1.9732	2.0025
250 (400.97)	v	0.0161	0.0166	0.0174	0.0186	2.1504	2.4662	2.6872	2.9410	3.1909	3.4382	3.6837	3.9278	4.1709	4.4131	4.6546
	h	68.66	168.63	270.05	375.10	1263.5	1319.0	1371.6	1423.4	1475.3	1527.6	1580.6	1634.4	1688.9	1744.2	1800.2
	s	0.1294	0.2937	0.4368	0.5667	1.5951	1.6502	1.6976	1.7405	1.7801	1.8173	1.8524	1.8858	1.9177	1.9482	1.9776
300 (417.35)	v	0.0161	0.0166	0.0174	0.0186	1.7665	2.0044	2.2263	2.4407	2.6509	2.8585	3.0643	3.2688	3.4721	3.6746	3.8764
	h	68.79	168.74	270.14	375.15	1257.7	1315.2	1368.9	1421.3	1473.6	1526.2	1579.4	1633.3	1688.0	1743.4	1799.6
	s	0.1294	0.2937	0.4307	0.5665	1.5703	1.6274	1.6758	1.7192	1.7591	1.7964	1.8317	1.8652	1.8972	1.9278	1.9572
350 (431.73)	v	0.0161	0.0166	0.0174	0.0186	1.4913	1.7028	1.8970	2.0832	2.2652	2.4445	2.6219	2.7980	2.9730	3.1471	3.3205
	h	68.92	168.85	270.24	375.21	1251.5	1311.4	1366.2	1419.2	1471.8	1524.7	1578.2	1632.3	1687.1	1742.6	1798.9
	v	0.1293	0.2936	0.4367	0.5664	1.5483	1.6077	1.6571	1.7009	1.7411	1.7787	1.8141	1.8477	1.8798	1.9105	1.9400
400 (444.60)	v	0.0161	0.0166	0.0174	0.0162	1.2841	1.4763	1.6499	1.8151	1.9759	2.1339	2.2901	2.4450	2.5987	2.7515	2.9037
	h	69.05	168.97	270.33	375.27	1245.1	1307.4	1363.4	1417.0	1470.1	1523.3	1576.9	1631.2	1686.2	1741.9	1798.2
	s	0.1253	0.2935	0.4366	0.5663	1.5282	1.5901	1.6406	1.6850	1.7255	1.7632	1.7983	1.8325	1.8647	1.8955	1.9250
500 (467.01)	v	0.0161	0.0166	0.0174	0.0186	0.9919	1.1584	1.3037	1.4397	1.5708	1.6992	1.8256	1.9507	2.0746	2.1977	2.3200
	h	69.32	169.19	270.51	375.38	1231.2	1299.1	1357.7	1412.7	1466.6	1520.3	1574.4	1629.1	1684.4	1740.3	1796.9
	s	0.1292	0.2934	0.4364	0.5660	1.4921	1.5595	1.6123	1.6578	1.6990	1.7371	1.7730	1.8069	1.8393	1.8702	1.8998

(Continued)

Temperature, F

Abs. press. lb/sq. in. (sat. temp)		Sat.	100	200	300	400	500	600	700	800	900	1000	1100	1200	1300	1400	1500
600 (486.20)	v	0.0166	0.0161	0.0186	0.0174	0.9456	1.0726	1.1892	1.3008	1.4093	1.5160	1.6211	1.7252	1.8284	1.9309		
	h		69.58	169.42	270.70	375.49	1215.9	1290.3	1351.8	1408.3	1463.0	1517.4	1571.9	1627.0	1682.6	1738.8	1795.6
	s		0.1292	0.2933	0.4362	0.5657	1.4590	1.5329	1.5844	1.6351	1.6769	1.7155	1.7517	1.7859	1.8184	1.8494	1.8792
700 (503.08)	v		0.0161	0.0166	0.0174	0.0186	0.0204	0.7928	0.9072	1.0102	1.1078	1.2023	1.2948	1.3858	1.4757	1.5647	1.6530
	h		69.84	169.65	270.89	375.61	487.93	1281.0	1345.6	1403.7	1459.4	1514.4	1569.4	1624.8	1680.7	1737.2	1794.3
	s		0.1291	0.2932	0.4360	0.5655	0.6889	1.5090	1.5673	1.6154	1.6580	1.6970	1.7335	1.7679	1.8006	1.8318	1.8617
800 (518.21)	v		0.0161	0.0166	0.0174	0.0186	0.0204	0.6774	0.7828	0.8759	0.9631	1.0470	1.1289	1.2093	1.2885	1.3669	1.4446
	h		70.11	169.88	271.07	375.73	487.88	1271.1	1339.2	1399.1	1455.8	1511.4	1566.9	1622.7	1678.9	1735.0	1792.9
	s		0.1290	0.2930	0.4358	0.5652	0.6885	1.4869	1.5484	1.5980	1.6413	1.6807	1.7175	1.7522	1.7851	1.8164	1.8464
900 (53195)	v		0.0161	0.0166	0.0174	0.0186	0.0204	0.5869	0.6858	0.7713	08504	0.9262	0.9998	1.0720	1.1430	1.2131	1.2825
	h		70.37	170.10	271.26	375.84	487.83	1260.6	1332.7	1394.4	1452.2	1508.5	1564.4	1620.6	1677.1	1734.1	1791.6
	s		0.1290	0.2929	0.4357	0.5649	0.6881	1.4659	1.5311	1.5822	1.6263	1.6662	1.7033	1.7382	1.7713	1.8028	1.8329
1000 (544.58)	v		0.0161	0.0166	0.0174	0.0186	0.0204	0.5137	0.6080	0.6875	0.7603	0.8295	0.8966	0.9622	1.0266	1.0901	1.1529
	h		70.63	170.33	271.44	375.96	487.79	1249.3	1325.9	1389.6	1448.5	1504.4	1561.9	1618.4	1675.3	1732.5	1790.3
	s		0.1289	0.2928	0.4355	0.5647	0.6876	1.4457	1.5149	1.5677	1.6126	1.6530	1.6905	1.7256	1.7589	1.7905	1.8207
1100 (556.28)	v		0.0161	0.0166	0.0174	0.0185	0.0203	0.4531	0.5440	0.6188	0.6865	0.7505	0.8121	0.8723	0.9313	0.9894	1.0468
	h		70.90	170.56	271.63	376.08	487.75	1237.3	1318.8	1384.7	1444.7	1502.4	1559.4	1616.3	1673.5	1731.0	1789.0
	s		0.1289	0.2927	0.4353	0.5644	0.6872	1.4259	1.4996	1.5542	1.6000	1.6410	1.6787	1.7141	1.7475	1.7793	1.8097
1200 (567.19)	v		0.0161	0.0166	0.0174	0.0185	0.0203	0.4016	0.4905	0.5615	0.6250	0.6845	0.7418	0.7974	0.8519	0.9055	0.9584
	h		71.16	170.78	271.82	376.20	487.72	1224.2	1311.5	1379.7	1440.9	1499.4	1556.9	1614.2	1671.6	1729.4	1787.6
	s		0.1288	0.2926	0.4351	0.5642	0.6868	1.4061	1.4851	1.5415	1.5883	1.6298	1.6679	1.7035	1.7371	1.7691	1.7996

Pressure (psia) (Tsat, °F)																
1400 (587.07)	v	0.0161	0.0166	0.0174	0.0185	0.0203	0.3176	0.4059	0.4712	0.5282	0.5809	0.6311	0.6798	0.7272	0.7737	0.8195
	h	71.68	171.24	272.19	376.44	487.65	1194.1	1296.1	1369.3	1433.2	1493.2	1551.8	1609.9	1668.0	1726.3	1785.0
	s	0.1287	0.2923	0.4348	0.5636	0.6859	1.3652	1.4575	1.5182	1.5570	1.6096	1.6484	1.6845	1.7185	1.7508	1.7815
1600 (604.87)	v	0.0161	0.0166	0.0173	0.0185	0.0202	0.0236	0.3415	0.4032	0.4555	0.5031	0.5482	0.5915	0.6336	0.6748	0.7153
	h	72.21	171.69	272.57	376.69	487.60	616.77	1279.4	1358.5	1425.2	1486.9	1546.6	1605.6	1664.3	1723.2	1782.3
	s	0.1286	0.2921	0.4344	0.5631	0.6851	0.8129	1.4312	1.4968	1.5478	1.5916	1.6312	1.6678	1.7022	1.7344	1.7657
1800 (621.02)	v	0.0160	0.0165	0.0173	0.0185	0.0202	0.0235	0.2906	0.3500	0.3988	0.4426	0.4836	0.5229	0.5609	0.5980	0.6343
	h	72.73	172.15	272.95	376.93	487.56	615.58	1261.1	1347.2	1417.1	1480.6	1541.1	1601.2	1660.7	1720.1	1779.7
	s	0.1284	0.2918	0.4341	0.5626	0.6843	0.8109	1.4054	1.4768	1.5302	1.5753	1.6156	1.6528	1.6876	1.7204	1.7516
2000 (635.80)	v	0.0160	0.0165	0.0173	0.0184	0.0201	0.0233	0.2488	0.3072	0.3534	0.3942	0.4320	0.4680	0.5027	0.5365	0.5695
	h	73.26	172.60	273.32	377.19	487.53	614.48	1240.9	1353.4	1408.7	1474.1	1536.2	1596.9	1657.0	1717.0	1777.1
	s	0.1283	0.2916	0.4337	0.5621	0.6834	0.8091	1.3794	1.4578	1.5138	1.5603	1.6014	1.6391	1.6743	1.7075	1.7389
2500 (668.11)	v	0.0160	0.0165	0.0173	0.0184	0.0200	0.0230	0.1681	0.2293	0.2712	0.3068	0.3390	0.3692	0.3980	0.4259	0.4529
	h	74.57	173.74	274.27	377.82	487.50	612.08	1176.7	1303.4	1386.7	1457.5	1522.9	1585.9	1647.8	1709.2	1770.4
	s	0.1280	0.2910	0.4329	0.5609	0.6815	0.8048	1.3076	1.4129	1.4766	1.5269	1.5703	1.6094	1.6456	1.6796	1.7116
3000 (695.33)	v	0.0160	0.0165	0.0172	0.0183	0.0200	0.0228	0.0982	0.1759	0.2161	0.2484	0.2770	0.3033	0.3282	0.3522	0.3753
	h	75.88	174.88	275.22	378.47	487.52	610.08	1060.5	1267.0	1363.2	1440.2	1509.4	1574.8	1638.5	1701.4	1761.8
	s	0.1177	0.2904	0.4320	0.5597	0.6796	0.8009	1.1966	1.3692	1.4429	1.4976	1.5434	1.5841	1.6214	1.6561	1.6888
3200 (705.08)	v	0.0160	0.0165	0.0162	0.0183	0.0199	0.0227	0.0335	0.1588	0.1987	0.2301	0.2576	0.2827	0.3065	0.3291	0.3510
	h	76.4	175.3	275.6	378.7	487.5	609.4	800.8	1250.9	1353.4	1433.1	1503.8	1570.3	1634.8	1698.3	1761.2
	s	0.1276	0.2902	0.4317	0.5592	0.6788	0.7994	0.9708	1.3515	1.4300	1.4866	1.5335	1.5749	1.6126	1.6477	1.6806
3500	v	0.0160	0.0164	0.0172	0.0183	0.0199	0.0225	0.0307	0.1364	0.1764	0.2066	0.2326	0.2563	0.2784	0.2995	0.3198
	h	77.2	176.0	276.2	379.1	487.6	608.4	779.4	1224.6	1338.2	1422.2	1495.5	1563.3	1629.2	1693.6	1757.2
	s	0.1274	0.2899	0.4312	0.5585	0.6777	0.7973	0.9508	1.3242	1.4112	1.4709	1.5194	1.5618	1.6002	1.6358	1.6691
4000	v	0.0159	0.0164	0.0172	0.0182	0.0198	0.0223	0.0287	0.1052	0.1463	0.1752	0.1994	0.2210	0.2411	0.2601	0.2783
	h	78.5	177.2	277.1	379.8	487.7	606.9	763.0	1174.3	1311.6	1403.6	1481.3	1552.2	1619.8	1685.7	1750.6
	s	0.1271	0.2893	0.4304	0.5573	0.6760	0.7940	0.9343	1.2754	1.3807	1.4461	1.4976	1.5417	1.5812	1.6177	1.6516

(Continued)

Table 8-3. Conclusion

Abs. press. lb/sq. in. (sat.temp)		(sat)	100	200	300	400	500	600	700	800	900	1000	1100	1200	1300	1400	1500
5000	v	0.0159	0.0164	0.0171	0.0181	0.0196	0.0219	0.0268	0.0591	0.1038	0.1312	0.1529	0.1718	0.1890	0.2050	0.2203	
	h		81.1	179.5	279.1	381.2	488.1	604.6	746.0	1042.9	1252.9	1364.6	1452.1	1529.1	1600.9	1670.0	1737.4
	s		0.1265	0.2881	0.4287	0.5550	0.6726	0.7880	0.9153	1.1593	1.3207	1.4001	1.4582	1.5061	1.5481	1.5863	1.6216
6000	v		0.0159	0.0163	0.0170	0.0180	0.0195	0.0216	0.0256	0.0397	0.0757	0.1020	0.1221	0.1391	0.1544	0.1684	0.1817
	h		83.7	181.7	281.0	382.7	488.6	602.9	736.1	945.1	1188.8	1323.6	1422.3	1505.9	1582.0	1654.2	1724.2
	s		0.1258	0.2870	0.4271	0.5528	0.6693	0.7826	0.9026	1.0176	1.2615	1.3574	1.4229	1.4748	1.5194	1.5593	1.5962
7000	v		0.0158	0.0163	0.0170	0.0180	0.0193	0.0213	0.0248	0.0334	0.0573	0.0816	0.1004	0.1160	0.1298	0.1424	0.1542
	h		86.2	184.4	283.0	384.2	489.3	601.7	729.3	901.8	1124.9	1281.7	1392.2	1482.6	1563.1	1638.6	1711.1
	s		0.1252	0.2859	0.4256	0.5507	0.6663	0.7777	0.8926	1.0350	1.2055	1.3171	1.3904	1.4466	1.4938	1.5355	1.5735

Temperature, F

Source: Abstracted from *Thermodynamic and Transport Properties of Steam*, (© 1967 by the American Society of Mechanical Engineers.

Table 8-4. Enthalpy of Air Under Atmospheric Pressure
(.013 lb moisture/lb dry air)

Temperature (°F)	Enthalpy (Btu/lb)	Temperature (°F)	Enthalpy (Btu/lb)
0	109.9	950	351.3
50	122.2	1000	364.8
100	134.4	1050	378.3
150	146.8	1100	391.9
200	159.1	1150	405.6
250	171.4	1200	419.3
300	183.8	1250	433.2
350	196.3	1300	447.1
400	208.8	1350	461.0
450	221.4	1400	475.1
500	234.0	1450	489.2
550	246.7	1500	503.3
600	259.5	1550	517.5
650	272.4	1600	531.8
700	285.3	1650	546.2
750	298.3	1700	560.5
800	311.5	1750	575.0
850	324.7	1800	589.5
900	338.0	1850	604.0
		1900	618.6

Source: Specific heat of water vapor taken from Reference 3. Enthalpy of dry air taken from Table I in *Gas Tables* by Joseph 11. Keenan and Joseph Kaye, ©1948 by John Wiley & Sons, New York.

8.1.3 Heat losses.

Enthalpy is a useful way of describing the amount of energy contained in steam or in steam condensate. Much of the energy lost from boilers, however, is radiated to the environment, and this loss must be calculated to complete the heat balance.

First, consider the problem of heat losses from hot pipes or boiler surfaces. These heat losses are associated with convection (usually treated as if it were the free convection associated with still air) and with radiation. This kind of heat loss must be considered whenever boilers or piping are contained within the system being analyzed, and it must be calculated for all surfaces. Useful formulas for these heat losses are given in Reference 2:

$$\text{radiative loss} = A \times .1714 \times 10^{-8} \times \left(T_S^4 - T_R^4\right) \qquad (8\text{-}1)$$

and

$$\text{convective loss} = A \times .18 \times (T_S - T_R)^{4/3} \qquad (8\text{-}2)$$

where

$$\begin{aligned}
A &= \text{surface area in ft}^2 \\
.1714 \times 10^{-8} &= \text{Stefan-Boltzmann constant in Btu/h} \cdot \text{ft}^2 \cdot {}^\circ R^4 \\
T_S, T_R &= \text{surface and room temperatures, respectively, in} \\
&\quad {}^\circ R \,({}^\circ F + 460)
\end{aligned}$$

Example 8-2: Calculate the amount of heat loss from the 140°F surface of a boiler into a room whose temperature is 80°F. The exposed surface of the boiler is 400 ft^2.

Solution: The radiative loss, per square foot per hour, is given by Equation 8-1 as:

$$\begin{aligned}
\text{Radiative loss} &= 1 \times (.1714 \times 10^{-8}) \times [(140 + 460)^4 - (80 + 460)^4] \\
&= 76.4 \text{ Btu/ft}^2/\text{h}
\end{aligned}$$

The convective loss, per square foot, is given by Equation 8-2 as:

$$\text{Convective loss} = (.18 \times 60^{4/3}) = 42.2 \text{ Btu/ft}^2/\text{h}$$

Thus, the total heat loss due to convection and radiation is found as:

$$\begin{aligned}
\text{Total loss} &= 400 \text{ ft}^2 \times (42.2 + 76.4) \text{ Btu/ft}^2/\text{h} \\
&= \underline{47{,}440 \text{ Btu/h}}
\end{aligned}$$

A second major heat loss occurs in the energy removed in the steam itself. The enthalpy is given in Tables 8-1, 8-2 and 8-3 for various steam temperatures and pressures; the flow rate and temperature can be observed as readings on gauges mounted on the steam lines. For example, using Table 8-1, the amount of energy flowing through a pipe with steam at 250 psig at a rate of 600 lb/h is 600 lb/h × 1201.1 Btu/lb, or 720,660 Btu/h.

Any other material leaving the system must be accounted for in the heat balance. For example, flue gas carries with it both latent and sensible heat, as was explained in Chapter 7, and the energy lost this way must be calculated as a significant factor in the heat balance. Any water used in blowdown contains heat that must also be counted as lost heat. Boiler blowdown is required periodically in order to remove some of the dirt, scale and other contaminants in the boiler water that build up over time. Some of the boiler water is drawn off and replaced with chemically purified makeup water. The blowdown process might be continuous, and if so, there would be a continuous loss of heat in this water.

The following example shows how a heat balance is calculated and used in the economic evaluation of a proposed boiler improvement.

8.1.4 Example of boiler improvement evaluation.

Consider a boiler with the data shown in Figure 8-1. This figure is a preliminary input-output diagram with the input and output quantities expressed in common units. A preliminary mass balance can be performed comparing the input quantities to the output quantities and correcting any discrepancy. (A discrepancy of 1-2% is considered acceptable.) Assume the fuel is a Virginia coal, with the ultimate analysis shown in Table 8-5.

Figure 8-1. Input-output diagram for boiler.

Table 8-5. Fuel Analyses for a Virginia Coal

Ultimate analysis		Flue gas analysis (dry)	
Constituent	% (weight)	Constituent	% (weight)
C	80.3%	CO_2	25.4%
H_2	4.9	SO_2	.1
S	.6	N_2	71.1
N_2	1.7	O_2	2.8
O_2	4.2	(15% excess air)	
H_2O	5.5		
Ash	2.8		

With these data, it is possible to calculate the amount of each heat energy flow in the boiler and to thereby choose the best place to look for possible energy cost savings.

Step One: First calculate the heat gains or heat inputs to the boiler system. Sources of this heat energy are fuel, combustion air, makeup water, and returned condensate.

Heat Gain from Fuel. The amount of heat energy available from the fuel is the fuel heating value multiplied by the amount of fuel used per hour. In this example, the energy input is [(101.3 tons/day)/(24 h/day)] × 2000 lb/ton × 14,210 Btu/lb, or 120 million Btu/h.

Heat Gain from Combustion Air. As shown in Figure 8-1, the temperature of the incoming combustion air is 80°F. Since the enthalpy of the flue gas was calculated relative to the temperature of the entering combustion air, the relative enthalpy of the combustion air is zero. This convention avoids counting the enthalpy of combustion air in both the input and the output computations.

Heat Gain from Makeup Feedwater. The makeup water enters the boiler at the ambient temperature of 80°F and thus does not represent a heat gain or a heat loss.

Heat Gain from Condensate Return. Condensate returned to the boiler accounts for a substantial amount of heat energy input. This input is 64,000 lb/h × (150°F − 80°F) Btu/lb, or 4,480,000 Btu/h.

Step Two: Now, calculate the heat losses as follows:

Radiative and Convective Losses (from Equations 8-1 and 8-2).

From the boiler:

radiative loss $= 1250 \times .1714 \times 10^{-8} \times [(215 + 460)^4 - (460 + 80)^4]$
                   $= 214.25 \times 10^{-8} \times (2.076 \times 1011 - 8.503 \times 10^{10})$
                   $= 262,600$ Btu/h

convective loss $= 1250 \times .18 \times (215 - 80)^{4/3}$
                   $= 155,800$ Btu/h

From the flue:

radiative loss $= 450 \times .1714 \times 10^{-8} \times [(450 + 460)4 - (50 + 460)^4]$
                   $= 476,700$ Btu/h

convective loss $= 450 \times .18 \times (450 - 50)^{4/3}$
                   $= 239,700$ Btu/h

Therefore, the total radiative and convective losses are approximately 1,134,800 Btu/h.

Heat Loss in Steam. The heat energy in the steam is calculated from the enthalpy given in Table 8-1. For 250-psig steam, the enthalpy is 1201.1 Btu/lb. Thus the heat carried from the boiler in the steam is 85,000 × 1201.1, or 102,093,500 Btu/h.

Heat Loss in Flue Gas. To estimate the amount of heat carried away in the flue gas, it is necessary first to estimate the amount of water vapor in the gas, then to combine this with the flue gas analysis to determine the weight percentages of each gas in the flue gas, and finally to use the specific heat of each gas to estimate the enthalpy increase per cubic foot of the escaping flue gas. (Note that this enthalpy is the *increase* from the enthalpy of the entering combustion air. As stated above, the enthalpy of the entering combustion air is counted as zero to avoid counting the initial enthalpy twice.) The details of calculating these three values are outlined below.

1. Calculate the amount of water vapor in the flue gas.
   a. Calculate the O_2 needed per pound of fuel.

Constituent and lb/lb fuel		lb O_2/lb constituent	Total lb O_2
C:	.803	2.264	1.818
H_2	.048	7.937	.389
S:	.006	998	.006
	.857 lb	Total:	2.213
		$-O_2$ in fuel:	.042
		O_2 needed per lb fuel:	2.171 lb

b. Calculate the air used in combustion

Minimum air needed = (2.171 lb O_2/lb fuel) × (100/23.15 lb air/lb O_2)

= 9.378 lb air/lb fuel

Air used for combustion = 9.378 × (1 + excess air percentage)

= 9.378 × 1.15

= 10.785 lb air/lb fuel

c. Calculate the total moisture in the flue gas

Moisture in flue gas = moisture from hydrogen in fuel + moisture in fuel + moisture in air (assumed .013 lb/lb dry air)

Moisture from hydrogen in fuel =

$$\frac{.049 \text{ lb } H_2/\text{lb fuel}}{2 \text{ lb } H_2/\text{lb•mole } H_2} \times 18 \text{ lb/lb} \cdot \text{mol } H_2O$$

= .441 lb mole H_2/lb fuel

Moisture in fuel = .055 lb/lb fuel (from fuel analysis)
Moisture in air = (.013 lb H_2O/lb air) × (10.785 lb air/lb fuel)

= .140 lb/lb fuel

Thus,

Total moisture in flue gas = (.441 + .055 + .140) lb/lb fuel

= .636 lb/lb fuel

$$= \frac{.636 \text{ lb/lb fuel}}{10.785 \text{ lb air/lb fuel} + .857 \text{ lb other/lb fuel}}$$

= .0546 lb/lb flue gas

2. Calculate composition of wet flue gas:

Component	lb/lb dry gas	lb/lb wet gas /	Molecular weight	=	lb • mol/lb wet gas
CO_2	.254	.241	44		.00548
N_2	.717	.680	28		.02429
O_2	.028	.027	32		.00084
SO_2	.001	.001	64		.00002
H_2O	.0546	.052	18		.00289

3. Calculate the enthalpy increase per cubic foot of the flue gas.
    a. Using the results from the table above, calculate the sensible heat of the flue gas. The specific heats are found in Figure 7-8 in Chapter 7.

Component	lb • mol/lb wet gas	X	Specific heat (Btu/lb • mol, °F)	X	700–80	=	Btu/h
CO_2	.00548		10.4		620		35.3
N_2	.02429		7.1		620		106.9
O_2	.00084		7.3		620		3.8
SO_2	.00002		10.9		620		.1
H_2O	.00289		8.3		620		14.9
	.03352				Sensible heat		161.0

Total heat (enthalpy) of wet flue gas

= sensible heat + latent heat

= 161.0 Btu + (.052 lb H_2O × 970 Btu/lb)

= 211.4 Btu/lb

Total heat (enthalpy)/ft^3 $= (Btu/lb) \times (lb/lb \bullet mol) \times (lb \bullet mol/ft^3)$

$$= \frac{211.4}{.03352} \times \frac{1}{359} \times \frac{460+80}{460+700}$$

$$= 8.18 \, Btu/ft^3$$

The flue-borne products of each pound of fuel weigh an amount equal to the combustion air + burned constituents + moisture, or, in this example, (10.785 +.857 +.636) = 12.278 lb. Since 101.3 tons/day are used and since each pound of wet flue gas takes with it 211.4 Btu, the loss per day from the flue is given by:

$$
\begin{aligned}
\text{Loss/day} \quad &= \quad (101.3 \text{ tons/day}) \times (2000 \text{ lb/ton}) \times \\
&\qquad (12.278 \text{ lb gas}) \times (211.4 \text{ Btu/lb gas}) \\
&= \quad 525{,}860{,}000 \text{ Btu/day} \\
&= \quad 21{,}910{,}000 \text{ Btu/h}
\end{aligned}
$$

Note that this represents a loss of 211.4 × 12.278 = 2596 Btu/lb. Since the heating value of this coal is 14,210 Btu/lb, the flue loss represents a loss of (2596/14,210) or about 18.3%. Note also that the heat carried by 33,200 cfm of flue gas is 33,200 × 60 × 8.18 = 16,290,000 Btu/h, somewhat off from that calculated from the fuel measurements alone.

Heat Loss in Boiler Blowdown. The enthalpy of water at 400°F is 375.1 Btu/lb, and hence the blowdown of 14,000 lb/h represents a loss of 5,251,000 Btu/h.

Step Three: The heat energy leaving a system must be equal to the heat energy entering the system. It is convenient to summarize the input and output heat energy flows as shown in Figure 8-2.

To check the accuracy of your estimates, the input and output enthalpies per hour should be within 1-2% of each other. Summing the input and output enthalpies from Figure 8-2 gives the following results:

Input enthalpies:

Coal	123,330,000 Btu/h
Makeup water	0 Btu/h
Condensate return	4,480,000 Btu/h
Input air	0 Btu/h
	127,810,000 Btu/h

Figure 8-2. Heat balance for boiler.

	127,810,000 Btu/h
Output enthalpies:	
Steam	102,100,000 Btu/h
Flue gas	21,910,000 Btu/h
Blowdown	5,251,000 Btu/h
Surface losses	<u>1,134,000 Btu/h</u>
	130,395,000 Btu/h

The heat gain and heat loss are within about 2% of each other, which is within the 1-2% that would be desired. Thus, the heat balance has probably identified the significant inputs and outputs properly.

The heat balance of Figure 8-2 serves several objectives. First, it is a clear way of presenting the energy input and output of a system, whether it is a boiler or an industrial process. This diagram can thus be used to communicate the energy flows, whether the communication is part of the data-checking process or whether it is part of a presentation being made in defense of a particular proposal. Second, preparing the diagram necessitates examining the system in detail in order to find and quantify the energy inputs and outputs. Many of these may not have been evaluated before, and their magnitudes are probably different from what they are thought to be. Third, the balance immediately and visibly suggests areas to examine in order to improve the system operation. It also clearly indicates the level of emphasis to place on the various opportunities. Some of the possibilities are discussed below.

8.2 WASTE HEAT RECOVERY

Waste heat is heat that goes into the atmosphere or to some other heat sink without providing any appreciable benefit to the user. Examples of such heat loss include flue gases, boiler blowdown exhausted to the air, heated air exhausted directly to the environment by vents, and heat lost from pipes that pass through unheated spaces. (Note that when these pipes provide protection against freezing, the heat radiated from the pipes is not waste heat.) Waste heat can be used to generate steam, to provide a source of energy for turbines, or to provide a source of free heat for incoming fluid streams. Any of these uses can improve the efficiency of energy utilization in a plant whether the source of the waste heat is a boiler, an industrial process, or an HVAC system. Benefits from waste heat recovery often include fuel savings and lower capital cost of heating and cooling equipment. Other benefits can include increased production capacity and, under some circumstances, revenue from the sales of recovered heat or energy [4].

8.2.1 Analyzing the Potential for Waste Heat Recovery

Waste heat sources and their uses can be conveniently categorized by the temperature at which the heat is exhausted, as shown in Table 8-6 [4].

The use of waste heat to power a turbine or pump should be considered if turbine work or pumping is needed and if enough energy is available to justify the cost involved. If the economics justify the production of electricity, cogeneration, discussed in Section 8.6, may be a possibility. The use of waste heat to heat a fluid stream should be considered if (1) the waste heat source is close enough to the fluid stream that the fluid temperature will still be high enough to be useful even after taking into account all heat lost in transporting the fluid from source to stream, (2) using waste heat from the source will not create problems at the source, and (3) the transfer of heat from the source to the stream is technically feasible.

The first step in analyzing an industrial process for possible waste heat recovery is the collection of data sufficient to describe the process with a heat balance as was done in Figure 8-2. The next step is to find all point sources of heat use or exhaust, to determine the annual Btu and mean temperature for each, and to show this information on an input-output diagram of the facility. Another way to summarize these data is shown in Figure 8-3. The most promising candidates for heat recovery are then examined in detail. Figure 8-4 shows the kind of data needed. If at all possible, the waste heat from a process should be used to improve the efficiency of that same process at its heat source. This practice avoids

Table 8-6. Sources of waste heat.

Temperature	Source	Use

High (1200+)	Exhausts from direct-fired industrial processes:
	    Cement kiln (dry), 1150-1350°F
	    Steel heating furnaces, 1700-1900°F
	    Glass melting furnaces, 1800-2800°F
	    Solid waste incinerators, 1200-1800°F
	    Fume incinerators, 1200-2600°F
		Cogeneration

Medium
(450-1200)
	Exhausts:
	    Steam boiler, 450-900°F
	    Gas turbine, 700-1000°F
	    Reciprocating engine, 450-1100°F
	Heat-treating furnaces, 800-1200°F
	Drying and baking ovens, 450-1100°F
	Annealing furnace cooling systems, 800-1200°F
		Steam generation

Low (90-450)	Process steam condensate, 130-190°F
	Cooling water from:
	    Furnace doors, 90-130°F
	    Bearings, 90-190°F
	    Welding machines, 90-190°F
	    Air compressors, 80-120°F
	    Internal combustion engines, 150-250°F
	Hot-processed liquids, 90-450°F
	Hot-processed solids, 200-450°F
		Supplemental
		heating
		Preheating

Source: Reference 4.

transportation losses and helps keep each process as independent from another as possible.

8.2.2 The Economics of Waste Heat Recovery

The benefits from waste heat recovery can be substantial; therefore, the benefits included in the economic analysis must be as complete as possible. Table 8-7 lists a number of benefits; this list should be amplified by the particular benefits of each specific project.

The analysis must also include complete details of the costs involved and the amount by which these costs are reduced by any tax benefits. Table 8-8 gives such a list.

8.2.3 Waste Heat Recovery Equipment

The factors that determine which equipment to select for waste heat recovery are the fluid temperature at the source, the intended use for the waste heat, and the distance the heated fluid (if any) must be transported. The most common equipment types and their uses are shown in Table 8-9.

Figure 8-3. Source-sink diagram for waste heat recovery. (*State of New Jersey Energy Audit Forms*, 1979. Reprinted with permission of the State of New Jersey Dept. of Energy)

Table 8-7. Economic benefits from waste heat recovery.

Benefits	How quantified
Reduced fuel costs	$/Btu × Btu/h × operating h/year
Increased production capacity	Marginal profit/unit × increase in capacity × no. h increased capacity will be used each year
Reduced overhaul costs	Cost per overhaul/mean time between overhauls
Reduced regular maintenance costs	Cost of maintenance per h × no. of h of maintenance per operating h
Lower capital costs of furnaces and heating and cooling equipment	Vendors' quotations
Sales of energy to utilities	*Firm* contracts with utilities, or regulations of state public service commission
Reduced pollution abatement equipment costs	Vendors' quotations

Figure 8-4. Data needed to do an in-depth analysis of a particular waste heat recovery measure.

Source date

     Location: _____

     Fluid used as a heat source: _____

     Flow rate: _____ lb/h

     Present temperature of fluid: _____ °F

     Present pressure of fluid: _____ psig

     Dew point of fluid: _____ °F

     Minimum allowable temperature: _____°F

Sink data

     Location: _____

     Fluid to be heated: _____

     Flow rate: _____ lb/h

     Present temperature of fluid: _____ °F

     Maximum allowable temperature of fluid: _____ °F

     Present pressure of fluid: _____ psig

     Maximum containable pressure of fluid: _____ psig

Heat transportation path data

     Horizontal distance: _____ ft

     Vertical distance: _____ ft

     Mass pumped: _____ lb/h

     Pumping cost: $ _____ /h

     Estimated heat loss per foot: _____ Btu/h

     Heat transported: _____ Btu/h

     Temperature drop: _____ °F

     Pressure drop: _____ psig

General data

     Building location: _____

     Heating cost, for planning purposes: $ _____ /Btu

     Electricity cost, for planning purposes: $ _____ /kWh

     Production: _____ (amount and units)/year

     Production cost savings: $ _____ /unit output

Table 8-8. Costs and cost reductions to be considered in waste heat recovery

Cost	Source of information
Capital cost of equipment	Vendors
Energy tax credits	Internal Revenue Service
Capital investment tax credits	Internal Revenue Service
State and local tax credits	State and local tax information sources
Increased property taxes	State and local tax information sources
Engineering design costs	Local consulting engineers (plan on 2-5% of total project costs)
Installation costs	Time and equipment cost estimates
Cost of production downtime installation	In-house production, time, and cost estimates
Operation and maintenance costs of new equipment	Vendors

Source: Taken in part from Reference 4.

Table 8-9. Waste heat recovery equipment.

Source fluid	Use	Equipment
Exhaust gases over 450°F	Preheat combustion air:	
	Boilers	Air preheaters
	Furnaces	Recuperators
	Ovens	Recuperators
	Gas turbines	Regenerators
Exhaust gases under 450°F	Preheat boiler feedwater or makeup water	Economizers
Condenser exhaust gases	Preheat liquid or solid feedstocks	Heat exchangers
Any waste heat source	Generate steam, power, or electricity	Waste heat boilers
Any waste heat source	Transfer heat	Run-around coils, heat wheels
Any waste heat source	Cooling	Absorption chillers

Source: Reference 4.

8.2.3.1 Recuperators.

A recuperator is a heat transfer device that passes gas to be heated through tubes that are surrounded by a gas that contains excess heat. The heat is transferred from the hot gas to the tubes and through the tube walls to the cool gas inside the tubes. Examples of two types of recuperators are shown in Figures 8-5 and 8-6.

8.2.3.2 Heat wheels.

A heat wheel is a large porous wheel that rotates between two adjacent ducts, one of which contains a hot gas and the other a cooler gas. In carrying heat from the hot gas to the cool gas, the heat wheel either captures some of the escaping heat in the hot gas, if that is desired, or cools the hot incoming gas, if that is desired. If the escaping gas is contaminated, the wheel can be constructed with a purge section so that the contami-

Figure 8-5. Radiation recuperator. (From Kenneth G. Kreider and Michael B. McNeil, *Waste Heat Management Guidebook*, National Bureau of Standards Handbook 121, 1977. Courtesy of the U.S. Department of Commerce.)

Figure 8-6. Convective-type recuperator. (From Kenneth G. Kreider and Michael B. McNeil, *Waste Heat Management Guidebook,* **National Bureau of Standards Handbook 121, 1977. Courtesy of the U.S. Department of Commerce.)**

nants are flushed from the wheel after they have given up a significant part of their heat. These wheels, illustrated in Figures 8-7 and 8-8, have high heat transfer efficiencies and can be used to significantly reduce either heating or cooling loads. It should be noted, moreover, that they can be designed to transfer latent heat, i.e., the heat in water vapor in the gas stream, as well as the usual sensible heat.

8.2.3.3 Air preheaters/Economizers.

In an air preheater or economizer, hot gas flowing through a series of closed channels transfers heat to cooler gas in adjacent channels. This kind of equipment allows the use of hot flue gas to preheat combustion air and reduces the amount of heat that must be supplied by the fuel.

8.2.3.4 Run-around coils.

A run-around coil heat exchanger consists of two heat exchanger coils connected by piping; a pump is also usually required. The heat is picked up by the heat exchange fluid in one coil. The fluid is then pumped to the other coil where the heat is removed and used. This heat transfer method makes waste heat recovery possible when the source and sink are somewhat separated.

8.2.3.5 Finned-tube heat exchangers.

This type of heat exchanger is a tube surrounded by perpendicular fins. The fins help to transfer heat from the tube to the surrounding air by

Figure 8-7. Simple heat wheel (Used with permission of the American Society of Heating, Refrigerating, and Air Conditioning Engineers, Inc., Atlanta, Ga.)

Figure 8-8. Heat wheel with purge section. (From Kenneth G. Kreider and Michael B. McNeil, *Waste Heat Management Guidebook,* National Bureau of Standards Handbook 121, 1977. Courtesy of the U.S. Department of Commerce.)

enlarging the heat transfer surface area. If the surrounding fluid has a higher temperature than the tube, the transfer works in the opposite way and transfers heat from the fluid to the material inside the tube. This type of heat exchanger is very common and is often used on boilers to recover some of the heat that would otherwise be lost in the stack gas. It is also used in baseboard heating and automobile radiators.

8.2.3.6 Heat pipe heat exchangers.

Heat pipes were described and illustrated in Section 6.6 of Chapter Six, and their application to moisture removal in air conditioning systems was discussed. A heat pipe is a gas-to-gas heat exchanger, and it has more widespread uses than just in air conditioning systems. They can be used as air preheaters for a boiler or furnace, as heat reclaimers from waste streams, and as heat exchangers in drying and curing ovens. Heat pipes have a variety of applications in industrial waste heat recovery because of their high efficiency and compact size.

8.2.3.7 Waste heat boilers.

If gas leaving some industrial process is sufficiently hot to vaporize water or some other working fluid, it may be possible to use a waste heat boiler. Such a boiler uses waste heat to produce vapor or steam which can be used directly in the industrial process or can be run through a turbine, pump, or generator to generate electricity or shaft power. Water has been the customary fluid for this purpose, but working fluids with lower boiling points are becoming more common.

8.3 BOILER EFFICIENCY IMPROVEMENT

In Chapter 7 we discussed combustion in boilers with particular attention to combustion air. In addition to monitoring combustion air closely, good boiler management requires close attention to the boiler control system and to balancing the load economically between boilers. There are also other opportunities for saving money and increasing production with boilers, and they are discussed in this section.

8.3.1 Inspecting the Boiler System

The boiler system often provides clear indications of opportunities to save energy and money. In a visual inspection of a boiler, look at the gauges first. If the gauges do not work, the boiler control system is probably not functioning correctly; this provides an opportunity for a substantial savings in fuel costs. If the boiler has not been professionally inspected and adjusted within the past two years, the same opportunity exists. If the boiler stack temperature exceeds the boiler water temperature by more than 150°F, the boiler is not operating as efficiently as it should; in that case, a professional adjustment is usually worthwhile. Look for rust in water gauges. This means that the pipes and tubes need to be cleaned for optimum operating efficiency. Check the boiler exhaust gas. If it is black, too little outside air is being used in combustion; if it is clear, too much outside air is being used.

A boiler can explode, so it is very important that all the safety features be in operating condition. Most building safety codes require that a boiler be inspected periodically; make sure that these inspection requirements have been complied with. Also make sure that outside air can get to the boiler to provide oxygen for combustion. Insufficient combustion air causes incomplete combustion and the generation of carbon monoxide. This is not only an inefficient use of energy, it is potentially dangerous.

8.3.2 Load Management

Boilers differ in efficiencies, and, in a system with several boilers, it makes sense to determine which boilers are the most efficient and to develop a method for allocating the load to them in order of decreasing efficiency. This allocation can be done either manually, by turning boilers down or off during off-peak seasons, or on a real-time basis through computer control. A boiler operates inefficiently with low loads, and it is usually worthwhile to operate one boiler at 90% capacity rather than two boilers at 45% capacity or three at 30%. The automatic controls needed are described in Reference 5, or information can be obtained from vendors.

8.3.3 Insulation

One common method for reducing heat loss from boilers is the addition of pipe and surface insulation. The choice of insulation type and thickness, together with the economics of insulation, is treated in detail in Chapter 11.

8.3.4 Components

Boiler efficiency can also be improved by replacing the burners and other components with more efficient models as they are developed. Besides burners, other components that should be examined are those where the heat balance indicates that waste heat may be successfully used, such as in units for preheating fuel oil or other fuels. As with other products, efficiency claims for boilers and their components are often over-stated; any new product should be evaluated carefully before deciding to buy it.

8.4 IMPROVING THE STEAM DISTRIBUTION SYSTEM

Steam distribution systems consist of piping from a boiler to points of use, together with the steam traps, condensate return lines, and any pumps needed for condensate return. The main function of the steam distribution system is to get the steam to where it is needed and to return the condensate to the boiler, doing both as efficiently as possible. Energy management can affect this system by improving the insulation, detecting and repairing steam and condensate leaks, maintaining the steam traps and condensate pumps, and providing water treatment. The steam distribution system should be inspected at the same time the boiler is inspected.

8.4.1 Insulation

Energy can be saved by insulating the pipes that carry either steam or condensate. The condensate return tanks should also be checked to see whether they need to be insulated. Chapter 11 describes how to determine the amount of insulation needed.

8.4.2 Steam Leaks

Steam leaks can be expensive if large amounts of steam are lost; condensate leaks represent lost heat and loss of treated water. In many environments, steam leaks can be detected by their hissing; in very noisy environments, it may be necessary to use an industrial stethoscope or an ultrasonic leak detector. Evidence of possible condensate leaks includes pools of hot water, dripping pipes, and rust spots on pipes. While looking for leaks, put your hand close to the pipes. If a pipe is too hot to touch, it probably should be insulated. The costs of steam leaks can be estimated by the following equation:

Cost/year = ($/Btu of steam) × (lb steam lost/h) × (Btu/lb) × (operating hours/year)

The pounds of steam lost can either be measured directly, estimated on the basis of experience, or calculated with a formula. Grashof's formula [3, pp. 4-47 and 4-48] gives the number of pounds of steam lost per month through an orifice of area A as:

$$\text{lb/h} = .70 \times .0165 \times 3600 \times A \times P_1{}^{.97} \tag{8-3}$$

where   .70 =   coefficient of discharge for hole
                    (for a perfectly round hole, this coefficient is 1)

.0165 = a constant in Grashof's formula
3600 = number of seconds per hour
A = area of hole in square inches
P_1 = pressure inside steam line in psia

Using this formula and Tables 8-1 and 8-2, it is possible to develop tables giving steam and Btu losses for any given leak size for a plant whose steam pressure is known. When the cost per million Btu is also known, it is then possible to estimate the cost of steam leaks directly from the size of the leaks. Table 8-10 gives an example of such calculations for a 600-psig system, assuming that the energy cost is $6.00/million Btu.

Another way to estimate steam losses is given by Waterland [6]:

A crude but frequently quite effective way of evaluating steam leaks is based on an arbitrary rating system. Tour a defined plant area at a weather condition or time of day when leaks are quite prominent, making a note of each steam leak and rating it as a wisp, moderate leak, or severe leak. Assign a value of 25 lb/hr for each wisp, 100 lb/hr for a moderate leak, and 500 to 1000 lb/hr for the severe leaks. When more than 20 leaks are evaluated, the total leakage rate determined in this way will usually be within 25% of the actual steam loss.

Table 8-10. Heat and dollar losses from leaks of 600-psig steam

Hole diameter	Steam loss (lb/h)	Heat loss (Btu/h)	Dollar loss ($/month)
1/16	74	89,000	$390
1/8	296	356,000	1,560
1/4	1183	1,420,000	6,230
3/8	2660	3,200,000	14,030
1/2	4730	5,690,000	24,942

When looking for steam leaks, the energy auditor should also examine the condensate return pumps for leaks and check to see whether the bearings are excessively noisy.

8.4.3 Steam Traps

The main function of *steam traps* is to drain condensed steam from the steam distribution network so that it can be returned to the boiler.

However, anything that reduces heat transfer from the steam to the pipe walls can be a source of inefficiency. In particular, air and dissolved gases act as insulators and should be removed, along with any condensed steam, as soon as possible within the steam distribution system. This removal is also a function of a steam trap. Two kinds of steam traps are shown in Figures 8-9 and 8-10.

The maintenance of steam traps is important in energy management because the condensate, dissolved air, and certain gases must be removed from the steam lines if the lines are to transmit steam. If condensate is not removed, the steam lines become lines carrying water but transmitting little heat. If dissolved air is not removed, the steam carries significantly less heat per pound since the pressure of the steam is reduced by the pressure of the air. If dissolved CO_2 is not removed, carbonic acid is formed, and this has a corrosive effect upon pipes. Any of these things can happen if steam traps fail closed. If steam traps fail open and they are open to the air, the effect is the same as if there were a large leak in a steam line. Thus, steam trap maintenance can be a very important source of energy cost savings.

Figure 8-9. Inverted bucket steam trap. (Courtesy of Armstrong Machine Works).

Figure 8-10. Float and thermostatic steam trap. (Courtesy of Armstrong Machine Works.)

Steam traps usually operate with a characteristic clicking and, depending on the installation, with a visible discharge of a small amount of flash steam-condensate that evaporates upon contact with air. The clicking can be detected with an industrial stethoscope or by putting one end of a long screwdriver on the steam trap and the other end next to your ear. If the steam trap is carrying a small amount of steam relative to its capacity, the interval between clicks may be 5-20 min. A sure sign that a steam trap is not working is the presence of live steam—steam that is under pressure—blowing into the air near a steam trap, where no other pipe or joint leak is present. If escaping steam is live downstream from a trap, the trap has failed open and is letting steam through along with the condensate; if the escaping steam is live at the condensate return tank, one or possibly all of the steam traps have failed. (The cost of steam trap failures is calculated the same way as the cost of steam leaks.)

Another problem caused by failed steam traps is water hammer. This phenomenon is caused when steam pushes a wave of condensate against a bend in the steam distribution system; it is generally accompanied by a loud noise. Water hammer usually occurs when the condensate is not being removed properly by the steam traps. Occasionally the steam distribution system was not designed correctly. However, this fault should not be suspected unless the steam traps have been checked and found to be working.

8.4.4 Water Treatment

Scaling also has an adverse effect on heat transfer. The more scale buildup, the less heat is transmitted through pipe walls. Reference 7 gives an example where a layer of $CaSO_4$ only .024 in. thick caused a temperature drop of 362°F in a boiler, leading to an outer tube temperature of 1004°F and to ultimate failure of the boiler tubes. This scaling can be prevented by proper *water treatment*, making water treatment one of the essential elements in boiler management. The amount of water treatment needed depends on the hardness of the water and the quantity of water used. Because the condensate is pure water, returning it to the boiler saves money on additional water treatment.

8.5 IMPROVING THE HOT WATER DISTRIBUTION SYSTEM

The hot water system is often a part of the overall boiler and steam distribution system where steam is used to produce the lower temperature hot water needed for process use, cleaning and bathroom use. This system can also provide a number of energy management opportunities. In performing an energy audit of this system, first see if more hot water is heated than is needed; second, determine whether the hot water can be heated to a lower temperature without adverse effects; and, third, assess the need for insulation.

Need for hot water. To estimate the need for hot water, look at the industrial cleaning uses first and then examine the personnel needs for washing. The amount and temperature of hot water needed in industrial processes will be defined by each process, and the auditor should make sure that neither the amount nor the required temperature are overstated. Each process should be examined to see whether cold water can be substituted for hot. In a laundry, for example, the unavailability of hot water can lead to the discovery that cold water works as well as hot, without any change in procedures except for a possible change in detergents.

The generally accepted figures for warm water for personal washing needs are 2-3 gal/day per person in offices, 20-25 gal/day per person in homes, and 40 gal/day per person in hospitals. If your facility exceeds these amounts, it may be worthwhile to valve off (carefully!) one or more hot water tanks and see if any complaints result. Another necessary step in this audit is to check hot water faucets for leaks. Hot water leaks are expensive; a continuous leak of 1 gal/hour of water at 155°F represents a loss of 7 million Btu/year, equivalent to about 7000 ft^3 of natural gas.

Temperature. To find out whether the temperature of the hot water can be lowered, first determine the present temperature. If the water is used for bathrooms, its delivered temperature can be reduced to 105°F by adjusting the thermostat on the hot water heater. Be sure to check the hot water thermostat. In one case, both the minimum and the maximum temperatures were set at 180°F, and one employee had complained about scalding. If the present temperature is too high, lower the hot water thermostat to the desired level, possibly in 5 degree increments to avoid complaints. Also consider eliminating the use of hot water completely. Most of us have used washrooms in gasoline stations or rest areas where only cold water was provided without any ill effects.

If the water is used in a kitchen, it must be hot enough to meet local health codes—usually at least 140°F. In many cases, this temperature is achieved by heating the water to an excessive temperature at the hot water tank; this increases the heat loss between the hot water tank and the kitchen. This loss can be avoided by having water temperatures of 105-110° in the kitchen with a booster hot water unit in the kitchen for the dishwasher. This unit should be on a timer so that water is heated only when needed.

Insulation. Where hot water is needed, it may be desirable to insulate the pipes, hot water tanks, and valves. If a pipe or tank feels hot to the touch, its temperature is probably 120°F or higher, and it is probably losing a significant amount of heat to the surrounding air. Valves and flanges have large heat transfer surfaces, and custom-made insulation for them may also be a worthwhile investment.

8.6 COGENERATION

Cogeneration is the process of sequentially producing both electricity and steam from a single fuel source. A cogeneration facility uses some of the thermal energy that a plant producing only electric power would otherwise reject to the environment. Thus, cogeneration can produce a given amount of electric power and thermal energy for 10 to 30% less fuel than a plant which produces the same amount of electricity alone. The specific fuel savings is highly dependent on both the cogeneration technology used and the quantity of thermal energy used. For many facilities, cogeneration offers a way to provide both low-cost electric power and the large amounts of thermal energy needed for thermal energy [8]. For example, a citrus processing plant in Florida installed a cogeneration plant to provide low-cost electricity and to provide the large quantities of steam needed to pasteurize orange juice and sterilize bottles that it is packed in.

Cogeneration has become more attractive because any cogenerated electricity that contributes to the total peak deliverable capacity of an electric utility enables the utility to avoid building an equal amount of new capacity, and the utility is obligated to buy such electricity from the cogenerator. (There are strict conditions on this purchase: The company must use some of the electricity it produces, the cogenerated electricity must be controlled and interconnected in such a way that it will not damage the utility network, and it must not force a net loss of revenues upon the utility.) Cogeneration also offers additional benefits to the utility, since the electric power is produced with less environmental pollution than if it came from a central station utility plant.

Cogeneration offers a number of benefits to the facility, particularly if it is an industry that has a source of waste fuel. Wood chips, black liquor, bagasse, garbage, and waste heat are all sources of fuel that would replace the need for a facility to purchase expensive oil, gas or coal. High-efficiency cogeneration helps industries and businesses compete in national and international markets, and helps keep more jobs in these industries and businesses. A facility can reduce its energy cost by lowering its need to buy expensive power from the electric utility company. It may also get some income from selling excess power to the utility.

8.6.1 Cogeneration Technologies

There are four major technologies available for cogenerating electric energy and thermal energy. Three of these are called topping cycles, where fuel is burned to generate electrical energy, and then the remaining heat or steam is used in an industrial process. The fourth technology is the bottoming cycle, where fuel is burned to produce industrial process heat, and then the remaining heat is used to generate electrical energy. These technological distinctions are important since they have a significant effect on how the efficiencies and environmental impacts are defined and evaluated.

In a steam turbine topping cycle application, fuel such as coal, gas or oil is burned in a boiler to produce high temperature and pressure steam, and that steam powers a turbine that drives an electrical generator. Only a portion of the energy in the steam is converted to electrical energy, and the remaining thermal energy is available for an industrial process such as drying or cooking.

In a gas turbine topping cycle application, gas or light oil is burned in a version of a jet engine, and the mechanical shaft power is used to drive an electrical generator. Then the waste heat from the gas turbine is captured and used directly for drying or other purposes, or is sent to a waste heat boiler where it produces steam for an industrial process.

In a diesel engine topping cycle application, gas or light oil is burned in a diesel engine—similar to a truck or marine diesel engine—and the mechanical shaft power is used to drive an electrical generator. Then the waste heat from the combustion exhaust, the jacket cooling water, and the oil cooling system can be used directly, or the waste heat from the exhaust can be sent to a waste heat boiler where it produces steam for an industrial process.

In a bottoming cycle application, fuel such as gas, oil or coal is burned to produce high temperatures for an industrial process such as drying cement in a kiln or melting glass in a furnace. The waste heat from the industrial process is then captured and sent to a waste heat boiler where it produces steam to drive a turbine generator for electrical energy production.

Many industries use the topping cycle for cogenerating electric energy and process heat or steam. Other industries use bottoming cycles with processes where large amounts of heat are generated in chemical reactions, and then this heat is captured and used in waste heat boilers and then for power generation—with additional fuel supplied. A particular example is in the phosphate industry where sulfur is burned to produce sulfur dioxide which is then reacted with water to produce sulfuric acid. The sulfuric acid is then used as a principal component of the process which yields the final product which is phosphoric acid. In this process, the sulfur is burned, and tremendous amounts of heat are produced. The heat is really a waste product, but when it is captured and sequentially used it has substantial value in its ability to reduce the amount of oil, gas or coal that is then needed to produce high temperature steam for generating electric energy.

8.6.2 Efficiency of Cogeneration

Whenever industrial steam or heat is generated, there is an opportunity for simultaneously generating high-efficiency electric power. If industrial steam is generated at a pressure and temperature above that required for the end use and the steam is then brought down to the desired pressure and temperature through a turbine generator, electric power can be produced with about half the fuel needed by a new central-station power plant. If industrial heat is produced for a production process, the remaining heat is a waste product, and any use of that heat represents a large overall efficiency improvement, and a reduced need for added fossil fuel.

Cogeneration, however, is a broad technical term, and there are substantial differences in the efficiency of different cogeneration technologies and different cogeneration facilities. Some facilities cogenerate in a highly efficient manner; others cogenerate only incidentally [9]. If, after generating electricity, a topping cycle cogenerator uses all the remaining thermal energy for a production process, then the plant cogenerates as efficiently as it can. Conversely, if a substantial part of the remaining heat is exhausted to cooling water or cooling air, then the efficiency may be quite low. For a cycle using waste heat, efficiency is somewhat difficult to define, since it uses energy that would otherwise be lost to the environment. However, any fuel use that is displaced by the available waste heat is an energy savings.

The best cogenerators are more than twice as efficient as new coal-fired power plants. For example, performance measures for generating electric power are associated with net heat rates; typical heat rates are 10,000-10,500 Btu/kWh for large, modern coal-fired, central station power plants and 4300-4700 Btu/kWh for highly-efficient, steam-topping cogeneration plants. Industrial size gas turbine topping cycles average around 5500 Btu/kWh and diesel engine topping cycles about 6500 Btu/kWh. Waste heat added from an industrial process may reduce the heat rate of an associated generating unit to the range of 5000 Btu/kWh.

8.6.3 Conditions for Successful Application

Cogeneration is not a new idea. In process industries where a great deal of heat is needed, many of the facilities were designed to incorporate cogeneration into the process. For example, one of the largest producers of magnesium takes the magnesium from magnesium chloride by electrolysis. The magnesium chloride comes from salt brines that have been dried. Drying the salt brines takes a large amount of heat; the electrolysis uses large quantities of electricity. The only way that this process can be carried out economically is to generate the electricity and use the waste heat to drive off the water from the brines. Another similar use of cogeneration appears in refineries.

In addition to industrial process use, cogeneration has also been used for district heating, where steam or hot water is supplied to a district by a utility whose primary product is electricity. Such use became less prevalent after World War II, when fuel was cheap and the economies of returned condensate were more attractive than those of wide-area distribution of steam and hot water. More recently, cogeneration is being used in large apartment buildings and in shopping centers as a less expensive source of both heating and electricity.

Cogeneration offers the promise of cheap electricity, with the electrical price under the control of the plant owner rather than of a utility company. But this promise may well be an illusion unless at least five conditions are met.

Condition 1. The need for heat must coincide with the need for electricity unless there are facilities for substantial heat or electrical energy storage. For example, one major university uses 400°F water for building heat and has an average electrical demand of 1.15 MW with peak demands of 2.3 MW. Although cogeneration seemed feasible in this situation, the highest electrical usage came from air conditioning in the summer at the same time that the demand for heat was at a minimum. If the university decided to use air conditioning based on a central chilled water facility with absorbing chillers, there might still be an economic opportunity for cogeneration. Hotels, motels and hospitals are other good candidates because they have a coincident need for electricity and hot water. Most industries also fit this condition.

Condition 2. Unless the economics are overwhelming, retrofitting for cogeneration should not be considered. Bottoming cycle equipment, in particular, is usually very heavy and can require substantial renovating of building foundations. To define these and other installation problems, it is a good idea to get a detailed feasibility study from a reputable engineering firm with experience in bottoming cycles. Furthermore, a facility designed to produce electricity in addition to the process heat now being provided (or vice versa) will necessarily be larger than the present facility. This increase in scale is usually expensive. Other problems can also arise. The authors have had experience with a lumber products operation where the boiler was operated completely on hogged fuel—chopped tree wastes. Changing this unit into a cogeneration facility would have required more energy than was available from the hogged fuel, necessitating either the purchase of gas or the development of some new fuel source. These mill operators were not enthusiastic about cogeneration. Alternatively, the authors have also seen many facilities where the boilers were greatly oversized for the existing loads. Such oversized boilers might well be economic to run at full capacity and cogenerate electricity while still providing all the steam the facility needs.

Condition 3. If the economic justification for a cogeneration project depends on the sale of electricity, this sale must be based on legally binding documents, and the conditions under which the electricity can be sold, particularly to utilities, must be spelled out, and their economic consequences evaluated. An electric utility is under no obligation to buy electricity in any way that would result in a net loss. In particular, an electric company is not obligated to pay premium rates for electricity delivered when the company has a surplus, and the company is not obligated to tie into a less controlled electric source having a phase or frequency difference that could harm the electrical system. The cost and constraints imposed by such controls should be considered and carefully evaluated— they can be great enough to discourage a sale of electricity in favor of using all the electricity in the cogenerating plant. Many facilities that cogenerate find that supplying their own loads provides significant savings without having to sell power to the utility.

Condition 4. The source of fuel for the cogeneration system should be considered carefully. When electricity is to be sold to a utility, the price is based in part on the reliability of the cogenerating source, and this places greater burdens on the cogenerator. If the source of fuel is some type of industrial waste, this may have to be supplemented with purchased fuel at an unacceptable cost. The lumber operation described above illustrates this problem. For those facilities supplying their own loads, the use of waste or alternative fuels provides an advantage, and can lead to the generation of very low-cost electric power compared to utility purchase.

Condition 5. The cogenerator must be prepared to purchase or provide backup heat and/or electricity for occasions when the cogeneration unit is out of service. This is an important economic consideration, since many utilities charge substantial rates to provide backup power. However, the reliability of most industrial cogeneration facilities is well over 90%, and this equipment is maintained well and repaired quickly.

Cogeneration is similar to other projects in that there are always a large number of considerations that influence the economics of the particular project. Overall analysis of the cost effectiveness of cogeneration is complex, and a specialized consultant should always be employed to help evaluate the various factors involved. However, the cogeneration and small power production area is growing rapidly, and this technology offers the potential for attractive savings in many facilities.

8.7 SUMMARY

In this chapter we have presented the concept of a heat balance and have provided examples of its application in boilers and waste heat utilization. We have also described the steam distribution system and hot water distribution system and provided some guidelines for maintaining and improving these systems. In addition, the general concept of cogeneration has been presented, along with some broad recommendations for its cost-effective application.

REFERENCES
1. Philip S. Schmidt, "Steam and Condensate Systems," Chap. 6 in *Energy Management Handbook, Second Edition*, Wayne C. Turner, Senior ed., Fairmont Press, Atlanta, GA, 1993.

2. Frank Kreith, *Principles of Heat Transfer*, Fourth Edition, Harper and Row, New York, NY, 1986.

3. Theodore Baumeister, Eugene A. Avallone, and Theodore Baumeister, III, *Marks' Standard Handbook for Mechanical Engineers*, McGraw-Hill, New York, 1978.

4. Kenneth G. Kreider and Michael B. McNeil, *Waste Heat Management Guidebook*, National Bureau of Standards Handbook 121, U.S. Department of Commerce, Washington, D.C., 1977.

5. Sam G. Dukelow, *The Control of Boilers*, Instrument Society of America Press, Research Triangle Park, NC, 1986.

6. Alfred F. Waterland, "Energy Auditing: A Systematic Search for Energy Savings Opportunities," in Chap. 3 *Energy Management Handbook*, Wayne C. Turner, Senior ed., Wiley Interscience, New York, 1982.

7. *Steam—Its Generation and Use*, 39th Edition, Babcock and Wilcox, New York, NY, 1978.

8. Barney L. Capehart and Lynne C. Capehart, "Efficiency in Industrial Cogeneration: The Regulatory Role," *Public Utilities Fortnightly*, March 15, 1990.

9. Lynne C. Capehart and Barney L. Capehart, "Efficiency Trends for Industrial Cogeneration," *Public Utilities Fortnightly*, April 1, 1991.

BIBLIOGRAPHY

Improving Steam Boiler Operating Efficiency, A Workshop Manual prepared by the Georgia Institute of Technology Industrial Experiment Station, Georgia Institute of Technology, Atlanta, GA, 1993.

Steam Conservation Guidelines for Condensate Drainage, Armstrong Machine Works, Three Rivers, Mich., 1976.

Total Energy Management, Third Edition, National Electrical Contractors Association, Washington, DC, 1986.

Turner, W.C., *Waste Heat Recovery Module*, Oklahoma State University Industrial Energy Conservation and Management Program, Stillwater, Okla., 1980.

Witte, Larry C., Schmidt, Philip S. and David R. Brown, *Industrial Energy Management and Utilization*, Hemisphere Publishing Corporation, Washington, DC, 1988.

Chapter 9

Control Systems and Computers

9.0 INTRODUCTION

Energy use can be controlled in order to reduce costs and maximize profits. The controls can be as simple as manually turning off a switch, but often automated controls ranging from simple clocks to sophisticated computers are required. Our view is that the control should be as simple and reliable as possible. Consequently, this chapter starts with manual controls and proceeds through timers, programmable controllers, and digital computers.

As one moves through this hierarchy of controls, each level of automation and complexity requires additional expenditure of capital. That is, the automated controls are more expensive, but they do more. Because choosing the proper type of control is often a difficult task, we will explore this decision process.

Computers can also help the energy manager in the analysis of proposed and present energy systems. Some excellent large-scale computer simulation programs have been written that enable the energy analyst to try alternative scenarios of energy equipment and controls, so in the last part of this chapter we discuss these computer programs and their use. BLAST 3.0 and DOE-2.1D are the two described in depth, but several others are mentioned.

9.1 WHY CONTROLS ARE NEEDED

Every piece of energy-consuming equipment has some form of control system associated with it. Lights have on-off wall switches or panel switches, and some have timers and dimmer controls. Motors have on-off switches, and some have variable speed controls. Air conditioners have thermostats and fan switches; they sometimes have night setback controls or timers. Large air conditioning systems have extensive controls consisting of several thermostats, valve and pump controls, motor speed controls, and possibly scheduling controls to optimize the operation of all of the components. Water and space heaters have thermostats and pump controls or fan motor controls. Large heating systems have modulating controls on the boilers and adjustable speed drives on pumps and variable air volume fans.

These controls are necessary for the basic safety of the equipment and the operators, as well as for the proper operation of the equipment and systems. Our interest is in the energy consumption and energy efficiency of this equipment and these systems, and the controls have a significant impact on both these areas. Controls allow unneeded equipment to be turned off, and allow equipment and systems to be operated in a manner that reduces energy costs. This may include reductions in the electric power and energy requirements of equipment, as well as the power and energy requirements associated with other forms of energy such as oil, gas and purchased steam.

9.2 TYPES OF CONTROLS

In this section, we present the different types of controls in order of increasing complexity and cost. In each subsection, the control discussed can perform the functions covered in that subsection as well as all those functions covered in the preceding subsections. For example, the functions discussed in the second subsection on timers can be performed very well by a timer or any of the succeeding types of controls (programmable controllers, microprocessors, and large computers) but not by a manual system.

9.2.1 Manual Systems

Manual control systems can be used to turn equipment off when it is not needed. Turning equipment off when not in use can lead to dramatic savings. For example, lights are often left on at night, but they should normally be turned off whenever possible. (Often a small series of lights is left on for security purposes.)

One of the best opportunities for manual control exists in the area of exhaust and makeup air fans. These fans are often located at the top of a high ceiling, and they are frequently left on unnecessarily because their running is undetectable without close scrutiny. The savings for turning off exhaust fans is twofold. First, electricity is no longer required to run the fan motor, and, second, conditioned air is no longer being exhausted. Consider the following example.

Example 9-1: Suppose that a fan is exhausting air at a rate of 10,000 ft^3/ min from a welding area. The fan is run by a 5-hp motor and is needed for two shifts (8:00 a.m. to 12:00 midnight) 5 days/ week. Previously, the fan has been left running all night and on weekends. If the space is not air-conditioned and is heated by a gas furnace that is 80% efficient, and the efficiency of the motor is 84%, what is the savings for turning the fan off at night and on the weekends? Gas costs $5.00/million Btu and electricity, $.08/kWh. (There will be no demand savings since peaking does not occur at night). Assume the outside temperature averages 30°F for the hours the fan can be shut off. (This would have been determined through weather data analyses, as discussed in Chapter 2.)

Solution:

1. Electricity savings:

The electric energy savings from turning the motor off during nights and weekends is found by multiplying the motor load in kW times the number of hours saved (number of hours the motor is not running). This energy savings in kWh is then multiplied by the energy cost to get the dollar cost savings.

Electric energy savings =

$$5\,hp \times \frac{.746\,kW}{hp} \times \frac{1}{.84} \times \left[\frac{5\,days}{week} \times \frac{8\,h}{day} + \frac{2\,days}{week} \times \frac{24\,h}{day}\right] \times \frac{52\,weeks}{year} \times \frac{\$.08}{kWh}$$

$$= \$1365.48$$

2. Heating savings: (see equation 6-13 in Chapter Six)

Heating cost savings =

$$\frac{10,000\,ft^3}{min} \times \frac{60\,min}{h} \times \left[\frac{5\,days}{week} \times \frac{8\,h}{day} + \frac{2\,days}{week} \times \frac{24\,h}{day}\right] \times \frac{52\,weeks}{year}$$

$$\times \frac{.075\,lbs}{ft^3} \times \frac{.24\,Btu}{lb\,°F} \times (65°F - 30°F) \times \frac{\$5.00}{10^6\,Btu} \times \frac{1}{.8}$$

$$= \$11,618.01$$

where the density of air =.075 lb/ft^3 and the specific heat of air =.24 Btu/ lb°F.

    3. Total annual savings   = <u>$12,983.49</u>

     In another example, a large office complex made a detailed study of building utilization and found that only a few tenants worked at nights or on the weekends. By making provisions for these few, the office complex was able to reduce lighting and space conditioning, saving about a third of its annual energy bill. For most industrial plants and many office buildings, the use of night setback to lower heating temperatures offers significant savings with little or no capital expenditure.

     The calculation procedure for determining this savings is relatively simple and involves heat loss calculations during the hours of setback. Bin weather data on outside temperatures and inside thermostat settings are required. The heat losses are calculated for the old thermostat setting and again for the revised setting. The difference is the heating savings in Btu. To simplify this procedure, or at least to give an approximation, a nomograph is given in Figure 9-1. The following example shows how to perform the calculation.

Example 9-2: A manufacturing company of 100,000 ft^2 is located in an area where heating demands are 4000 degree days. The company keeps its thermostats set at 70°F all the time even though it works only one shift. Presently, the company figures it consumes 240×10^3 Btu/ft^2 of gas for heating. (Normally, this can be estimated from gas bills.) If the company pays $4.50/10^6$ Btu for its natural gas and the heaters are 75% efficient, what would the savings be for turning the thermostats back to 55°F at night when the building is not occupied?

Solution: To use the nomograph, follow the heavy black lines in Figure 9-1. The savings are approximately 125×10^3 Btu/ft^2. Total savings then are found by first determining the actual fuel savings and then finding the dollar savings.

Fuel savings in Btu   = $(125 \times 10^3$ Btu/ft^2) (100,000 ft^2) $\times 1/.75$
                                  = <u>$16,700 \times 10^6$ Btu</u>

Savings in dollars   = $(16,700 \times 10^6$ Btu) ($4.50/10^6$ Btu)
                                  = <u>$75,150/year</u>

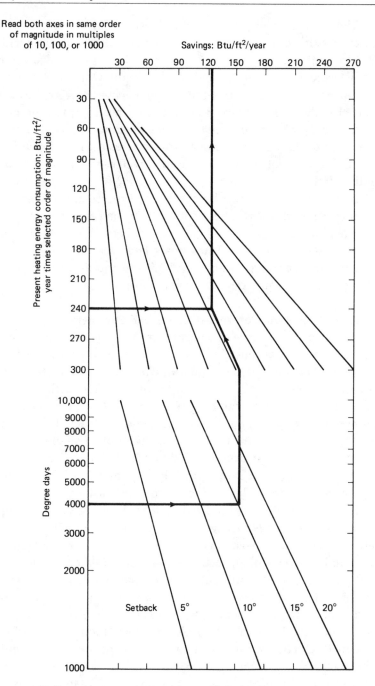

Read both axes in same order
of magnitude in multiples
of 10, 100, or 1000

Figure 9-1 Estimation of savings via thermostat night setback. (*Identifying Retrofit Projects for Buildings*, FEA/D-76-467, Sept. 1976.)

The savings for thermostat setback can be substantial, as shown in the example. In warm climates similar savings can result from turning off the air conditioners at night and on weekends, but the dollar amount is usually less since heating demands peak at night, while cooling demands peak during the day. Of course, the energy manager needs to be careful to ensure that the night setback does not cause heating plant problems or cause process problems, e.g., changing the tolerances on large metal parts or affecting the hardening rate of thermocure resins. This depends on the particular equipment and controls used as well as the thermal load of the building itself.

Night setback can also be applied to process areas. For example, large furnaces such as brick kilns should be turned off when possible, but often the preheat time required and/or the thermal wear on the furnace walls makes this impossible. Many times, however, the thermostat can be adjusted downward significantly without causing these problems. Trial and error may be required to determine the optimum setting.

9.2.2 Basic Automatic Controls—Timers and Dimmers

The next step in level of control complexity is the use of automatic controls such as timers and dimmers. Timers can range from very simple clocks to fairly complicated central time clocks with multiple channels for controlling numerous pieces of equipment on different time schedules. Automatic timer controls can range from simple thermostats each with a built-in time clock (costing somewhere around $100 each) to a central time clock that overrides all the thermostats. An installed single-channel central time clock will cost around $1000, but it can control numerous thermostats if all are on the same schedule. Different setback schedules require multiple channels, increasing the cost somewhat.

Some companies have utilized time clocks to duty cycle equipment such as exhaust fans. For example, a large open manufacturing area will likely have several exhaust fans. If there are six fans, then a central time clock could turn one fan off each 10 minutes and rotate so that each fan is off 10 minutes of each hour, but no more than one fan is ever off at the same time. This saves on electrical consumption (kWh) to run the fan, electrical demand (kW, since one fan is off at any time), and heating (since less conditioned air is exhausted). General ventilation over a wide area is maintained. Of course, care must be taken to ensure that no ventilation problems develop.

The use of timers allows a company to start-stop equipment at exactly the correct time. It is not necessary to wait for maintenance people to make their rounds, turning off equipment and adjusting thermostats.

However, although timers don't forget to do their job, they do suffer from other problems. For example, power outages may require timers to be reset unless a battery backup is used. Also, arrival or termination of daylight savings time requires all timers to be set up an hour in the spring and back an hour in the fall. Finally, the clocks must be maintained and replaced as they wear out.

The authors had an opportunity to audit a plant that had sophisticated time clock controls on its equipment, but management was not maintaining the clocks. The 7-day time clocks allowed for night and weekend setbacks. The audit was done on a Thursday, but the time clocks read Saturday. Consequently, the thermostats were on night setback, and the employees were cold. To remedy this, maintenance had purchased several additional portable heaters. If they had come in on a Saturday, when the clock read Monday, the plant would have been nice and warm. In this case, the poorly maintained clocks were costing the company a great deal of money. Timers and any other type of control system must be maintained.

Another type of control that has some attractive savings potential is a light dimmer. Dimmers can be automatically controlled depending on time, and on natural lighting levels if photocell sensors are used. It is important to be sure the dimming system chosen actually reduces electrical consumption and is not simply a rheostat (variable resistor) that consumes the same amount of energy regardless of the amount of light delivered. Supermarkets can often use relatively sophisticated dimming systems. For example, supermarkets might:

1. Use photocells to detect natural light and dim the window lights as appropriate.

2. Use photocells that turn parking and security lights off at dawn and on at dusk.

3. Use photocells to determine dusk so that interior lighting can be reduced. (Studies have shown that people coming from a dark street to a brightly lit room are actually uncomfortable. Lower lighting levels are preferred.)

As with timers, photocells and dimmers must be maintained. They sometimes fail, and if undetected, the failure can cause other more severe problems. A regular maintenance schedule of checking photocells and dimmers should be used. Photosensors were also discussed in Chapter 5.

9.2.3. Programmable Controllers

A programmable controller is a control device that has logic potential but is not powerful enough to be called a computer. As might be expected, it fills a need for systems requiring more than a timer but less than a computer. It can do all that timers can do and considerably more but at a cost significantly less than that of a computer. Fixed logic devices —such as timers—are primarily useful in buildings under 50,000 ft^2 with maybe a dozen control points. (A control point is a switch, a thermostat, or a control actuator.) Computerized systems would be applicable in buildings of 100,000 ft^2 with more with 100 control points. The middle group would be suitable for programmable controllers. Some control system selection guidelines are given in reference 1.

A programmable controller adds logic capability to control systems. Demand shedding is a prime example. When the controller senses that the electrical demand is approaching a critical programmed level, the unit then shuts off equipment and/or lights to keep the demand from passing that critical level. As shown in Chapter 3, demand can be a large part of an electrical bill, so the savings can be significant.

Another example is excess air control for a boiler or any larger combustion unit. By sensing CO_2 or O_2 and perhaps CO levels in exhaust, the controller can adjust the combustion air intake to yield optimum combustion efficiency. As shown in Chapter 7, this can be a real money saver. Continuous control through the use of a programmable controller allows the air intake to be adjustable according to the heating demand on the unit.

Programmable controllers can also be used to control outside air for heating, ventilating and air conditioning systems. In air conditioning, if the outside air is more comfortable than the inside air, the outside air should be used rather than returning the inside air. In fact, sometimes the air conditioning units can be turned off completely and outside air used for cooling. Programmable controllers can sense the difference between outside and inside enthalpy and determine the optimum damper setting. The same controller can shut off outside air completely for early morning start-up and nighttime operation during heating seasons. Outside air control is discussed further in the next section.

9.2.4 Computerized Systems

Most computerized energy management control systems (EMCSs) sold today are microprocessor-based. These capabilities run from a few control points up to several thousand, with the larger ones often performing fire-safety functions, equipment maintenance status monitoring and report generation as well as energy management [2]. The technology is changing very rapidly and there are many vendors in the field, each introducing new equipment. A potential user should consult several vendors and be well prepared to discuss the facility's needs. Some general approaches to the selection and specification of computer control systems for energy management are given in references 3 and 4.

EMCS users must recognize the need for feedback. The computer may have sent a signal to turn off a load, but was the load actually turned off? A sensor is required to feed the control status back to the computer. Also, it is helpful if the computer maintains a record of when control was exercised. With these records, histograms can be developed periodically to show how frequently any given load is being shed.

Additional control options are available to the EMCS user, but mostly they are some combination of the techniques discussed in the previous sections. Figure 9-2 is useful in summarizing these techniques. In Figure 9-2(a), the original electrical demand profile is shown before control is applied. Note that the area under these curves is the integration of demand over time and thus is the kWh consumption. In Figure 9-2(b), demand control is applied. Here, a peak demand is determined, and loads are shed once that peak is approached. Shedding requires predetermining what loads can be shed and in what priority. For example, display lighting would likely be shed before office lighting.

Many systems also "remember" loads previously shed. They will rotate among shedable loads, and will obey preset maximum shed times. As Figure 9-2(b) shows, shed loads must sometimes be recovered and sometimes not. For example, shedding refrigeration saves demand, but sooner or later the unit must catch up. Shedding lights, on the other hand, saves energy because lighting cannot be recovered. As shown, then, some consumption is shifted but usually not as much as was shed. Demand savings remain the predominant goal.

In Figure 9-2(c), a fixed start-stop schedule is utilized. Now units are turned on and off at exactly the same time each day. No longer are personnel required to make rounds, turning equipment on and off. In Figure 9-2(d), an optimized start-stop schedule is employed. The precise time of need is determined each day, and the equipment is turned on at that time. For example, if the outside and inside temperatures are both warm, the heating units do not have to be turned on as early as they would be if the respective temperatures were quite cold.

Figure 9-2(e) shows what happens when the use of outside air is optimally controlled. In this case, the plant requires air conditioning. The fans—but maybe not the compressors—are turned on early in the morn-

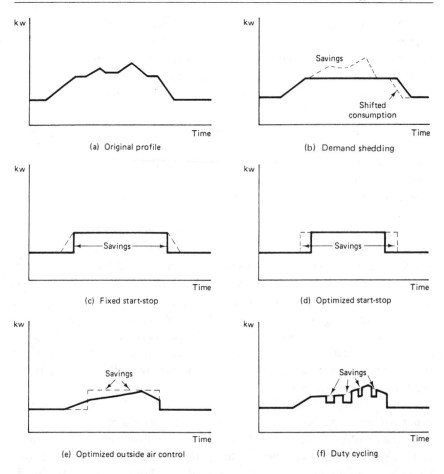

Figure 9-2. EMCS control techniques. (From Dick Foley, "Reducing Waste Energy with Load Controls," *Industrial Engineering,* July 1979, p. 24. Extracted with permission from Institute of Industrial Engineers, Inc., 25 Technology Park/Atlanta, Norcross, Ga., 30092, ©1979.)

ing to draw in cool outside air for precooling. As the daytime temperatures warm, less outside air is used, and the compressors have to run longer; at the peak air conditioning time a minimum of outside air is used. Toward the end of the operating time, it may again become profitable to utilize outside air.

Duty cycling is depicted in Figure 9-2(f). Here loads are selected to be turned off on a predetermined schedule. For example, in our earlier discussion exhaust fans were turned off 10 minutes out of each hour. Duty cycling will produce a savings in kWh since the units are operating less. If these schedules are determined so that some piece of equipment is being cycled off at any time that peaks are likely to occur, a demand savings will also result.

All the techniques shown in Figure 9-2 deal with electrical consumption. Other techniques cannot be easily demonstrated in such a figure because of their nature and/or because they affect fuel consumption rather than electricity. Examples include the following:

1. Light dimming, as discussed in Section 9.1.2.

2. Combustion air control for furnaces. This affects fuel consumption.

3. Night setback for heating. This normally affects fuel consumption instead of electricity.

4. Surge protection. If power outages occur, EMCSs can be programmed to start turning loads off to prevent an extremely large surge of power once the service is reconnected.

5. Temperature reset. Here the temperature of supply air or water is modified to meet actual demand. For a heating system, the supply air temperature may be reduced by 10 to 20°F when heating demands are small. This could save substantial fuel.

There are several different generic forms of computer EMCS configurations available. In a centrally controlled system, control is vested in one central unit—a microprocessor or microcomputer. The control points are accessed directly (a *star network*) or through common wiring (a *common data bus network*). These two configurations are shown in Figure 9-3. In the star network, control is more direct but installation is considerably more expensive. So, its use is limited to facilities with few control points. The common bus design allows for common use of wiring, so its installation cost is less for facilities with a large number of control points. Some of the current activities in the design of EMCSs involve the movement to standardized communication methods called protocols [5]. Standardized systems should be cheaper and easier to maintain and expand.

Instead of using a centrally controlled EMCS, many newer EMCS systems use a distributed configuration where remote processing units using microprocessors or microcomputers perform the actual control functions. A central unit is still used, but primarily for coordination and report generation. This is illustrated in Figure 9-4 using a star network.

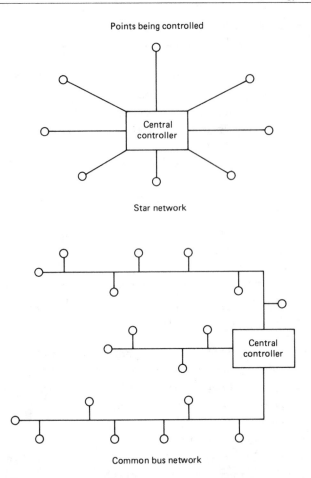

Figure 9-3. Centrally controlled EMCS star network and common bus network.

The remote control units vary in capability from a simple transfer function to complete control. In a true distributed control system, the remote controllers can function with or without the central unit—at least for a period of time. Intelligent remote controllers are only slightly more expensive than unintelligent ones. Most designers predict that future EMCSs will have more distributed control as the cost of remote controllers is reduced. Of course, systems can be hybrids so that some star networking is used along with common bus designs. Also, some remote controllers may be intelligent and others not, and some points may be directly controlled by the central controller even in a distributed control basic design.

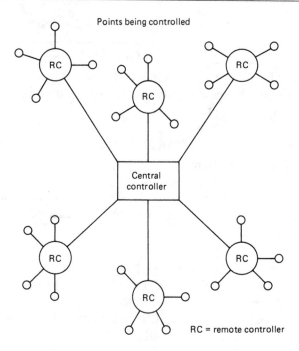

Figure 9-4. Distributed control EMCS star network.

An important factor to consider in the selection and cost of EMCSs is the type of control needed at the points being controlled. The simplest and cheapest is a digital point which is a simple on-off control. Examples include switches for fans, lights, and motors. Analog controls, on the other hand, are more complex and therefore more expensive. Analog controls are needed when signals of varying intensity are required. For example, control of outside air for air conditioning systems requires an analog signal to adjust the position of a supply vent damper. As the outside air cools, the EMCS will continue to open the damper, allowing more outside air to enter. A significant number of analog points will run up the cost of an EMCS rather rapidly, but the savings should also increase. Careful studies should be conducted in order to determine the optimum number of control points.

EMCSs sometimes fail to provide the energy cost savings that have been predicted. These reasons usually fall into one of the categories shown in Table 9-1. The prudent energy manager will consider these potential pitfalls and plan accordingly.

Table 9-1. Why EMCS units fail to produce the desired results.

1. *Simple things not done first.* The EMCS can do many things, but it should not be asked to do the things that should have already been done. For example, the EMCS cannot turn a standard lighting system into an energy-efficient one.

2. *Simple alternatives overlooked.* An EMCS may not be necessary. Manual or time control may suffice.

3. *Requirements not carefully defined.* This is the most important reason. The buyer must define the requirements before choosing the system. Considerable planning is necessary.

4. *Inadequate buyer commitment/inadequate seller backing.* All too often, buyers seem to expect the EMCS units to install, program, and maintain themselves. Sometimes the seller misrepresents the amount of work necessary to get the EMCS operational. The energy manager should insure that the seller will back the product and provide the necessary technical aid.

5. *Poor vendor assessment.* The energy manager should screen the vendors carefully. Ask for reference letters and check with other energy managers. Be wary.

In summary, EMCSs can do many things, but they are only machines. The energy manager must be aware of their limitations as well as their strengths and design their applications accordingly. The major part of the design should occur before selection of the equipment—not afterwards.

9.3 COMPUTER UTILIZATION

Computers have become so inexpensive and so powerful that they are used everywhere today, including use in a wide variety of tasks in energy management and energy analysis. The energy manager must be aware of computing capabilities and applications, and must carefully integrate computers into his or her environment. In addition to the direct EMCS applications, there are many other energy management uses for computers. Personal computers can be programmed to perform cash flow analyses, waste heat recovery studies, excess air control studies, and a myriad of other aids. Some desktop computers are capable of running large building simulations or equipment design programs requiring significant data storage and lengthy computing times.

Energy Engineering, the journal of the Association of Energy Engineers, has started publishing an annual Directory of Software for Energy Managers and Engineers. In the 1991 directory there were one hundred and thirty-three different software products for energy applications that were produced by forty-four companies [6]. The editor of this journal commented that there was more interest in the Directory Issue than any other issue produced [7]. The Directory is compiled by companies voluntarily sending in information, so it is not a complete listing of all energy-related computer programs. However, the listings in the 1991 Directory are more than twice as long as the listings in the 1990 Directory, which indicates a growing interest among both the users and the suppliers of these programs.

It is impossible to summarize or list all of the possible energy-related uses for computers, but one does stand out from the rest. That is the use of computers to perform building energy use analysis and simulation studies. In the rest of this chapter we will examine some of the programs that are most commonly available for this task. Building energy use analysis and simulation studies require the input of weather data, operating times, and other energy-consuming parameters such as number and type of lights and equipment, efficiency of various devices, etc. as well as the parameters of the building shell such as wall construction, insulation levels, amount of window area, etc. The computer will then simulate a year of building operation (or whatever cycle is chosen), and develop energy consumption and energy bills. Thus various scenarios involving energy efficiency improvements to the building shell or to building equipment can be fed into the computer and the likely savings identified and estimated. Most of the programs available also contain a financial analysis subroutine that provides the economic decision measures needed to help select the most cost-effective EMOs. Thus, the complete energy management study can be done by the computer.

Some of the better known programs include BLAST 3.0, DOE-2.1D, AXCESS, ASEAM, TRACE, and ECUBE, although there are many others. A call to your local utility, university, and/or energy management consultant can identify which are available in your area. We will discuss the capabilities of BLAST 3.0 and DOE-2.1D. Brief discussions of AXCESS, ASEAM, TRACE, and ECUBE, as well as where to get additional information on those programs, can be found in the Software Directory from reference [6].

9.3.1 BLAST 3.0

BLAST can be used to investigate the energy performance of new and retrofit building design options of almost any type and size [8]. BLAST is the acronym for the Building Loads Analysis and Systems Thermodynamics family of programs. Not only can BLAST calculate the building peak loads using design day criteria, which is necessary for mechanical equipment design, it can also estimate the annual energy performance of a facility. This is essential for the design of solar and cogeneration systems and for determining the building's compliance with a design energy budget.

Apart from its comprehensiveness, the BLAST system differs in three key aspects from other similar programs. First, BLAST uses extremely rigorous and detailed algorithms to compute loads, to simulate fan systems, and to simulate boiler and chiller plants. Second, the program has its own user-oriented input language and is accompanied by a library which contains the properties of all materials, wall, roof, and floor sections. Third, BLAST's execution time is short enough to allow many alternatives to be studied economically. In this way, efficient designs can be separated from the inefficient, and proper equipment type, size and control can be determined.

The BLAST Energy Analysis Program contains three major subprograms. First, the Space Load Predicting subprogram computes hourly space loads in a building based on weather data and user inputs detailing the building construction and operation. Next, the Air Distribution System Simulation subprogram uses the computer space loads, weather data, and user inputs describing the building air-handling system to calculate hot water, steam, gas, chilled water, and electric demands of the building and air-handling system. Finally, the Central Plant Simulation subprogram uses the weather data, the results of the air distribution system simulation, and user input describing the central plant to simulate boilers, chillers, on-site power-generating equipment, and solar energy systems; and then computes monthly and annual fuel and electrical power consumption.

Early versions of BLAST were considered difficult to use, but several new methods have been developed to communicate with BLAST. Available methods include BTEXT and Drawing Navigator for Autocad. BTEXT is a text-based scrolling menu program to solicit information about the building model. It builds a special file of building information and can generate BLAST input files. Drawing Navigator for Autocad uses the graphic information accessible in drawings to generate the necessary building geometric data; it passes information into BTEXT for eventual BLAST input file creation. Both preprocessors simplify input file creation [8].

MICRO BLAST 3.0 is available to run on any PC compatible 386 or 486 computer with 20-100 MB of hard drive memory and 4 MB of RAM. Users may obtain access to the BLAST program, and additional information about it from the BLAST Support Office, University of Illinois, 1206 West Green Street, Urbana, IL 61801.

9.3.2 DOE-2.1D

The DOE-2.1D Computer Simulation Program was developed for the Department of Energy (DOE) to perform energy analysis and simulation of plants and buildings [9]. It calculates the hour-by-hour energy use of a building and its life-cycle cost of operation using information on the building's location, construction, operation, and heating, ventilating, and air conditioning system. This program is used to design energy efficient new buildings, analyze energy conservation measures in existing buildings and calculate building design energy budgets. The program is divided into five major subprograms: (1) Building Description Language (BDL), (2) LOADS, (3) SYSTEMS, (4) PLANT, and (5) ECONOMICS.

The Building Description Language (BDL) subprogram allows the user to enter key building design information. The program uses a library of properties of all materials, walls, roof, and floor sections. The user also inputs a description of the HVAC systems, occupancy, equipment, lighting schedules, and other parameters.

The LOADS subprogram computes hourly space loads resulting from transmission gains and losses through walls, roofs, floors, doors, and windows; internal gains from occupants, lighting, and equipment; and infiltration gains and losses caused by pressure differences across openings. The LOADS calculations are based on ASHRAE algorithms, including the response factor technique for calculating transient heat flow through walls and roofs and the weighting factor techniques for calculating heating and cooling loads.

After the building loads are calculated, the program begins the SYSTEMS analysis. The SYSTEMS subprogram takes the hourly space loads, along with characteristics of secondary HVAC equipment, the component and control features, and the thermal characteristics of the zone, and determines the actual room temperature and heat extraction or addition rates using ASHRAE algorithms.

The PLANT subprogram uses the building thermal energy load data determined by SYSTEMS and various other user-input operating parameters of the plant equipment to allocate available equipment and simulate their operation. The PLANT program simulates conventional central plants, solar heating and cooling systems, and plants with on-site generation and waste heat recovery. It also permits load management of plant equipment and energy storage. It calculates the monthly and annual cost and consumption of each type of fuel used, the daily electrical load profile, and the energy consumption at the site and at the source.

The ECONOMICS subprogram uses the life-cycle costing methodology derived from DOE guidelines. Life-cycle costing method investment statistics such as cost savings, savings-to-investment ratio, energy savings, energy savings-to-investment ratio, and discounted payback period are calculated to provide a measure for comparing the cost effectiveness of each case against a reference case.

The MICRO DOE-2.1D program is available for a PC compatible 386 or 486 computer with 2 MB of RAM and at least 20 MB of hard disk space. A newer version, DOE-2.1E, has recently been made available, and a greatly expanded version named DOE-3 is currently under development. The DOE-2 program is used extensively by electric utilities, government agencies, national laboratories, architect/engineering firms, universities, and many private organizations. A version of DOE-2 is used by the State of California to determine building code compliance.

Users may obtain access to the DOE-2.1D program or the MICRO DOE-2.1D program from various commercial vendors. Some of these vendors also offer user support and training for the system. Additional information on DOE-2.1D is available through the Building Energy Analysis Group, Energy and Environment Division, Lawrence Berkeley Laboratory, Berkeley, CA 94720. Information is also available through the Department of Energy, Office of Building Systems Technology, Architectural and Energy Systems Branch, 1000 Independence Avenue, SW, Washington, DC, 20008.

9.4 SUMMARY

In this chapter we have examined control systems and computer applications for energy management. We began with a discussion of the types of controls, including manual, timer, programmable controllers, and computers. Then we discussed each level of control, giving advantages and limitations. Basically, the simpler controls are the least expensive and least robust. The more expensive controls (such as EMCSs) are more robust in that more control activities can be utilized. Computers can be used in other areas of energy management also. Data manipulation, data summary, and large-scale modeling or simulation are among some of the examples of other areas where computers can be utilized. Large-scale computer simulation models of energy systems are available and are quite useful in simulation of system operation, various scenarios of new equipment selection, or use of revised control schemes. BLAST 3.0 and DOE-2.1D are two programs discussed in some depth in the chapter.

REFERENCES

1. "Programmable Controllers: A Viable Alternative," *Plant Energy Management*, March 1982, pp. 44-45.

2. Bynum, Harris D., "What's New in Building Automation: Network Applications Add New Levels of Synergy," *Strategic Planning and Energy Management*, Vol. 9, No. 2, 1989.

3. Payne, F. William, *Energy Management and Control Systems Handbook*, The Fairmont Press, Atlanta, GA, 1984.

4. "Guidelines for the Design and Purchase of Energy Management and Control Systems for New and Retrofit Applications," American Consulting Engineers Council, Washington, DC, 1984.

5. Fisher, David M., "Open Protocol: Today's Options," *Strategic Planning and Energy Management*, Vol. 9, No. 2, 1989, pp. 51-56.

6. "Directory of Software for Energy Managers and Engineers," *Energy Engineering*, Vol. 88, No. 2, 1991, pp. 6-79.

7. Williams, Anna Fay, "Editorial," *Energy Engineering*, Vol. 88, No. 2, 1991, p. 5.

8. Lawrie, Linda K., "Day-to-Day Use of Energy Analysis Software," *Energy Engineering*, Vol. 89, No. 5, 1992, pp. 41-51.

9. Curtis, R.B., et. al., "The DOE-2 Building Energy Analysis Program," Technical Report LBL-18046, Applied Science Division, Lawrence Berkeley Laboratory, University of California, Berkeley, CA, April 1984.

BIBLIOGRAPHY

Blake, Frederick H., "New Energy Management Controls for Supermarkets," *Energy Engineering*, Aug.-Sept. 1981 PP. 49-54.

Lawrie, Linda K. and T. Bahnfeth, "Guidelines for Evaluating Energy Analysis Software," USACERL Technical Report E89-15, September 1989.

McDonald, J. Michael and Michael B. Gettings, "Military EMCS: Implications for Utilities, Cities, and Energy Services," *Energy Engineering*, Vol. 88, No. 4, 1991, pp 20-35.

Peterson, Kent W., and John R. Soska, "Control Strategies Utilizing Direct Digital Control," *Energy Engineering*, Vol. 87, No. 4, 1990, pp 30-35.

Stewart, Alan F., "Modeling and analysis of HVAC Systems with an EMCS," *Energy Engineering*, Vol. 87, No. 2, 1990, pp 6-17.

Thumann, Albert, Editor, *The Energy Management Systems Sourcebook*, The Fairmont Press, Atlanta, GA, 1985.

Turner, Wayne C., Senior ed., *Energy Management Handbook, Second Edition*, The Fairmont Press, Atlanta, GA, 1993.

Chapter 10

Maintenance

10.0 INTRODUCTION

The quality of maintenance influences energy use in a facility. In almost every energy-using system, a lack of maintenance is associated with increased energy consumption. Boilers with well maintained controls use 20-30 percent less fuel than boilers with controls that have not been taken care of for 2 or more years. HVAC systems with clogged filters use more energy and provide less clean air than those where the filters are changed when they get dirty. Controls drift out of calibration, motors wear and shift on their mounts, and all the air conducting components get dirty. In hot water systems, washers get worn, and faucets start to leak. In industrial processes, the heat transfer surfaces on waste heat recovery units get fouled, control systems drift out of calibration, and oven gaskets get worn. Steam traps that have failed shut do not provide the process protection they were designed for and can allow equipment to freeze; steam traps that have failed open can allow steam to escape, which also costs money. Lighting systems that are maintained with periodic relamping and cleaning of reflective surfaces are more effective than those where dirt is allowed to accumulate.

In general, maintenance can significantly reduce energy consumption, can increase the availability of production equipment and thus lower production costs, and can make a more comfortable and safe environment for plant personnel. We have two objectives in this chapter: to describe some good maintenance procedures for the major energy using systems and to show how a good routine maintenance program can improve the energy use in a facility. The first step is to assess current needs for repairs and for routine maintenance. Then some suggestions are presented for determining how often routine maintenance should be performed, and determining how long each maintenance action should take. We also examine record-keeping and use of computerized maintenance systems. The use of outside sources for maintenance work is also discussed.

10.1 DETERMINING MAINTENANCE NEEDS

10.1.1 Determining Basic Repair Needs

For companies who do not have a comprehensive maintenance program for each major energy-system, the first thing the energy auditor should do is to determine what repairs are needed to bring each system to an acceptable operating condition. When performing the energy audits described in Chapter 2, the maintenance aspects of each of the major energy-using systems must also be examined in detail. For example when the authors examined the HVAC system in one facility, we saw a circulating pump where the pump motor had been removed for repair two years before we arrived and had not yet been replaced. Meanwhile, rooms on the periphery of the building had 45°F temperatures, and rooms on the interior had 95°F temperatures. This company did not keep good enough maintenance records to follow through on the initial maintenance action.

Poor boiler maintenance can result in noticeable loss in energy efficiency. Fuel-handling equipment gets worn, and burners and controls get out of adjustment, with a resulting loss of efficiency if they are not inspected and maintained regularly. As a rule of thumb, if a boiler has not been maintained for two years, a 20-30 percent gain of efficiency is immediately possible with maintenance.

Similar situations occur frequently, and one of the functions of the energy audit is to identify them. As these systems are inspected, the auditor should take notes on what is wrong with each system and identify corrective measures. The company must then prioritize these measures, obtain necessary spare parts, schedule personnel and perform the necessary maintenance actions.

10.1.2 Routine Maintenance

In addition to the repair needs shown by the energy audit, the energy auditor must determine what routine or preventive measures should be scheduled on a regular basis. In Chapter 5 we pointed out three causes of lighting quality deterioration—dirt, lamp dimming, and lamp wear out. (Maintaining the lighting system was discussed in detail in Section 5.3 of Chapter 5.) Analogies to these can be found in most systems. A preventive maintenance program should include: (1) a monitoring/ metering system which can detect deteriorating conditions using such equipment as a light meter; (2) a schedule for performing housekeeping and routine tasks such as lubrication or cleaning light fixtures; (3) a program for replacing worn parts like bearings or lights; and (4) a regular assessment of the general condition of the system—for example, noting that a motor is overheating or that a ballast is beginning to smoke. Preventive maintenance helps to lengthen the life of equipment so that major repairs are not necessary as often. It also helps to detect major problems early enough that repairs can be scheduled at a convenient time.

What preventive maintenance tasks should be done? Vendors are a good source for this information. So are books such as that by Sack [1] and the Department of Defense facilities maintenance manual [2]. These books provide a guide as to the time, skill level, and equipment needed for routine maintenance. In addition, the manuals that come with the facility's equipment should be kept and their instructions followed. Special procedures can be used for some items such as bearings; the Tornberg procedure for bearings [3] is an example. Other sources for preventive maintenance information are the professional societies, such as the Institute of Industrial Engineers or the American Institute of Plant Engineers, and trade journals. One example of a typical list of maintenance tasks and solutions is shown in Figure 10-1 [4].

As maintenance procedures are developed, a database record should be prepared on every piece of equipment showing specifications such as bearing numbers, equipment serial numbers, fan belt sizes and serial numbers, and other information unique to the particular item of equipment. These database records are invaluable when something breaks down or when spare parts must be ordered. A microcomputer with special purpose software to record the dates and times maintenance is performed as well as the specific actions taken is often the most effective system for providing this information. An alternative to a computerized record system is a system using file cards. For more information on computer database systems, see Section 10.4.

Maintenance needs differ from company to company, depending on how heavily the equipment has been used, how well it has been maintained in the past, and what environmental conditions it is exposed to (dust in the atmosphere, temperature swings, etc.) However, no matter what facility you are examining, the objective should be the same: increased production and lower energy bills through wise maintenance.

10.2 DETERMINING MAINTENANCE INTERVALS

When choosing preventive maintenance intervals, the plant maintenance team should understand the following three points. First, the more often a piece of equipment is maintained, the higher the direct cost of maintenance labor and the greater the likelihood that something will be

Figure 10-1. Maintenance for the major energy systems.

Problem	Solution	Cost/benefit[a]
Building envelope (outside walls and roof):		
Doors: Loose fitting	Weatherstripping	VL/LA
Do not close	Correct fit; check pressures	L/LA
Windows: Air leakage	Weatherstrip, caulk	VL/LA
Broken	Replace panes	L/MA
Walls: Cracks	Patch or seal	M/H
Roofs: Holes	Patch or cover	M/M
Boiler and steam distribution system:		
Boiler: Safety problems	Correct now	S
Inoperable gauges	Overhaul	VH/VH
Stack gas clear or black	Check and adjust	H/VH
Scale deposits	Fix water treatment	H/H
Steam traps leak	Repair or replace	L/H
Steam leaks	Repair	H/VH
Steam line insulation ragged or missing	Repair or replace	L/M
Condensate return system uninsulated but hot	Insulate	L/M
Heating, ventilating, and air conditioning system:		
Filters very dirty	Replace or clean	L/LA
Dampers blocked or disconnected	Repair linkage and check controls	L/H
Ductwork blocked inside	Examine and repair	L/LA
Grillwork blocked or dirty	Remove blockage; clean	VL/MA

Fan motor not connected	Replace belts or disconnect motor	VL/M
Thermostat reading not accurate	Calibrate and correct	VL/LA
Electrical system:		
Transformers:		
Leak	Replace immediately	S
Heat transfer surfaces fouled	Clean carefully	L/L
Contacts show burned spots	Repair immediately	S
Frayed wire	Tape carefully	S
Switches arc	Replace	L/M
Lights, windows, and reflective surfaces:		
Unused space illuminated	Remove lights or install better switches	VL/L
Ballasts buzz	Adjust	L/LA
Lights flicker	Replace or eliminate	VL/LA
Walls dark	Use light colored paint; clean walls	VL/LA
Windows dirty	Clean or paint over	L/L
Hot water distribution system:		
Water too hot	Lower thermostat setting	VL/M
Faucets or lines leak	Repair or replace	L/M
Water heated when not needed (weekends)	Install or reset time clocks	L/H
Scale buildup	Check water treatment system	M/MA

(Continued)

Figure 10-1. (*Continued*).

Problem	Solution	Cost/benefit[a]
Air compressors and compressed air distribution system:		
Compressor air leaks	Overhaul	M/L
Compressor oil leaks	Overhaul	M/H
Oil in pneumatic control system	Overhaul compressor	M/VH
Air line leaks	Repair	VL/M
Manufacturing system—processes:		
Controls not working	Overhaul	VH/VH
Working temperatures too high	Study and correct	H/VH
Motors noisy	Check bearings	L/H
Motors hot	Adjust voltages on legs of three-phase input; correct belt-tension	M/M
Oven refractory problems	Repair or replace	M/H
Oven gaskets leak	Replace	VL/M
Material handling systems:		
Drafts	Check envelope, pressure balance (inside-outside)	L/MA
Uncontrolled fuel use	Institute accounting of fuel per vehicle	L/MA

Excessive tire wear	Check tire pressure	L/M
Excessive fuel use	Keep vehicles tuned up	L/MA
MH system runs when nothing is being carried	Check load switches regularly	VL/H
Excess use of diesel fuel	Follow manufacturer's recommended procedures; if timed switches limit idling check starters often	

aKey:

Costs:
VL = less than $25
L = $25 - $100
M = $100 - $300
H = $300 - $1000
VH = more than $1000

Benefits:
L = less than $25/year
M = $25-$150
H = more than $150
A = substantial noncost benefits (comfort, appearance, etc.)
S = safety-related

(W. J. Kennedy, Jr., "Apply a Cost-Benefit Approach in Building an Energy System Maintenance Program," *Industrial Engineering*, Dec. 1981, pp. 62-63. Reprinted with permission from the Institute of Industrial Engineers, Inc., 25 Technology Park/Atlanta, Norcross, Ga. 30092, (©1981.)

adjusted incorrectly. Second, the more often a piece of equipment is in-spected, the more likely it is that equipment defects will be detected in time to schedule repairs at a convenient time (unless the inspection itself contributes to the problem). Third, if preventive maintenance is perceived as a detriment to production, it will be postponed or it may not be performed at all.

10.2.1 Measurements

The choice of a maintenance interval is easier to defend if it is based on some kind of measurement. For example, if the water gauge pressure across a filter drops one inch when the filter is dirty, a manometer can be installed and the pressure drop across the filter can be recorded every week or month. Then a preventive maintenance interval can be chosen so that the pressure drop never exceeds one inch water gauge.

Maintenance intervals can also be based on recognition of increased energy consumption if the change of energy consumption with time can be measured. For example, if a steam system is old and has not been maintained very well, then the number of steam leaks that appear be-tween inspections may be related to the length of the inspection interval, and the steam loss will increase in some proportion to the length of this interval. Therefore, the steam system maintenance interval may be set based on counting the number of steam leaks. Once the number of leaks exceeds a pre-set value, it is time to perform maintenance on the equip-ment.

10.2.2 Performance degradation

It would be helpful if curves showing how equipment performance degrades over time were available for each piece of equipment. Then maintenance could be scheduled so that the performance never drops below a prescribed level. This can be done to a certain extent with lighting, as shown in Chapter 5. In most situations, however, the interval between inspections and the time to perform regular preventive maintenance must be determined on the basis of experience. In a dusty environment, for example, air filters may need to be changed once each month. In a clean environment, it may be possible to check filters once each year and only replace them every 2 years. Some criteria must be established to decide how often it is necessary to check to see if a filter is dirty, and setting these criteria requires experience.

Another point that should be considered when determining mainte-nance intervals is the likelihood of performance loss in the energy-con-suming systems. A steam leak costs money in lost energy, and it also costs money in degraded performance in the process which uses the steam. In one case, in a plywood plant where steam was used to dry plywood veneer, the energy cost of the steam loss was very small because the steam was generated by waste wood products. However, the loss of steam was directly associated with a loss of productive capacity so the actual cost of the steam loss was much greater than the energy cost of the leaks. In a compressed air distribution system, leaks cause the compressor to work more and use more energy. They also may cause problems with the control system using the compressed air. Thus performance degradation should be a factor in choosing maintenance intervals.

In many systems, performance degradation is gradual, and the optimum interval may cover a large range, for example, 4-8 weeks. In that case, the maintenance staff can choose the most convenient maintenance interval. They can group similar maintenance tasks and perform them at the same time. That way they can take advantage of the same economies that make group relamping attractive. If maintenance is performed by company personnel, it is usually desirable to choose maintenance intervals so that the load on personnel is constant.

Sometimes the degradation of a system can create safety problems. These systems should be inspected on a regular basis regardless of other considerations. Safety is a primary responsibility of management and must be a paramount concern to any professional.

10.2.3 Keeping track of the maintenance schedule

Once a maintenance schedule has been chosen, it should be communicated to anyone affected by the schedule. Someone should also have responsibility for making sure the schedule is followed. The schedule can be checked either through a computerized system with wide access to the communications data base or through a tickler file with one group of cards for each day. Either way, management should know what is supposed to be done and what has been done. Keeping a record of the schedule will allow the company to decide whether the maintenance interval should be lengthened or shortened.

10.3 STAFFING AND SUPPORT

It is difficult to determine staffing needs for maintenance. Three questions must be answered: How many people are needed, what skill levels should be employed, and how much should be invested in support equipment.

10.3.1 Estimating the Number of People Needed

The usual method for determining the number of staff members required for a manufacturing operation is to (1) calculate the number of operations that need to be performed, (2) estimate the standard time (including factors for personal time, fatigue, machine-caused delay, and transportation) for each operation, (3) add overhead, and (4) use the previous steps to estimate the number of man-years required per year. Then staff members are selected accordingly. For routine maintenance, this is possible, and References 1 and 2 give good guidance. In many systems, however, the major problem is repair rather than preventive maintenance, and a different technique must be used.

Factors which affect the times for repair maintenance include the experience of people doing the repair, the availability of necessary spare parts, the availability of diagnostic test equipment, and the possibility of having the work done by outside contractors. Generally, a time estimate should be made by the person who is responsible for the work, with an evaluation of this estimate based on the maintenance person's past performance.

In addition to obtaining time estimates for repair maintenance, it is important to obtain cost estimates for repair equipment, spare parts, and outside expertise. It is equally important to estimate the costs of consequences if repairs are not performed. Again, *safety should be considered first.*

Once benefits and costs have been tabulated for each repair, the repairs can be prioritized and worked into the schedule of regular maintenance personnel. If the repair load is very large, it may be necessary to hire additional people, depending on the urgency of the repairs and their potential benefits.

There is a great temptation to forgo preventive maintenance so as to have more time for repair. This is understandable but does not really make much sense. Preventive maintenance may increase the short-term work load, but it decreases the long-term work load. If such time-consuming items as bearing replacement can be regularly scheduled, then the number of in-plant bearing repairs, and the incidence of production stoppages due to bearing failures, will decline, and the load of unexpected repairs will also decline.

10.3.2 Skill Requirements

It pays to hire the best possible people for maintenance and to pay them enough to keep them. Maintenance personnel are usually the ones who lubricate and repair processes and equipment, and often are the ones that monitor the equipment operating records in order to predict the need for preventive actions. A journeyman will have enough experience to perform non-routine repair as well as to do routine cleaning, lubrication, and maintenance; an apprentice may have the enthusiasm but will probably not have the skill. If employees are given specific training in energy management, they can add these tasks as well and, in the authors' experience, will do so gladly.

The psychology of dealing with maintenance personnel is important. If the most visible results of an energy auditor's visit are criticism of maintenance procedures and an increase in work load, the auditor's next visit will not be greeted with enthusiasm. Many energy management measures, however, reduce the maintenance workload, and such reductions should be pointed out.

Special attention should be given to suggestions from the maintenance people; often their suggestions have been made before but management has ignored them on the philosophy that (1) the person making them is in maintenance and therefore not worth listening to or, (2) the person is working in our plant and therefore cannot be very smart. When good suggestions from maintenance personnel are given support by the energy management team with full credit to their source, the team gains credibility and further assures the cooperation of the maintenance staff.

10.3.3 Support Equipment

For energy management maintenance to be effective, the energy manager must pay attention to support equipment. This specifically includes manuals, diagnostic equipment, tools needed for repairs, spare parts, and instruments used in routine inspections. In addition there should be a good maintenance documentation program, and a good instrument calibration program.

10.3.3.1 Manuals and Documentation.

Vendors will usually provide manuals for most of their equipment items. One copy of each manual should be kept in a central location with an index system designed to facilitate its use. Other copies should be available where they will be needed. These manuals are a good source of basic information, including maintenance schedules and equipment. Documentation with up-to-date prints and schematics for the facility should also be kept near these manuals.

10.3.3.2 Diagnostic equipment.

Good diagnostic equipment is a great aid in identifying the exact cause of a problem. What constitutes "good," however, depends on the

quality of the equipment, the nature of the documentation, and the skill of
the user. Information on diagnostic equipment can be obtained from
vendors, training seminars, and basic experience of maintenance personnel. All these should be consulted, both formally and informally. It is also
important to have a program of maintaining and calibrating the diagnostic equipment, so that it is in good working order, and provides accurate
readings.

10.3.3.3 Spare parts inventories.

In-house inventories of spare parts must be sufficient to assure that
maintenance can be performed either when planned or when immediate
repairs are needed. But some parts are too expensive to stock routinely,
and other parts never break down, so it is not necessary to store every part
for every piece of equipment.

A thorough job of maintenance management includes keeping and
analyzing records on all equipment breakdowns and on the needs for
spare parts. It is also crucial to know who the vendors are and what the
lead times are for materials, under both routine and emergency conditions. The objectives of these analyses are to ensure that all parts are on
hand when maintenance is scheduled, to keep inventories of maintenance
parts to a small dollar value but to have enough parts on hand for critical
breakdowns, and to keep a ready supply of cooperative vendors available
so that emergency parts can be obtained when needed.

10.3.3.4 Instruments for routine inspections.

The authors consider it essential to have on hand a number of
instruments for routine inspection of energy management systems. These
instruments, their uses, and their approximate costs are shown in Figure
10-2.

10.4 USE OF MICROCOMPUTERS

10.4.1 Record-keeping systems

Microcomputer-based record-keeping systems are capable of storing
a large amount of data that can be readily accessed. They can keep current
track of an inventory with 10,000 different items and can generate orders
for spare parts. However, they will not solve problems associated with
sloppy maintenance planning; they cannot take the place of well-trained
people; and they will probably not decrease the amount of personnel
money spent on maintenance.

Figure 10-2.
Instruments for maintenance inspections in energy management.

Instrument	Cost	Function
Digital thermometer with surface and probe attachments, 50-2000°F	$200-350	Check stack, process, and room temperatures
Vane anemometer	300-400	Determine velocity of air in ducts and through envelope gaps
Pitot tube anemometer	200-300	Measure velocity of flue gas or fluids in ducts
Industrial light meter	40-60	Measure light conditions
Combustion gas analyzer	700-900	Analyze composition of flue gas
Industrial stethoscope	300-400	Check bearings on motors
Clamp-on ammeter with recorder	600-800	Measure current over chart time; determine electrical peak demand times
Clamp-on volt-ohm-milliammeter	80-200	Measure voltage, current where needed; check for short circuits

For maintenance software to be effective, it must be used; for it to be used, it must be understood and be seen as useful. There are some important things to remember:

(1) Some people are still afraid of computers. They are therefore unwilling to trust a computer program they do not understand to do something they can do equally well with a system they do understand, namely a file of cards.

(2) Personnel must be trained in the use of any software, under conditions of use, and this training must be carried out for a substantial period of time after the software purchase. A certain amount of handholding is necessary, from having the vendor staff available when the new software is put in, to having a reliable phone number where the vendor will answer questions that come up about the software.

(3) When a new software system is installed, the old system must be kept running in parallel with the new one until the bugs have been worked out of the new system. Otherwise, production and customers can be lost.

(4) When software is purchased, hardware must also be upgraded to provide the means for a backup of the data, and personnel must be trained to make a daily backup. This can save much time and confusion after an accident to the facility or to the computer. In some cases, it may be desirable to keep two backup copies, one on-site, one off-site.

10.4.2 Use of expert systems

Whenever a major system component fails, the expertise of on-site maintenance people may be exceeded. In this case, a company usually brings in outside vendors and repair personnel, thus incurring substantial cost and the downtime necessary for these experts to come to the site from their home bases. However, an alternative is currently being developed in the form of an "expert" system. An expert system is a unit of software that asks a user questions, responds to his/her questions on the basis of a series of rules and a base of system knowledge, and guides the user toward a solution to the problem. Some systems now available give diagrams, pages of the appropriate manuals, and even sounds as clues to what kind of repair is needed and how to do it. Many of these systems are included with the price of the equipment. Such systems are usually very easy to use and can reduce production downtime substantially.

10.5 DETERMINING THE NEED FOR OUTSIDE VENDOR SERVICES

Most companies do a mix of in-house repair and contract repair [5]. Some operations are a part of the routine preventive maintenance performed regularly by plant personnel. Other operations require equipment or skills that the plant personnel do not possess, and these are routinely handled by contracts. There are some significant advantages to a company to perform a maintenance operation rather than to have the operation performed by an outside contractor. Doing the operation in-house avoids the time necessary for a contractor to travel to the company or for the company to transport the item to be repaired to the contractor and back. Doing the operation in-house also enables the kind of detailed analysis of surrounding conditions that may lead to eliminating the source of the problem completely, an opportunity which may not be present if the repair is done by a contractor. Another significant advantage is that in-house repair enables the company to benefit from the learning process that occurs as it continually improve its own methods for each repair.

But there may be significant disadvantages to performing a maintenance operation in-house. First, it may require the purchase of new repair or diagnostic equipment, and this may be uneconomical. Second, it takes the time of maintenance personnel. If the present maintenance personnel are unfamiliar with the operation, they may lose time which could be better spent doing other maintenance tasks.

Some companies have purchased all of their maintenance services from an outside vendor. Advantages claimed for this are (1) the vendor pays the salaries and fringe benefits to employees, so the company does not need to incur such expenses as health insurance and retirement, (2) the vendor can develop a larger pool of expertise than any single company for trouble-shooting tasks, and (3) the vendor can take advantage of price breaks on spare parts and other consumable materials that would not generally be available to the company because of the smaller volume of such products it uses.

10.6 SUMMARY

In this chapter we have briefly discussed maintenance as a part of an energy management program. An effective maintenance program should include some elements aimed at reducing energy costs and/or increasing the amount of production available from a given amount of energy. Serious thought should be given to the people involved in maintenance: how many, who, and how they should be treated. Sufficient support equipment must be provided as part of regular maintenance costs. Microcomputers should be considered, both for data storage and retrieval and for use with expert systems for diagnosis of complex equipment.

REFERENCES
1. Thomas F. Sack, *A Complete Guide to Building and Plant Maintenance*, Prentice-Hall, Englewood Cliffs, N.J., 1971.

2. Unit Price Standards Handbook, NAVFAC P-716 (or ARMY TB 420-33 or AIR FORCE 85-56), Nov. 1977.

3. Wayne C. Turner, Senior ed., *Energy Management Handbook*, 2nd edition, Wiley-Interscience, New York, 1993.

4. W.J. Kennedy, Jr., "Apply a Cost-Benefit Approach in Building an Energy System Maintenance Program," *Industrial Engineering*, Dec. 1981, pp 60-65.

5. W.J. Kennedy, "Modeling In-house vs. Contract Maintenance, with Learning Effects" paper submitted to the *International Journal of Production Economics*, 1993.

BIBLIOGRAPHY

Criswell, John W., *Maintenance Time Management*, Fairmont Press, Atlanta, GA, 1991.

Criswell, John W., *Planned Maintenance for Productivity and Energy Conservation*, Fairmont Press, Atlanta, GA, 1990.

Engineers Digest, a monthly publication on maintenance, Intertec Publishing Corporation, Overland Park, KS.

Energy User News, a monthly publication on energy that often has articles on maintenance, Chilton Business News, Radnor, PA.

"Maintenance Management and Control," Chapter 73, in *Handbook of Industrial Engineering*, Second Edition, Gavriel Salvendy, Editor, Wiley, New York, NY, 1992.

Petrocelly, K.L., *Maintenance Computerization Handbook*, Fairmont Press, Atlanta, GA, 1993.

Power, a monthly publication which often has articles on maintenance in the utility industry, McGraw-Hill Publishing Company, New York, NY.

The Maintenance Journal, a bi-monthly publication of the International Maintenance Institute, Madisen Publishing Division, Appleton, WI.

Chapter 11

Insulation

11.0 INTRODUCTION

Unwanted heat loss or gain through the walls or roofs of buildings, and heat loss from the pipes, tanks and other equipment in buildings or plants can significantly increase energy use and energy costs. Thermal insulation plays an important role in reducing these energy costs in many situations. Good engineering design of insulation systems will reduce undesirable heat loss or gain by at least 90% in most applications and will often improve environmental conditions at the same time. Consequently, it is highly beneficial to understand insulation theory and applications, and to recognize when cost-effective insulation EMOs can be implemented.

In the first part of this chapter we discuss the theory behind the calculations involving heat transfer and its relation to the amounts or types of insulation used. The two basic areas of insulation applications are buildings and process equipment. In the rest of the chapter we discuss the various insulation types available for buildings and for process applications, and show examples of energy cost savings for several typical insulation EMOs.

Solar energy causes substantial heat gain in buildings, and even some heat gain in process equipment. Unwanted solar heat gains can often be controlled by suitable insulation, but sometimes the solar heat gain is a benefit in reducing use of fuel or electricity for heating. Usually a combination of insulation and architectural features such as overhangs or other blocking systems, films, and sometimes landscaping, is used to control this solar heat gain. The insulation aspects are covered in this chapter, but the direct use of solar heat in buildings and processes will be covered in Chapter 13 on alternative energy sources.

11.1 INSULATION THEORY

In our discussion of the theory of insulation, we cover the topics of heat transfer, thermal conductivity, and thermal resistance. The discussion of heat transfer explains the three ways that heat is moved into or out of a material. Thermal conductivity relates to the material itself and the ease with which heat moves through the material. Thermal resistance is the inverse of thermal conductivity, and relates to how well a given material will block or retard the movement of heat through it. We also develop the heat transfer calculations for flat surfaces and for cylindrical surfaces such as pipes.

11.1.1 Heat Transfer

There are three basic modes of heat transfer. They are conduction, convection, and radiation. These modes are defined as follows:

- *Conduction* is the transfer of heat from a hot side to a cooler side through a dividing medium. The hot side heats the molecules in the dividing medium and causes them to move rapidly, heating the adjacent molecules until the cool side is heated. The transfer stops when the temperature of the hot side equals that of the cool side.

- *Convection* is the transfer of heat between a moving liquid or gas and some conducting surface. Usually the heated fluid rises, causing cooler fluid to come in contact with the conducting surface, which is then heated and rises, etc.

- *Radiation* is the transfer of heat based on the properties of electromagnetic waves so no transfer medium is necessary. For example, the sun heats by radiation.

11.1.2 Thermal Conductivity

The *thermal conductivity* (K) of a material is a physical property that describes the ability of the material to conduct heat. K is measured by the amount of energy (Btus) per hour that can pass through one square foot of surface one inch thick* for a one-degree F temperature difference between the two environments being separated. The units in which K is measured are $(Btu \cdot in)/(ft^2 \cdot h \cdot °F)$. Tables 11-1 and 11-2 present K values for various materials at room temperatures. Table 11-1 contains conductivity values

---

*Be careful; sometimes tables show K values for one-foot thicknesses.

for materials that are good heat conductors, and Table 11-2 has values for good insulating materials. Thermal conductivity will vary with temperature, which can be important for process applications, as will be seen. For K values of common insulating materials at varying temperatures, see Figure 11-1.

The rate of heat transfer is directly proportional to the temperature difference and the thermal conductivity as shown in Equation 11-1:

$$Q \propto K \Delta t \qquad (11\text{-}1)$$

where Q = rate of heat transfer per ft^2 of surface
             Δt = temperature difference
             K = thermal conductivity

Table 11-1. Thermal conductivity values for various materials at room temperature.

Description	$K \dfrac{\text{Btu} \bullet \text{in.}}{\text{ft}^2 \bullet \text{h} \bullet {}^\circ\text{F})}$
Aluminum (alloy 1100)	1536
Aluminum bronze (76% Cu, 22% Zn, 2% Al)	696
Brass:	
Red (85% Cu, 15% Zn)	1044
Yellow (65% Cu, 35% Zn)	828
Bronze	204
Copper (electrolytic)	2724
Gold	2064
Iron:	
Cast	331
Wrought	418.8
Nickel	412.8
Platinum	478.8
Silver	2940
Steel (mild)	314.4
Zinc:	
Cast	780
Hot-rolled	744

Source: Albert Thumann, *The Plant Engineer's and Manager's Guide to Energy Conservation*, ©1977. Reprinted with permission of Van Nostrand Reinhold Co., New York.

11.1.3 Thermal resistance

The insulating property of a material is generally specified in terms of the *thermal resistance* (R) of the material, also called the R-value. Thermal resistance is related to the K value as follows:

$$R = \frac{d}{K} \tag{11-2}$$

where d = thickness of material
　　　 K = thermal conductivity

Table 11-2.

Material	Description	Conductivity K^a	Conductance $C^{b,c}$
Building boards	Asbestos-cement board	4.0	
	Gypsum or plaster board...1/2 in.		2.25
	Plywood	0.80	
	Plywood...3/4 in.		1.07
	Sheathing (impregnated or coated)	0.38	
	Sheathing (impregnated or coated) 25/32 in.		0.49
	Wood fiber—hardboard type	1.40	
Insulating materials	Blanket and Batt:		
	Mineral wool fibers (rock, slag, or glass)	0.27	
	Wood fiber	0.25	
	Boards and slabs:		
	Cellular glass	0.39	
	Corkboard	0.27	
	Glass fiber	0.25	
	Insulating roof deck...2 in.		0.18
Masonry materials	Loose fill:		
	Mineral wool (glass, slag, or rock)	0.27	
	Vermiculite (expanded)	0.46	
	Concrete:		
	Cement mortar	5.0	
	Lightweight aggregates, expanded shale, clay, slate, slags; cinder; pumice; perlite; vermiculite	1.7	
	Sand and gravel or stone aggregate	12.0	
	Stucco	5.0	
	Brick, tile, block, and stone:		
	Brick, common	5.0	
	Brick, face	9.0	
	Tile, hollow clay, 1 cell deep, 4 in.		0.90
	Tile, hollow clay, 2 cells, 6 in.		0.54

Table 11-2. (*Continued*)

Material	Description	Conductivity K[a]	Conductance C[b,c]
Masonry materials (cont'd)	Block, concrete, 3 oval core:		
	Sand & gravel aggregate .... 4 in.		1.40
	Sand & gravel aggregate .... 6 in.		0.90
	Cinder aggregate .... 4 in.		0.90
	Cinder aggregate .... 6 in.		0.58
	Stone, lime or sand	12.50	
Plastering materials	Cement plaster, sand aggregate	5.0	
	Gypsum plaster:		
	Lightweight aggregate...1/2 in.		3.12
	Lt. wt. agg. on metal lath...3/4 in.		2.13
	Perlite aggregate	1.5	
	Sand aggregate	5.6	
	Sand aggregate on metal lath 3/4 in.		7.70
	Vermiculite aggregate	1.7	
Roofing	Asphalt roll roofing		6.50
	Built-up roofing...3/8 in.		3.00
Siding materials	Asbestos-cement, 1/4 in. lapped		4.76
	Asphalt insulating (1/2 in. board)		0.69
	Wood, bevel, 1/2 × 6, lapped		1.23
Woods	Maple, oak, and similar hardwoods	1.10	
	Fir, pine, and similar softwoods	0.80	
	Fir, pine & sim. softwoods 25/32 in.		1.02

[c]Same as U value.
[a]Conductivity given in Btu•in./h•ft^2•°F
[b]Conductance given in Btu/h•ft^2•°F
Source: Extracted with permission from *ASHRAE Guide and Data Book*, 1965. Reprinted with permission from the Trane Co., La Crosse, WI.

To determine the thermal resistance of something composed of several materials, the total thermal resistance (R_{total}) is simply calculated as the sum of the individual components:

$$R_{total} = R_1 + R_2 + ... + R_N \qquad (11\text{-}3)$$

where R_i = the thermal resistance of the ith component,
         i = 1, 2, 3,..., N.

11.1.4 Conductance
The insulating property of a material is often measured in terms of conductance rather than resistance. *Conductance* (U) is the reciprocal of resistance.

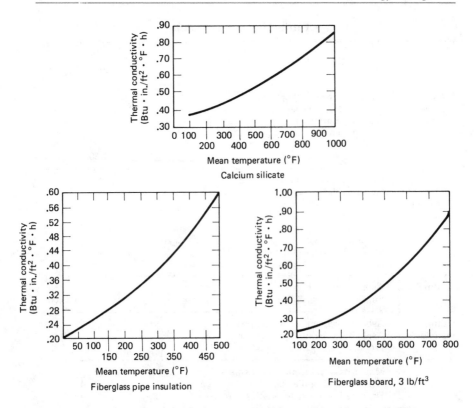

Figure 11-1. Thermal conductivities at varying temperatures.

$$U = \frac{1}{R} \tag{11-4}$$

The overall conductance of a total structure is

$$U_{total} = \frac{1}{R_{total}} \tag{11-5}$$

It is important to note that while the resistances are additive, the conductances are not.

11.1.5 Heat Transfer Calculations

Using the information developed so far, the final heat transfer equations can be constructed as follows:

$$Q = \frac{1}{R_1 + R_2 + \ldots + R_N} \Delta t \qquad (11\text{-}6)$$

$$Q = \frac{1}{R_{total}} \Delta t \qquad (11\text{-}7)$$

$$Q = U \Delta t \qquad (11\text{-}8)$$

$$Q_{total} = UA \Delta t \qquad (11\text{-}9)$$

where $\quad Q_{total} \quad$ = rate of heat transfer for total surface area involved
$\qquad\quad A \qquad$ = area of heat transfer surface

11.1.5.1 Heat Transfer Calculations for Flat Surfaces

The *surface film coefficient* is the amount of heat transferred from a surface to air or from air to a surface per square foot of surface for each degree of temperature difference. Surface film coefficients are usually specified in terms of the surface resistance, as shown in Table 11-3.

Example 11-1:

Assume we have a 1/2-inch-thick uninsulated mild steel tank storing a hot fluid as shown in Figure 11-2. (The K value for mild steel is found in Table 11-1.) The fluid is heated to 200°F, while ambient air is 70°F. What

Figure 11-2 Tank storage of hot fluid.

is the heat loss for the uninsulated tank? Ignore the heat loss to the ground. (Actually, transfer to the ground is a rather significant heat loss, but we will ignore it here.) The tank is 10 feet in diameter and 10 feet tall. Although this cylindrical tank is not truly a flat surface, it is large enough that it can be reasonably approximated as a flat surface. Smaller diameter cylinders such as pipes must be treated in a different way which is explained in the next section.

There are two primary sources of heat loss shown in Figure 11-2. Q_1 is the loss through the walls of the tank, and Q_2 is the loss through the roof. Although the surface film coefficient will vary somewhat for vertical heat flow versus horizontal heat flow, we will ignore that here, and assume that $R_{surface\ coefficient}$ is equal for the roof and walls:

$$R_{total} = R_{tank} + R_{surface\ coefficient}$$

The thermal resistance of the tank itself is quite small and is usually ignored, but we include it below for demonstration purposes; note how small it is compared to the surface resistance:

$$R_{tank} = \frac{d}{k} = \frac{.5}{314.4} = .0016 \frac{h \bullet ft^2 \bullet °F}{Btu}$$

From Table 11-2, assuming that the surface is dull metal and the surface temperature is 120°F*

$$R_{surface\ coefficient} = .50$$

$$U = \frac{1}{R_{surface} + R_{tank}} = \frac{1}{.5016} = 1.99 \frac{Btu}{ft^2 \bullet °F \bullet h}$$

$$A = area = \pi DH + \pi r^2 = \pi(10\ ft)\ (10ft) + \pi\ (5\ ft)^2 = 392.7\ ft^2$$

$$Q_{total} = UA\ \Delta t = \frac{1.99\ Btu}{ft^2 \bullet °F \bullet h}\left(392.7\ ft^2\right)(200° - 70°F)$$

$$= 101,591 \frac{Btu}{h}$$

---

*We will show you how to check this assumption shortly.

To see the impact that insulation can have on these calculations, assume that aluminum-jacketed fiberglass insulation 1 inch thick is added so that it covers the tank. The new heat loss would be:

$$R_{insulation} = \frac{d}{k} = \frac{1}{.25} = 4$$

Using Figure 11-1, K is found for fiberglass at a mean temperature of around 140°F.

$$U = \frac{1}{R_{tank} + R_{insulation} + R_{surface}} = \frac{1}{.0016 + 4 + .88}$$

$R_{surface}$ is found for aluminum, assuming a surface temperature of around 95°F.

$$U = \frac{1}{4.8816} = .205$$

$$Q_{total} = UA\,\Delta t = \frac{.205\ \text{Btu}}{\text{ft}^2 \bullet °F}\left(392.7\ \text{ft}^2\right)(130°F) = 10{,}465\,\frac{\text{Btu}}{\text{h}}$$

$$\text{savings} = 101{,}796 - 10{,}465 = 91{,}331$$

$$\% \text{ savings} = \frac{91{,}331}{101{,}796} = \underline{89.7\%}$$

Note that this is almost a 90% saving for only 1 inch of insulation. For 2 inches of insulation, the heat loss would be:

$$Q_{total} = UA\,\Delta t = \frac{1}{8.8816}(392.7)(130) = 5748\,\frac{\text{Btu}}{\text{h}}$$

$$\text{savings} = 101{,}796 - 5748 = 96{,}048\,\frac{\text{Btu}}{\text{h}}$$

$$\% \text{ savings} = \frac{96{,}048}{101{,}796} = \underline{94.4\%}$$

Table 11-3. Surface film coefficients, R_s Values[a] (h•ft^2•°F/Btu)

$t_s - t_a$ (°F)[b]	Still air		
	Plain fabric, dull metal, $\varepsilon = .95$	Aluminum, $\varepsilon = .2$	Stainless steel, $\varepsilon = .4$
10	.53	.90	.81
25	.52	.88	.79
50	.50	.86	.76
75	.48	.84	.75
100	.46	.80	.72

With wind velocities

Wind velocity (mph)			
5	.35	.41	.40
10	.30	.35	.34
20	.24	.28	.27

[a]For heat loss calculations, the effect of R_s is small compared to R_1, so the accuracy of R_s is not critical. For surface temperature calculations, R_s is the controlling factor and is therefore quite critical. The values presented are commonly used values for piping and flat surfaces.

[b]Note that t_s = surface temperature. Knowing the surface temperature requires measurement or calculation through the concept of thermal equilibrium, which will be discussed.

Source: Courtesy of Manville Corp.

Example 11-1 demonstrates a characteristic of the cost-effectiveness of insulation. The first increment of insulation thickness yields by far the largest energy savings. Each additional increment of thickness increases the energy savings, but at a rapidly decreasing rate. At some point, then, it becomes uneconomical to add any additional insulation. This will be demonstrated later in this chapter.

11.1.5.2 Heat Transfer Calculations for Pipes

For determining the effects of insulating pipes, the calculations become a bit more difficult because the heat flow is in a radial direction away from the pipe through the insulation to a larger surface area. The effect of dispersing the heat over a larger surface is to increase the insulation thickness. This effect is manifested in the calculations through a concept known as *equivalent thickness* d':

$$d' = r_2 \ln \frac{r_2}{r_1}$$ (11-10)

where   d' = equivalent thickness
           r_1 = outside radius of pipe
           r_2 = outside radius of pipe plus insulation

Table 11-4 shows the outside radius of a pipe if the nominal size—or inside diameter—is given.

Table 11-4. Nominal pipe size vs. outside radius.

Nominal pipe size (inches)	Outside radius (inches)	Nominal pipe size (inches)	Outside radius (inches)
1/2	.420	7	3.813
3/4	.525	8	4.313
1	.658	9	4.813
1-1/4	.830	10	5.375
1-1/2	.950	11	5.875
2	1.188	12	6.375
2-1/2	1.438	14	7.000
3	1.750	16	8.000
3-1/2	2.000	18	9.000
4	2.250	20	10.000
4-1/2	2.500	24	12.000
5	2.781	30	15.000
6	3.313		

Example 11-2:
A 6-inch nominal size pipe (meaning a pipe with a six-inch inside diameter) with 2 inches of insulation would have an equivalent thickness of insulation of

$$d' = r_2 \ln \frac{r_2}{r_1}$$

where r_1 = 3.313 from Table 11-4
         r_2 = 3.313 + 2.000 = 5.313

$$d' = 5.313 \ln \frac{5.313}{5.313} = 2.51$$

The equivalent thickness can now be used in the normal equations to determine the effective R value, etc.

Example 11-3:
    The same 6-inch pipe carrying fluid at 200°F and insulated by 2 inches of aluminum-jacketed fiberglass in a 70°F ambient area would have a heat loss of:

$$R_{insulation} = \frac{d'}{K} = \frac{2.51}{.25} = 10.04$$

$$Q_{total} = UA \, \Delta t$$

$$= \left(\frac{1}{10.04 + .88}\right) A \,(200 - 70)$$

$$= \frac{11.9 \, Btu}{ft^2 \bullet h}$$

Each linear foot of the 6-inch uninsulated pipe has a surface area of:

$$\frac{[3.313(2) \, in.] \, \pi(1 \, ft)}{12 \, in./ft} = 1.735 \, ft^2$$

To determine the savings from insulating the pipe we must calculate the heat loss for the uninsulated pipe. The heat loss per one-foot length of insulated pipe is:

$$\frac{(11.9 \, Btu)}{ft^2 \bullet h} \frac{\left(1.735 \, ft^2\right)}{ft} = 20.646 \, \frac{Btu}{ft \bullet h}$$

11.1.5.3 Thermal equilibrium and its applications

The concept of thermal equilibrium is important in many types of calculations including the checking of surface temperature assumptions demonstrated earlier. Thermal equilibrium simply says that the total heat flow through a system is equal to the heat flow through any part of the system. For example, a system as shown in Figure 11-3 consists of a wall, two layers of insulation, and an outside surface film.

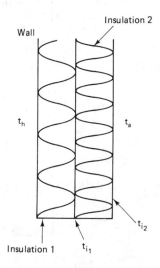

Figure 11-3. Insulation system.

Thermal equilibrium states that[*]

$$Q_{total} = \frac{t_n - t_a}{R_{i1} + R_{i2} + R_s} = \frac{t_n - t_{i1}}{R_{i1}} = \frac{t_{i1} - t_{i2}}{R_{i2}} = \frac{t_{i2} - t_a}{R_s}$$

$$= \frac{t_{i1} - t_a}{R_{i2} + R_s}$$

(11-11)

By using these equations, any surface temperature or other unknown quantity can be calculated if the heat flow and all other quantities are known.

---

[*]For purposes here, the inside surface film resistance is assumed to be zero. If not, there would be another set of expressions involving R_s.

Example 11-4:

The tank example demonstrated in Figure 11-2 assumed a surface temperature of 95°F. If we know the total heat loss, we can check this assumption by setting the total heat loss equal to one of the expressions shown in Equation 11-11. That is:

$$\frac{Q_{total}}{ft^2} = \left(\frac{10{,}500\,Btu}{h}\right)\left(\frac{1}{392.7\,ft^2}\right) = \frac{t_h - t_s}{R_{tank} + R_{insul}}$$

$$= \frac{200 - t_s}{.0016 + 4}$$

$$t_s = 93.0°F$$

As a further check:

$$Q_{total} = \left(\frac{t_s - t_a}{R_s}\right)\left(392.7\,ft^2\right)$$

$$= \left(\frac{93.0°F - 70°F}{.88}\right)\left(392.7\,ft^2\right)$$

$$= 10{,}300\,\frac{Btu}{h}$$

We assumed that the surface temperature was 95°F. This temperature results in a total heat loss of 10,300 Btu/hour. This value is close enough to the original value of 10,500 Btu/hour to reasonably validate our assumption.

These calculations can be used for many different objectives. For example, suppose the objective of insulation were to protect personnel and the surface temperature could not be higher than, say, 110°F. The energy manager could take the appropriate expressions as in Equation 11-11 and set the surface temperature equal to 110°F. By back-calculating, the required amount of insulation thickness could be determined.

Suppose the purpose were to prevent condensation from forming on cold pipes. Then the energy manager could determine the dew point of ambient air, set the surface temperature equal to that, and back-solve again for the insulation thickness.

In both cases, the energy manager would probably round off to the next highest insulation thickness or even increment up one thickness as a safety measure.

For energy management purposes, it is usually necessary to calculate the heat loss or gain over a year instead of for only an hour as already shown. To do this, it is necessary to sum the hourly heat losses over the number of hours in a year.

Example 11-5:

Suppose the pipe in Example 11-3 carried hot fluid 24 hours/day for 365 days/year. Further, assume that the unit generating the hot fluid is 80 percent efficient using natural gas costing $5.00/$10^6$ Btu. Find the energy cost savings per foot of pipe from insulating it with 2 inches of aluminum jacketed fiberglass. Assume the pipe is made from mild steel.

The total heat loss for a year for the uninsulated system is:

$$Q_{total} = UA \, \Delta t \; \left(\text{assume } T_s \approx 120°F\right)$$

$$= \left(\frac{1}{.50}\right)(1)(130)$$

$$= 260 \, \frac{Btu}{ft^2 \bullet h}$$

$$Q_{total(\$)} = \left(\frac{260 \, Btu}{ft^2 \bullet h}\right)\left(\frac{1}{.8}\right)\left(\frac{\$5.00}{10^6 \, Btu}\right)\left(\frac{8760 \, h}{year}\right)$$

$$= \$14.24/ft^2$$

Savings for insulating the pipe would be

$$\left(\frac{260 \, Btu}{ft^2 \bullet h} - \frac{11.9 \, Btu}{ft^2 \bullet h}\right)\left(\frac{1}{.8}\right)\left(\frac{\$5.00}{10^6 \, Btu}\right)\left(\frac{8760 \, h}{year}\right) = \$13.58/ft^2$$

Savings per linear foot of pipe would be

$$\left(\frac{\$13.58}{ft^2}\right)\left(\frac{1.735\ ft^2}{ft}\right) = \underline{\$23.57/ft}$$

Consequently, insulation can be a real money saver.

In dealing with systems exposed to outside conditions, it is often helpful to use the degree day or degree hour concept developed in Chapter Two. Degree days are used in the example below.

Example 11-6:
Assume a building wall has an R value of 3. The temperature inside is kept at 65°F during the winter and 75°F during the summer. The plant operates 365 days/year, 24 hours/day. Assuming a heating plant efficiency of 0.80 and a cooling coefficient of performance (COP) of 2.5, what is the cost of energy loss through the wall? Electricity costs \$.08/kWh and gas \$6.00/10^6 Btu. The plant experiences 4000°F heating days and 2000°F cooling days. The total wall area is 100 ft^2.

$$heat\ lost\ (\$) = \left(\frac{4000°F\ days}{year}\right)\left(\frac{1\ Btu}{3\ ft^2 \cdot h \cdot °F}\right)\left(\frac{24\ h}{day}\right)(100\ ft^2)$$

$$\times\left(\frac{1}{.8}\right)\left(\frac{\$6.00}{10^6\ Btu}\right)$$

$$= \underline{\$24.00/year}$$

$$cooling\ gain\ (\$) = \left(\frac{2000°F\ days}{year}\right)\left(\frac{1\ Btu}{3\ ft^3 \cdot h \cdot °F}\right)\left(\frac{24\ h}{day}\right)(100\ ft^2)$$

$$\times\left(\frac{1}{2.5}\right)\left(\frac{kWh}{3412\ Btu}\right)\left(\frac{\$.08}{kWh}\right)$$

$$= \underline{\$15.00/year}$$

total energy loss = \$15.00 + \$24.00 = \$39.00/year

If the plant described above had a night setback program, then you would have to use degree hours calculated through manipulation of the base corresponding to the hours of setback. For example, if the temperature were setback to 55°F, the heat loss would be calculated using heating degree hours determined from a 55°F base for the respective hours of setback rather than the standard 65°F base for all heating hours.

11.2 INSULATION TYPE

Before an energy manager can select the proper type of insulation for a particular application, he or she must know the properties of various kinds of insulating materials. In this section, the properties of insulation are discussed first, followed by a discussion of the different types of insulation.

11.2.1 Properties of materials used for insulation

Some of the more important properties of materials that would be used to provide insulation properties include the following:

- *Cell structure.* Cell structures are either open or closed. A closed cell is relatively impervious to moisture, especially in a moderate environment, so insulation with a closed cell structure may not need any additional moisture barrier. Open cells pass moisture freely and therefore probably require vapor barriers. For extremely cold applications where a lot of condensation occurs, a vapor barrier is probably required regardless of cell structure.

- *Temperature use.* Different insulating materials react to extreme temperatures in different ways. In some cases, high temperatures can destroy the binders and render the insulation useless. All insulation materials have temperature ranges for which they are recommended. Usually the restriction occurs at the warmest end rather than the coldest end.

- *Thermal conductivity (K).* As mentioned earlier, K values vary with the temperature—sometimes significantly. The energy manager must be familiar with the different types of insulation, their K values, and how the temperature affects the K values. In all cases, the K value chosen to reflect the appropriate conductivity should be that for the mean temperature (t_m) experienced by the insulation:

$$t_m = \frac{t_h + t_s}{2}$$

(11-12)

- *Fire hazard.* Fire hazard ratings measure a product's contribution to flame spread and smoke development in a fire. The rating is measured on a flame spread-smoke spread scale where 100/100 is the rating for red oak.

- *Forms.* Insulation is available in a number of different forms. Flexible blankets, batts, rigid board, blocks, and pipe half sections are some of the more popular ones. Insulation is also available in a number of sizes and thicknesses. For example, fiberglass batt insulation with a Kraft paper vapor barrier is available in 15-inch and 23-inch widths for thicknesses of 3.5 and 6 inches. Thicknesses of 12 inches are readily available in many parts of the country.

11.2.2 Common Insulating Materials

Some of the more popular types of materials with a discussion of some of their specific properties are given here. Details are summarized in Table 11-5.

- *Mineral fiber-rock wool.* Mineral fiber insulation is made from molten rock. It is fairly impervious to heat and so can be used in relatively high temperatures (see Table 11-5).

- *Fiberglass.* Probably the most popular type of insulation, fiberglass can be obtained in blankets, batts, boards, and pipe covering. Although organic binders are frequently used which limit temperature ranges somewhat, cell structure is such that the limitations can sometimes be exceeded and still have acceptable results.

- *Foams.* Several types of foam insulation are available; some types have problems meeting fire hazard classifications but have very good K values. Others meet the fire hazard requirements but do not offer very good K values. Foams are particularly applicable to cold applications.

- *Calcium silicate.* A very popular type of insulation for high-temperature use, calcium silicate is spun from lime and silica. It is extremely durable and offers a high thermal resistance.

Table 11-5. Industrial insulation.

Insulation type and form[a]	Temp. range (°F)	Thermal conductivity [Btu·in./h·ft²·°F at T_m (°F)]			Compressive strength (psi) at % deformation	Fire hazard classification or flame spread-smoke developed	Cell structure (permeability and moisture absorption)
		75	200	500			
Calcium silicate blocks, shapes, and P/C	To 1500	.37	.41	.53	100-250 at 5%	Noncombustible	Open cell
Glass fiber blankets	To 1200	.24-.31	.32-.49	.43-.73	.02-3.5 at 10%	Noncombustible to 25/50	Open cell
Glass fiber boards	To 1000	.22	.28	.51-.61			
Glass fiber pipe covering	To 850	.23	.30	.62			
Mineral fiber blocks and P/C	To 1900	.23-.34	.28-.39	.45-.82	1-18 at 10%	Noncombustible to 25/50	Open cell
Cellular glass blocks and P/C	-450 to 900	.38	.45	.72	100 at 5%	Noncombustible	Closed cell
Expanded perlite blocks, shapes, P/C	To 1500	—	.46	.63	90 at 5%	Noncombustible	Open cell
Urethane foam blocks and P/C	(-100 to -450) to 225	.16-.18	—	—	16-75 at 10%	25-75 to 140-400	95% closed cell
Isocyanurate foam blocks and P/C	To 350	.15	—	—	17-25 at 10%	25-55 to 100	93% closed cell
Phenolic foam P/C	-40 to 250	.23	—	—	13-22 at 10%	25/50	Open cell
Elastomeric closed cell sheets and P/C	-40 to 220	.25-.27	—	—	40 at 10%	25-75 to 115-490	Closed cell
MIN-K blocks and blankets	To 1800	.19-.21	.20-.23	.21-.24	100-190 at 8%	Noncombustible	Open cell
Ceramic fiber blankets	To 2600	—	—	.38-.54	.5-1 at 10%	Noncombustible	Open cell

[a]P/C means pipe covering.
Source: Wayne C. Turner, Senior Ed., *Energy Management Handbook*, ©1982. Reproduced by permission of Wiley-Interscience, New York.

- *Refractories-ceramic fiber.* An alumina-silica product, ceramic fibers are available in blankets or felts that can be used alone or added to existing fire brick.

- *Refractories-fire brick.* Fire bricks are made for high-temperature applications. Made of a refractory clay with organic binders which are burned out during manufacture, they offer good thermal resistance and low storage of heat.

- *Others.* Other types of insulation include cellular glass, perlite, and diatomaceous earth. Each has advantages and disadvantages with which the energy manager must become familiar.

11.3 ECONOMIC THICKNESSES

As mentioned in Example 11-1, insulation has an optimum thickness that can be calculated using the principles of engineering economy discussed in Chapter 4. Consider Figure 11-4.

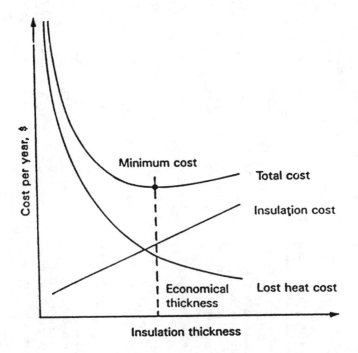

Figure 11-4. Illustration of the economic thickness of insulation.

As thickness of insulation is increased, the cost of material and installation goes up. The cost of lost energy, on the other hand, goes down, but at a decreasing rate. Said in other terms, the energy cost savings also goes up, but at a slower rate of increase than the cost of materials and installation. At some point, then, the total cost, which is the sum of the lost energy cost and the material cost, reaches a minimum point. That amount of insulation is called the *economic thickness*.

As Figure 11-4 shows, the total cost curve is relatively flat in the immediate neighborhood of the economic thickness. This means that the energy manager does not have to use the exact optimum amount of insulation. A small deviation either way will not affect the resulting annual cost very much.

To determine this economic thickness, the energy manager needs to construct cash flow diagrams for the different alternative thicknesses and calculate the annual equivalent cost for each increment. Since the cash flows include future fuel costs, careful handling of inflation is required.

Many simplified charts, graphs, tables, and computer programs have been developed to help determine optimum thicknesses. The Bibliography includes some of these, but it is fairly simple to develop a program that will quickly calculate equivalent annual cost for each increment.

Figure 11-5 is a printout resulting from a program written at Oklahoma State University. From this printout you can see that the economic thickness for this particular application is 3 inches. However, the total cost for 2 inches and 4 inches of insulation is about equal and is only slightly higher than the total cost for the economic thickness.

11.4 SUMMARY

Reducing heat loss or heat gain by adding insulation is often a cost-effective Energy Management Opportunity. An energy manager must know the different types of insulation available, and their respective advantages and disadvantages in specific applications. Different kinds of insulation are needed for walls and roofs of buildings, as compared to insulation needed for high temperature steam lines. Knowing how to perform heat loss-gain calculations is necessary to properly evaluate the effects of adding insulation, and to calculate the benefits of different levels of insulation. Finally, understanding insulation economics is important for good energy management since the first increments of added insulation produce very cost-effective benefits, while there is a point of diminishing returns of savings for increasing the levels of insulation. The calculation of the optimum economic thickness of insulation allows the deter-

Figure 11-5.
Computer printout of insulation economic thickness problem.

ECONOMIC THICKNESS DETERMINATION.

DEPARTMENT OF INDUSTRIAL ENGINEERING AND MANAGEMENT
OKLAHOMA STATE UNIVERSITY

INPUT PARAMETERS

FIRM: PLASTICS, INC.		CONTACT: S.A. PARKER	
SYSTEM: PLANT WALLS		DATE: OCTOBER 10, 198-	
INSULATION: FIBERGLASS BATTS		K-VALUE: 0.3600 BTU·IN/H·FT^2·°F	
AFTER-TAX MARR:	18.0%	GENERAL INFLATION RATE: 10.0%	
FUEL ESCALATION RATE:	20.0%	PRESENT R-VALUE: 1.427 HR·FT^2·°F/BTU	
HEATING DEGREE-HOURS: 112392.0		COOLING DEGREE-HOURS: 153600.0	
HEAT PLANT EFFICIENCY:	75.0%	COOLING PLANT EFFICIENCY: 210.0%	
COST PER MMBTU OF HEAT: $3700		COST PER MMBTU OF COOLING: $4900	
INCREMENTAL TAX RATE:	48.0%	AVAILABLE TAX CREDIT: 0.0%	
PROJECT LIFE: 10 YEARS		DEPRECIATION LIFE: 5 YEARS	

THICKNESS CALCULATIONS

INSULATION THICKNESS (IN.)	INSTALLED COST ($/FT^2)	ANNUAL ENERGY LOSS OR GAIN (BTU/FT^2)	NPV OF SAVINGS ($/FT^2)	ANNUALIZE COST ($/FT^2)
.00	.00	156271.1	.00	.493
1.00	1.31	53034.6	.52	.378
2.00	1.53	31936.6	.64	.351
3.00	1.65	22847.5	.68	.342
4.00	1.84	17785.7	.62	.355
5.00	2.01	14560.0	.54	.372
6.00	2.15	12324.7	.48	.387
7.00	3.27	10684.4	−.31	.563
8.00	3.43	9429.4	−.41	.584
9.00	3.57	8438.3	−.52	.608

mination of the greatest amount of insulation to add, while still producing the largest savings in energy cost.

BIBLIOGRAPHY

ASHRAE Handbook of Fundamentals, Chapter 20, Thermal Insulation, American Society of Heating, Refrigerating and Air Conditioning Engineers, Atlanta, GA, 1993.

Brand, Ronald E., *Architectural Details for Insulated Buildings*, Van Nostrand Reinhold, New York, NY, 1990.

Diament, R.M.E., *Thermal and Acoustic Insulation*, Butterworth, Guilford, Surrey, England, 1986.

Donnely, R.G. et al, *Industrial Thermal Insulation*, Oak Ridge National Laboratory, ORNL/TM-5515, 1976.

Graves, R.S. and D.C. Wysocki, *Insulation Materials, Testing and Applications*, Volume 2, American Society for Testing Materials, Philadelphia, PA, 1991.

McElroy, D.L. and J.F. Kimpflen, *Insulation Materials, Testing and Applications*, Volume 1, American Society for Testing Materials, Philadelphia, PA, 1990.

Parker, Steven A., Insulate, working paper, Oklahoma State University, Stillwater, Okla., 1982.

Silvers, J.P. et al, A Survey of Building Envelope Thermal Anomalies and Assessment of Thermal Break Materials for Anomaly Correction, prepared for Oak Ridge National Laboratory, 1985.

Stamper, Eugene, *National Program Plan for the Thermal Performance of Building Envelope Systems and Materials*, Third Edition, Building Thermal Envelope Council, Washington, DC, 1988.

Thumann, Albert, *The Plant Engineer's and Manager's Guide to Energy Conservation*, Fifth Edition, Fairmont Press, Atlanta, GA, 1990.

Turner, Wayne C., Senior ed., *Energy Management Handbook*, Second Edition, Fairmont Press, Atlanta, 1992.

Chapter 12

Process Energy Management

12.0 INTRODUCTION

In many facilities, energy management is simply a matter of managing the energy required for lighting and space conditioning. In many others, however, energy management is much more complex and involves large motors and controls, industrial insulation, complex combustion monitoring, unique steam distribution problems, significant amounts of waste heat, etc. Typical facilities offering large energy management opportunities include industrial facilities, large office and commercial operations, and government institutions such as schools, hospitals and prisons. Such facilities generally have specialized industrial, commercial or institutional processes that incorporate many of the concepts covered in other chapters. These processes require thorough analytical evaluations to determine the appropriate energy-saving measures. This chapter provides some examples.

The energy manager must be careful in process energy management. Processes can be quite complex, so a full understanding of the entire process is necessary. Examining only one area of a process and making energy use changes to that part may have adverse effects in another area. For example, small changes in heat treatment temperatures or atmosphere can sometimes dramatically decrease the product quality or subsequent workability.

In this chapter we first present a suggested procedure for process energy improvement. Then, motors and controls are discussed since they form an integral part of most processes. Next, some sample case studies of process energy management opportunities are provided. Finally, we outline some common process activities where better energy management can be practiced. Air compressors are also discussed.

12.1 STEPS FOR PROCESS IMPROVEMENT

Readers who have studied work simplification and improvement may remember the suggested order of changes as (1) eliminate; (2) combine; (3) change equipment, person, place, or sequence; and (4) improve [1].

The same order of change is appropriate for process energy management, but as mentioned earlier, the analyst must understand the entire system and the cascading impacts that changes might effect. In terms of energy management, examples of the preceding changes include the following:

- *Eliminate.* Does that cooling water really need to be there? Sometimes process cooling water is not really necessary; eliminating it saves pumping and chilling costs. Is the paint oven really necessary? Some newer paints will air dry quite well, and paint oven costs can be substantial.

- *Combine.* Machining operations can often be combined with jig and fixture modifications or changes in equipment. This saves the energy used by the additional machines; it also reduces material handling and may save process storage energy. Sometimes combining processes also saves the energy necessary to bring the material back to a required workability.

- *Change equipment, person, place, or sequence.* Equipment changes can offer substantial energy savings as the newer equipment may be more energy efficient. For example, new electric welders are considerably more energy efficient than older ones. Changing persons, place, or sequences can offer energy savings as the person may be more skillful, the place more appropriate, and the sequence better in terms of energy consumption. For example, bringing rework back to the person with the skill and to the place with the correct equipment can save energy.

- *Improve.* Most energy management work today involves improvement in how energy is used in the process because the capital expenditure required is often minimized. Examples include reducing excess air for combustion to a minimum, reducing temperatures to the minimum required (don't forget chilling—maybe the freezer temperature can be increased a few degrees), and removing excess lighting. Improving does sometimes require large amounts of capital. For example, insulation improvements can be expensive, but energy savings can be large, and there can be improved product quality.

12.2 MOTORS AND ADJUSTABLE SPEED DRIVES

Motor energy use represents well over half of all the electric energy consumed by industrial, commercial and institutional facilities. Motors are found on almost every piece of equipment used to perform a process in manufacturing, mining and agriculture. Even pieces of equipment that perform special functions often have motors as their principal part—for example, chillers and air compressors.

Because of the widespread application of electric motors in almost every facility, they are excellent candidates for improvements to their efficiencies and improvements in their utilization in machines and processes. Just a small improvement in electric motor efficiency can produce significant savings in the energy cost of operating a piece of equipment. The annual cost of operating a motor can often be five to ten times the original purchase price of the motor.

Many motor applications require variable speeds, or should use variable speeds to match the actual loads, and thus the area of motor controls is also very important. Adjustable speed drives—or variable speed drives—are motor control systems that reduce the energy input to a motor when it is not fully loaded. These ASDs or VSDs can produce substantial savings in the operational costs of motors, and can often improve the operation of the system that previously used a motor without a speed control.

12.2.1 High Efficiency Motors

Electric motors account for about three-quarters of all the electric energy used by industry (Figure 1.5, Chapter One), and almost half of all electric use by commercial facilities. Energy efficient motors are now readily available that are two to eight percent more efficient than the standard motors they would replace. Table 12-1 provides data on the efficiencies and cost premiums for motors in size ranges from 0.75 horsepower to 250 horsepower [2]. Since typical motors last over twenty years, using high efficiency motors offers business and industry substantial energy and dollar savings.

Table 12-1. Motor data.

hp Rating	Standard Efficiency	High Efficiency	Premium ($)
.75	0.740	0.817	35
1	0.768	0.840	39
1.5	0.780	0.852	48
2	0.791	0.864	56
3	0.814	0.888	73
5	0.839	0.890	69
7.5	0.846	0.902	97
10	0.864	0.910	111
15	0.875	0.916	149
20	0.886	0.923	186
25	0.897	0.929	224
30	0.901	0.931	273
40	0.908	0.934	371
50	0.915	0.938	469
60	0.916	0.940	553
75	0.917	0.944	678
100	0.919	0.950	887
125	0.924	0.952	1,172
150	0.930	0.953	1,457
200	0.940	0.956	2,027
250	0.943	0.956	2,159

Motor efficiency is a measure of the effectiveness with which electrical energy is converted to mechanical energy. Motor losses occur in five major areas: core losses, stator losses, rotor losses, stray load losses, and windage and friction losses. High efficiency motors are designed and manufactured to reduce these losses. In addition to having lower losses, high efficiency motors also have higher power factors during operation. Cost premiums for high efficiency motors range from 10% to 30%, but since a motor may use 75 times its initial cost in electric energy over its lifetime, the savings potential is great [3]. Many motors in commercial facilities, industries and institutions run 6000 to 8000 hours per year, so very cost-effective paybacks can be achieved.

Example 12-1: Ace Industries has a 50 hp air compressor that operates at full-load, all day for 365 days per year. If the motor for the air compressor cost $1400, the motor efficiency is 90%, and electricity costs $7.00/kW/month and $0.05/kWh, how much does it cost to operate the air compressor for one year? How much money will be spent to operate the air compressor over a ten-year period?

Solution: If a motor were 100% efficient, there would be an electric load of 0.746 kW/hp, or 37.3 kW. Since the motor is only 90% efficient, its electric load is:

Electric load   = (50 hp) × (0.746 kW/hp)/(0.9)
                    = 41.44 kW

The annual demand charge is then:

Demand charge   = (41.44 kW) × ($7.00/kW/month) ×
                       (12 months/year)
                       = $3481/yr

The annual energy charge is:

Energy charge   = (41.44 kW) × (8760 h/yr) × ($.05/kWh)
                    = $18,151/yr

The total electric cost of operating the motor is then:

Total cost = $3841 + $18,151 = $21,992/yr

Since the purchase cost for the motor was $1400, the annual electric cost for operating the motor/air compressor is over fifteen times the initial purchase price.

If the air compressor operates for ten years, the total operating cost will be 10 × $21,992 or $219,920. This is over 150 times the initial purchase price of the motor.

As this example shows, the original purchase cost of a motor can be a small part of the life-cycle cost. Thus, it is important to consider other factors besides the initial cost when buying a new motor for a piece of equipment.

Example 12-2: ACE Industries has been experiencing a period of rapid growth in the success of their products, and they plan to expand their production capacity by building a second plant nearby. They determine that they need another 50 hp air compressor which will also run continuously. They can purchase either the Standard or the Deluxe Model air compressor with the difference being that the Deluxe Model has a high efficiency motor. The motor efficiency for the Standard Model is 91.5%, and for the Deluxe Model it is 93.8%. The additional cost for the Deluxe Model is $470. Is this a good investment for ACE Industries?

ACE Industries also wants to know what kind of "cushion" they have on this decision, since their forecast for new business could be too optimistic. If the new air compressor is only run for two shifts a day, for a total of 5000 hours per year, is the additional investment still worthwhile?

Solution: The electric load for each model of the motor is found, and the difference will be used to compute the savings. The reduction in electric demand is:

Standard Model demand	$= (50 \text{ hp}) \times (0.746 \text{ kW/hp})/(.915)$ $= 40.77 \text{ kW}$
Deluxe Model demand	$= (50 \text{ hp}) \times (0.746 \text{ hp})/(.938)$ $= 39.77 \text{ kW}$
Demand savings = 1 kW	
Annual demand cost savings	$= (1 \text{ kW}) \times (\$7.00/\text{kW month})$ $\times (12 \text{ months/yr})$ $= \$84/\text{yr}$
Annual energy cost savings	$= (1 \text{ kW}) \times (8760 \text{ h/yr}) \times$ $(\$.05/\text{kWh})$ $= \$438/\text{yr}$
Total electric cost savings	$= \$522/\text{yr}$

We can now compute several economic decision criteria to evaluate how good an investment the additional $470 would be. Using equations 4-1 and 4-2, and the rewritten form of Equation 4-5 in Example 4-3 from Chapter Four gives:

Simple Payback Period = \$470/\$522/yr = <u>0.9 years</u>

If the motor has a lifetime of ten years running full-time, we can find the ROI as:

ROI = [\$522 − (\$470/10)]/\$470 = <u>101%/year</u>

If ACE Industries has an investment rate of 12%, we can find the Benefit-Cost Ratio as:

$$B/C \ = \$522 \times [(1 + I)^n - 1]/[I \times (1+I)^n]/\$470$$
$$= 1.11 \times [(1 +.12)^{10} - 1]/[.12 \times (1 +.12)^{10}]$$
$$= 1.11 \times 5.6502 = \underline{6.27}$$

By any one of these three economic evaluation measures, the decision to buy the Deluxe Model with the high efficiency motor is an excellent investment if the motor runs 8760 hours each year. To see how sensitive this result is to the use of the motor, we need to recompute the savings if the motor is only used 5000 hours each year.

The demand savings is the same. The new energy cost savings is:

Energy cost savings = (1 kW) × (5000 hours/yr) × (\$0.05/kWh)

= \$250/year

Total cost savings = \$84 + \$250 = \$334

The new values of the economic measures are:

SPP = <u>1.41 years</u>

ROI = <u>61.1%</u>

B/C = <u>4.02</u>

These are still excellent values for the three economic decision measures. Most companies would still find the investment in the more expensive motor highly attractive, even with the reduced hours of operation of the air compressor.

A number of computer programs have been written to help perform the economic analysis involved in motor selection. One of the best is the MotorMaster program written and made available by the Washington State Energy Office [4]. This program has an extensive data base on motor prices and efficiencies, and is updated twice a year.

12.2.2 Rewinding electric motors

There is at least one other important factor in motor replacement selection, and that is the potential for rewinding a motor that has failed. There are three options available for a facility that has just experienced a motor failure. One, they can buy a standard efficiency motor to replace the failed one. Two, they can buy a high efficiency replacement. Three, they can send the failed motor out to be repaired, and potentially rewound. Example 12-2 illustrated how cost-effective it can be to buy a more expensive, high-efficiency motor for a piece of equipment that is used heavily.

The cost of rewinding a motor is often substantially less than the cost of purchasing a new motor—whether it is a standard-efficiency model or a high-efficiency model. However, it is fairly common for motors to be damaged during the rewinding process, and to suffer losses in efficiency of 1-2% [5]. Example 12-2 showed the economic impact of a 2.3 percentage point difference in motor efficiency. Often, a 1 percentage point loss in efficiency will result in the cost of the additional electricity being greater than the total cost of rewinding. Thus it is important to consider this factor when replacing a motor. Not all rewinding operations damage motors, but the loss of efficiency is quite common.

12.2.3 Motor drives and controls

In addition to improving the efficiency of the motor itself, there are many other opportunities for energy savings in the complete motor system. One of these opportunities is in the use of solid-state, electronic controls that can provide soft-starts, speed control and power factor correction.

For large motors that have variable loads, the addition of electronic speed controllers—or adjustable speed drives (ASDs)—can be very cost effective. ASDs are electronic devices that vary the speed of a motor to match that of the load being put on the motor. The size of a motor is usually based on maximum load, even though normal design conditions seldom require this full load size. An ASD will reduce the speed of the motor by adjusting the frequency, voltage, or current of the motor input so that the motor performance exactly matches the present load. ASDs are also called Variable Speed Drives (VSDs) or Variable Frequency Drives (VFDs).

Fans and pumps are typical applications where ASDs can improve motor performance. The cube law for fans discussed in Chapter Six (Equation 6-14) showed the potential for substantial energy savings when the speed of a fan is reduced. Since the energy used in many fan and pump applications is proportional to the cube of the flow rate, then small reductions in the required flow rates translate to large savings in energy needed. In addition, many motors are purposefully over-sized to have a safety factor in handling the required load. This over-sizing is not beneficial to energy efficiency, and results in many motors running at conditions that unnecessarily waste energy. Properly sizing motors is probably one of the most cost-effective EMOs that a facility could implement.

ASD costs vary with the size of the motor. Equipment costs range from $150 to $450 per horsepower for ASDs smaller than 50 horsepower, and $100 to $150 per horsepower for larger units [6]. General estimates are that ASDs can save an average of 20-30 percent of the energy used for a typical motor.

Example 12-3: In one industrial application, three cyclone fans used for ventilating stack gases were replaced with one larger fan motor having an ASD. The cost of the new fan, motor, and installation of new ductwork was $18,250. The cost of the ASD was $20,000. The electrical consumption from the new system dropped 500,000 kWh/yr from that of the old fan system. If the facility paid $0.061/kWh, determine the cost-effectiveness of this EMO.

Solution: The electrical cost savings from the new system is found as:

Annual savings   = (500,000 kWh/yr) × ($0.061/kWh)
                 = $30,500

The total cost of implementation is the sum of the costs for the new fan, motor, ductwork, installation and the ASD:

Total implementation cost   = $18,250 + $20,000
                            = $38,250

The SPP for this EMO is found using Equation 4-1 from Chapter Four:

SPP = $38,250/$30,500/yr = <u>1.25 years</u>

If the new system had an expected lifetime of 15 years, the ROI can be found using Equation 4-2 from Chapter Four:

ROI = [30,500 – (38,250/15)]/38,250
      = <u>73.1%</u>

These two measures demonstrate that this was an excellent EMO investment.

12.2.3 Other factors in motor system efficiency

In addition to the efficiency of the motor itself and the use of adjustable speed drives, there are still several other factors that affect the overall motor system efficiency. One of these factors is the mechanism for the transmission of power from the motor to the load. In particular, care should be taken to insure that efficient pulleys, drives, belts and gears are used to couple the motor to the load.

Belts. Many motors are coupled to their loads with belt drives. There are several types of belts, including V-belts, cogged V-belts, and synchronous belts. Standard V-belts are the most common type of drive belt, and have transmission losses that occur because of flexing and slippage of the belt. The cogged V-belts and the synchronous belts are more efficient because they do not allow the same amount of flexing and slippage as the standard V-belts. Motor system performance can be improved by 2-4% with the use of these more efficient belts.

Lubrication. Motor lubrication is also a factor in motor efficiency. There are synthetic lubricants available that reduce the friction losses in motor-driven equipment. Savings of 1-2% are common, and much larger savings are possible for some equipment. Often, however, manufacturers recommend that these synthetic lubricants be used only in new pieces of equipment, so that there is no contamination of the synthetic lubricant.

Maintenance. Motor maintenance is also a factor. This area was discussed in some detail in Chapter Ten on Maintenance. The operating temperature of a motor should be checked periodically, as well as its mechanical and electrical condition. Each motor in a facility should be inspected periodically to determine the condition of its bearings and its pulley and belt alignment if it uses a belt drive. The bearing condition of a motor can be checked with an industrial stethoscope. In this procedure, the stethoscope is used to measure the noise at both ends of a motor that uses chains or belts to drive equipment. In a normal motor, the active end will read 2-3 db louder than the fixed end. If there is no difference or if the fixed end is louder than the active end, the bearings at the fixed end are worn; if the active end is 4-6 db louder than the fixed end, the bearings are badly worn; if the active end is 7-8 db louder, the bearing housing is turned inside the motor; and if the active end is 9 db or more louder than the fixed end, the motor should be replaced immediately. This procedure, developed at Safeway Stores by Mr. Harold Tornberg, has proved practical in industry.

12.2.4 Utility rebates for motors and drives

A general discussion of electrical utility company rebates and incentives was given in Chapter Three. However, since most utilities offer rebates and incentives for electric motor system improvements, it is worth mentioning again. There are generally two forms that the rebates and incentives take for motor systems: an incentive based on kW savings, or an incentive based on the horsepower of the motor involved. The level of the rebate or incentive for a peak load reduction due to a motor system improvement depends greatly on the individual utility. If the utility is working hard to limit its peak demand so that it will not have to add new power generation capability, the incentives may be quite large. Incentives of a few hundred dollars per kW of motor load reduced are quite common.

The incentive may also be related to the horsepower of the motor replaced or the horsepower of the motor that an adjustable speed drive was added onto. For a high efficiency motor replacement, the utility usually has a list of minimum efficiencies that qualify for rebates. If a customer replaces a motor with one that meets the minimum efficiency level, then the customer is automatically eligible for the rebate. This may be in the range of $6 – $15 per hp. Rebates for ASDs generally would qualify for levels of $100 to $300 per hp.

Example 12-4: In Example 12-2, ACE Industries would receive a $6/hp rebate from its electric utility if the company purchases the high efficiency 50 hp motor for the air compressor. How does this rebate influence the cost effectiveness of the purchase decision for ACE Industries?

Solution: The rebate of $6/hp reduces the cost of the motor by:

Cost reduction = ($6/hp) × (50 hp) = $300

Thus, the cost to ACE Industries is now only:

$$\text{Motor cost} = \text{price differential—utility rebate}$$
$$= \$470 - \$300$$
$$= \$170$$

The new SPP for operating the motor 8760 hours is:

$$\text{SPP} = \$170/\$522/\text{yr} = \underline{0.33 \text{ years}}$$

For 5000 hours of operation, the SPP is:

$$\text{SPP} = \$170/\$334/\text{yr} = \underline{0.51 \text{ years}}$$

The utility rebate significantly improves the attractiveness of this EMO. A simple payback period of four to six months for this motor purchase would be considered a highly cost-effective investment by most companies.

12.3 AIR COMPRESSORS

Many facilities use substantial quantities of compressed air to power machinery, process operations and control systems. This becomes a large energy cost, and the compressed air system should be designed and operated so that it is as energy efficient as possible. Selecting the best types and sizes of air compressor units is important in achieving this goal. A high efficiency motor should be specified for any air compressor, unless it is one that will only be used for short periods of time.

12.3.1 Types of air compressors

There are two major classifications of air compressors: positive displacement compressors and dynamic compressors [7,8]. A reciprocating compressor is an example of a positive displacement compressor. In this type of compressor, successive volumes of air are trapped in a closed space, and the pressure is increased as the piston moves toward the top of the cylinder and reduces the size of the closed space. Reciprocating air compressors have good energy efficiency characteristics at both part-load and full-load. It is more difficult to capture the waste heat from a reciprocating compressor than from a rotary screw or centrifugal air compressor since the pistons in the reciprocating unit are exposed to the open air around the machine.

The rotary screw compressor is another example of a positive displacement air compressor. In this unit, air enters the inlet and is trapped between mating male and female rotors and compressed to the required discharge pressure. Rotary screw compressors have excellent efficiencies at full-load conditions, and average efficiencies at part-load conditions. Heat recovery is easiest from the rotary screw compressors since the entire compressor section is enclosed.

The centrifugal air compressor is an example of a dynamic compressor, where air is compressed by the dynamic action of rotating impellers or vanes imparting velocity and pressure to the air. Almost all large air compressors (greater than 3000 CFM and 50-psig) are centrifugal compressors. The efficiency of these centrifugal air compressors is lower than either the reciprocating or rotary screw models, so smaller air compressors are almost never the centrifugal type. Heat recovery is also somewhat difficult from these compressors.

12.3.2 Designing the compressed air system

Most facilities have a number of air compressors that can be operated in combinations to satisfy the demand for compressed air at any time. Large units are needed for periods of high demand, and smaller units are needed to supply reduced amounts of compressed air during weekends or periods of slack production. A centralized air compressor facility together with a common-header and a computer control system for managing the load should provide efficient, cost-effective operation for most plants. In some cases there are large distances between parts of a facility that need compressed air. In those cases it might be more efficient to provide a small compressor at each location to minimize the energy lost in transmitting the compressed air through long pipelines.

Air compressors operate most efficiently when they use cool air for the intake. One of the factors that should be considered in designing the compressed air system is the access to cool intake air. Air compressors are often located in parts of a facility which are quite warm, and the air intakes use this warm air. In these cases, an outside air intake vent should be installed to allow the compressors to use cooler air.

12.3.3 Waste heat recovery from air compressors

Nearly 90% of the energy that goes into an air compressor becomes waste heat. Thus, it is important to recognize the value of air compressors as waste heat sources. Warm air can be recovered for space heat or process drying, or water can be heated for washing parts, cleaning equipment or for bathroom use. Air compressors can often be located next to the place where their waste heat will be utilized. This can save on costs of ducting and piping that would have been needed to move the waste heat from a compressor located some distance away.

The use of waste heat from a compressor to heat a facility is illustrated in Section 12.4.2. In this example, it is very cost-effective to install the ducting to transfer the hot exhaust air from the compressor to the warehouse where it can reduce the need for using gas to provide space heat. Since the heat is only needed a few months of the year, it must be vented outside during the other months.

12.3.4 Improving the operation of the compressed air system

One way to determine how much energy is being used in the compressed air distribution system is to attach a recording ammeter to one leg of the motor driving the air compressor. This gives both the time and the amount of energy consumption and can point to excess usage or to leaks as possible problems. The assignment of dollar values to air leaks of various sizes is possible, and the amount of air lost to leakage can be a significant fraction of the total air used. In facilities where the air leaks are small, more attention should probably be paid to the ways in which the air is used. In any case, monitoring the electrical consumption of the compressor motor can determine whether this use of energy is worth the attention of the auditor. Replacing standard-efficiency motors with high-efficiency motors for air compressors is usually a very cost-effective EMO.

While considering the compressed air system, think of replacing air-powered equipment by equipment powered by electricity. An air-powered hoist, for example, uses 5 hp in the air compressor for every hp that would be used if the hoist were electrical. When the main function of the air compressor is to control HVAC equipment, water or oil in the control air can wreck the controls on the equipment. Most air compressors have water or oil traps at the bottom, and they can be inspected to see if they are acting as the manufacturer intended. Obtain the operating manual before attempting the inspection; otherwise you may manipulate the safety valve rather than the water trap, with hazardous results. If water or oil is not removed at the compressor, it may travel to the thermostats and impede their operation. Each thermostat operates differently, however, and inspecting a thermostat for oil or water is a task best left to a vendor or to a trained service person.

In addition, clean dry air is so important for the proper operation and longevity of air-powered controls and tools, that most facilities have an air dryer that is placed in the air supply line after the compressor. This may be a mechanical vapor-compression cycle dryer (similar to an air conditioner), or it can be a desiccant-type dryer where the desiccant material is periodically reconditioned on an automatic basis.

Electric motors transmit power to compressors through belts, and a misalignment of the belt pulleys can cause severe damage and early failure of both bearings and belts. Other problems that can occur with compressors include inoperable switches, gauges that do not work, loose or frayed wiring, and leaks. Look for these problems, and listen for sounds of escaping air. If compressor problems persist, the cost savings of a working control system more than pays for the help of a trained technician.

12.4 EXAMPLES OF PROCESS ENERGY IMPROVEMENTS

In this section, some examples of abbreviated studies of process energy improvements are presented. They are intended as illustrative examples only and should not be used as general calculation guidelines. Individual circumstances will vary from these examples, and calculations should be tailored to fit the specific conditions of the facility being studied.

12.4.1 Recuperator for a Large Brick Kiln
Summary:
The two drying kilns at a large brick manufacturing company presently use ambient air for the combustion air. The air intake for the kiln burner could be modified to draw air from the cooling section of the kiln, thus serving as a recuperator. (A recuperator is a device that preheats the combustion air for a boiler, furnace or oven. Preheating the combustion air increases the system efficiency.) This air has the full oxygen content, yet has been heated as it has cooled the hot bricks.

A simple insulated duct could connect the cooling section to the combustion air intake duct. It is recommended that the air drawn from the cooling section be no more than 800°F and the duct be insulated. Usually the higher the temperature of the combustion air, the more efficient the combustion process. Unfortunately, when dealing with temperatures above 800°F, there is a risk of the burners becoming too hot, resulting in a shorter burner life. By using air at 800°F and taking into account the heat loss in the duct, the risk of harming the burners is reduced. Additional controls may also be needed on the air intake motor due to higher air temperatures. Also, a filtering system may be needed if the new air source has unwanted dust particles.

The cost-effectiveness analysis of this EMO follows.

Data:

Present air intake temperature: 90°F

Proposed air intake temperature: 800°F

Airflow*: 3548 cfm/kiln

Specific heat of air (Cp) at 800°F: 0.259 Btu/lb °F

Specific heat of air (Cp) at 90°F: 0.240 Btu/lb °F

Density of air (p) at 800°F: 0.03 lb/ft^3

Density of air (p) at 90°F: 0.075 lb/ft^3

Heat loss through ductwork: 30% (using 3-inch hot
   pipe insulation)

Operating hours: 8760 h/year

Natural gas cost: $3.30/Mcf

Calculations:

In raising the combustion air temperature, the air mass flow rate must remain constant in order to maintain a correct air-fuel mix:

* **Energy Savings:**

Air mass flow rate $= m$

$= (3548 \text{ ft}^3/\text{min}) \times (.075 \text{ lb/ft}^3)$

$= 266.1 \text{ lb/min}$

Savings in Btus $= (\text{air mass flow}) \times (T_2 - T_1) \times (\text{average specific heat}) \times (1 - \text{total heat loss})$

$= (266.1 \text{ lb/min}) \times (800°F - 90°F) \times (.250 \text{ Btu/lb°F}) \times (60 \text{ min/h}) \times (8760 \text{ h/year}) \times (2 \text{ kilns}) \times (1 - .30)$

$= 34{,}756 \times 10^6 \text{ Btu/year}$

Gas savings in Mcf $= (34{,}756 \times 10^6 \text{ Btu/yr}) \times (1 \text{ Mcf}/10^6 \text{ Btu})$

$= 34{,}756 \text{ Mcf/year}$

---

*The airflow is calculated from the proposed burner consumption after excess air control is optimized.

(Note: This is a conservative estimate for the savings in gas because it does not include the efficiency of the gas heater.)

- **Cost Savings:**

  savings in $ = (34,756 Mcf/year) × ($3.30/Mcf)
  = $114,695/year

<u>**Implementation Cost Data:**</u>

  Length of duct: 75 ft/kiln
  Ductwork: 304 stainless steel at $5.45/lb
     (24-in.-diameter duct of 14-gauge steel: 20 lb/ft)
  Insulation for ductwork: $10.00/ft
  Engineering design cost: $5000
  Labor and contingency: $5000

- **Implementation Cost:**

  Implementation cost = [(75 ft) × ($5.45/lb) × (20 lb/ft) +
     (75 ft) × ($10.00/ft)] × (2 kilns) +
     $5000 + $5000
  = $27,850

- **Simple Payback Period:**

$$SPP = \frac{implementation\,cost}{annual\,savings}$$

$$= \frac{\$27,850}{\$114,695/year}$$

$$= 0.24\ year$$

Thus, this EMO is highly cost-effective.

12.4.2 Heat Recovery from Compressors to Space Heat a Warehouse
Summary:

An industrial warehouse is heated with natural gas. The shop next to the warehouse has one 75-hp and one 100-hp air compressor. By installing ductwork from the air compressors to the warehouse, the hot air from the compressors can be used to heat the warehouse. Two dampers are required so that air may be exhausted in the summertime. For this system, automatic dampers might be required so that the temperature of the air for space heating will not be too high. The system can then mix cool

outside air with the hot compressor exhaust air if needed.
The cost-effectiveness analysis of this EMO follows.

Data:

Compressor size: 100 and 75 hp
Average air temperature before compressor (T_1): 90°F
Average air temperature after compressor (T_2): 110°F
Hot air flow rate from 100-hp compressor: 10,000 cfm
Operation hours: 992 h/year (24 h/day, 5 days/week; 4 h/day,
    1 day/week; 8 weeks/year)
Natural gas cost: $3.30/Mcf
Efficiency of gas heater: 0.80
Percent load on compressor: 75%
Ductwork length: 35 ft each (70 ft for both)

Calculations:

The Btu savings as well as dollar savings from using the hot air from these compressors can be calculated.

• Energy Savings:

Btu savings = (air flow rate) × (density of air) ×
              (specific heat of air) × (temperature difference) ×
              (load factor)

For the 100-hp compressor,

Btu savings = (10,000 ft^3/min) × (.075 lb/ft^3) × (.24 Btu/lb°F)
              × (110-90°F) × (60 min/h) × (992 h/year) × (.75 load)
            = 160.7 × 10^6 Btu/year

For the 75-hp compressor, we'll assume that the savings will be about 75% of savings for the 100-hp compressor, so we have

Btu savings   = (1607 × 10^6 Btu/year)(.75)
              = 120.5 × 10^6 Btu/year

Total Btus saved   = (160.7 × 10^6 + 120.5 × 10^6 Btu/year)
                   = 281.2 × 10^6 Btu/year

Total Mcf saved   = (281.2 × 10^6 Btu/year)(1 Mcf/10^6 Btu) × (1/.80)
                  = 351.5 Mcf/year

Total $ saved   = (351.5 Mcf/year) × ($3.30/Mcf)
                = $1,160/year

Implementation Cost:
Material cost (for the air intake ductwork):
   Insulated flexible duct with vinyl-coated spring steel
   (or aluminum): $2.90/linear foot
   Two dampers: $30 each

Labor cost (for installation of the duct):
   Two persons: $20/h each
   Time: 8 hours

Total cost   = material cost + labor cost
             = (duct cost/linear ft) × (total linear ft)
                + (2 dampers) × ($30/damper) + (number of
                laborers) × (number of h worked) × (wage/h)

             = ($2.90/linear ft) × (70 ft) + ($30/damper) ×
                (2 dampers) + (2 laborers) × (8 h) × ($20/h)

             = $583

Simple Payback Period:

$$ SPP = \frac{implementation\,cost}{annual\,savings} = \frac{\$583}{\$1160/year} = 0.5 \text{ year} $$

This EMO is also highly cost-effective.

12.4.3 Installation of an Economizer for a Plastic Plant
Summary:
This EMO considers the installation of an economizer and an associated control system in a plastics plant. To remove the heat being generated by a large injection molder, the plant currently air-conditions 9 months a year (March through November), three shifts a day. The economizer control will read the dry and wet bulb temperatures to determine if the outside air conditions (temperature and humidity) are more desirable than the present return air. When outside air is more desirable (less enthalpy), the economizer cycle will allow the use of cooler outside air to replace the need for additional conditioned air so that the least amount of

energy will be expended to get the air to the desired temperature and humidity.

It was assumed that the return air was at 78°F and 50% relative humidity (a conservative estimate), which corresponds to an enthalpy reading of 30 Btu/lb (air). Btu savings occur when the enthalpy of the outside air is less than the return air. The savings can be easily calculated in this manner by using bin data, which give the number of hours of weather experience in a month at a given temperature range for the geographic area where the facility is located. (See Table 12-2 at the end of this chapter for bin data.)

The cost-effectiveness analysis of this EMO follows.

Data:

Electrical energy cost: $.034/kWh
Return air: 78°F, 50% relative humidity
Air flow rate of unit: 7120 ft^3/min
Operation time: 3 shifts (24 h/day Monday-Friday;
    4 h/day Saturday) COP of air conditioner = 3.0

Calculations:
• Energy Savings

Savings in Btus =   (total savings of Btu•h/lb air•year)
       × (cfm of unit) × (60 min/h)
       × (density of air) × (1/COP)

(See Table 12-3 at the end of this chapter for the calculation of the total savings of Btu•h/lb air•year)

$$= \quad (25{,}858.5 \text{ Btu•h/lb air•year}) \times (7120 \text{ ft}^3/\text{min}) \times (60 \text{ min/h}) \times (.075 \text{ lb/ft}^3 \text{ air})(1/3)$$

$$= \quad 276.2 \times 10^6 \text{ Btu/year}$$

Savings in dollars

$$= \quad (\text{Btu/year}) \times (\text{kWh/3412 Btu}) \times (\text{cost/kWh})$$

$$= \quad (276.2 \times 10^6 \text{ Btu/year}) \times (1 \text{ kWh/3412 Btu}) \times (\$.034/\text{kWh})$$

$$= \quad \$2752.29/\text{yr}$$

Implementation Cost Data:

Two 7.5-ton economizer units: $1216 each
Installation cost: $600

$$\text{Implementation cost} = 2(\$1216) + \$600$$
$$= \$3032$$

Simple Payback Period:

$$SPP = \frac{\text{implementation cost}}{\text{annual savings}} = \frac{\$3032}{\$2752/\text{year}} = 1.1 \text{ years}$$

This EMO is also quite cost-effective.

12.4.4 HPS Relamp of Refrigerated Storage
Summary:

This EMO recommends a major relamping of a refrigerated storage area at a meat packing company. The storage area presently uses 150-watt incandescent lamps. Relamping would replace the existing lamps with 70-watt high pressure sodium (HPS) lamps and maintain the same level of illumination.

High-pressure sodium lighting is one of the most efficient high-intensity discharge (HID) sources and has excellent lumen maintenance over the lifetime of the lamps. The yellow color of HPS lamps has been a limiting factor for many interior applications, but improvements in color rendition have made its application for interior lighting acceptable even for color-critical areas.

The cost-effectiveness analysis of this EMO follows.

Data:

Present lighting:
    Type: incandescent   Initial lumens: 2350
    Size: 150 W   Life: 2500 h
           (150-W input to fixture)
    Quantity: 68   Lamp cost: $1.10
Proposed lighting data:
    Type: high-pressure sodium   Initial lumens: 5400
    Size: 70 W   Life: 20,000 h
           (88-W input to fixture)
   Spacing-to-mounting-   Fixture cost: $91.55
       height ratio: 2.0   Lamp cost: $39.69

Electrical energy cost: $0.034/kWh
Demand cost: $4.97 (June-October)
               $3.29 (November-May)
               $3.99 (average year-round)

Total area: 9800 ft^2
Mounting height: 8.5 ft
Hours of operation: 4576 h/year (16 h/day, 5.5 days/week,
                      52 weeks/year)
Cooling unit coefficient of performance (COP): 2.5

Calculations:

Present illumination   = (68 lamps)(2350 lumens/lamp)
                       = 159,800 lumens

$$\text{number of HPS lamps required} = \frac{159,800\,\text{lumens}}{5400\,\text{lumens/lamp}}$$

$$= 30\,\text{lamps}$$

$$\text{area of lighting/HPS lamp} = \frac{9800\,\text{ft}^2}{30\,\text{lamps}}$$

$$= 326.67\,\text{ft}^2/\text{lamp}$$

$$\text{spacing requirements} = \left(326.67\,\text{ft}^2\right)^{1/2}$$

$$= 18.07\,\text{ft/lamp}$$

$$\text{space–to–mounting–height ratio} = \frac{18.07\,\text{ft}}{8.5\,\text{ft}}$$

$$= 2.1$$

(This is extremely close to the recommended spacing-to-mounting-height ratio of 2.0 obtained from the manufacturer's catalog.)

- **Energy savings:**
  Present kWh   =   (68 lamps) × (150 W/lamp) × (4576 h/year)
                    (1 kW/1000 W)
                =   46,675.20 kWh/year

$$\text{Proposed kWh} = (30 \text{ lamps}) \times (88 \text{ W/lamp}) \times (4576 \text{ h/year})$$
$$\times (1 \text{ kW/1000 W})$$
$$= 12{,}080.64 \text{ kWh/year}$$

$$\text{Savings in kWh} = (46{,}675.20 \text{ kWh/year} - 12{,}080.64 \text{ kWh/year})$$
$$= 34{,}594.56 \text{ kWh/year}$$
$$= (34{,}594.56 \text{ kWh/year}) \times (3412 \text{ Btu/kWh})$$
$$= 118.04 \times 10^6 \text{ Btu/year}$$

• **Demand savings:**

$$\text{Present kW (demand)} = (68 \text{ lamps}) \times (150 \text{ W/lamp-month})$$
$$\times (1 \text{ kW/1000 W})$$
$$= 10.20 \text{ kW/month}$$

$$\text{Proposed kW (demand)} = (30 \text{ lamps}) \times (88 \text{ W/lamp-month})$$
$$\times (1 \text{ kW/1000 W})$$
$$\text{Savings in kW (demand)} = 10.20 \text{ kW/month} - 2.64 \text{ kW/month}$$
$$= 7.56 \text{ kW/month}$$

• **Replacement cost savings:**

$$\text{Present replacement cost} = (68 \text{ lamps}) \times (\$1.10/\text{lamp}) \times$$
$$(1/2500 \text{ h}) \times (94567 \text{ h/year})$$
$$= \$136.91/\text{year}$$

$$\text{Proposed replacement cost} = (30 \text{ lamps}) \times (\$39.69/\text{lamp})$$
$$\times (1/20{,}000 \text{ h}) \times (4576 \text{ h/year})$$
$$= \$272.43/\text{year}$$

$$\text{Savings in replacement cost} = \$136.91/\text{year} - \$272.43/\text{year}$$
$$= -\$135.52/\text{year}$$

• **Savings in reduced refrigeration needs:**

(The reduction in wattage from HPS relamping will also reduce the amount of heat generated by the lamps that must be removed by the refrigeration system. This translates to a savings in energy (kWh) but not a savings in demand (kW) since the air conditioner still operates at its rated kW.)

$$\text{Present cost of heat removal} = (46{,}675.20 \text{ kWh/year}) \times$$
$$(1/2.5) \times (\$.034/\text{kWh})$$
$$= \$634.78/\text{year}$$

Cost of heat removal with proposed lighting
$$= \quad (12{,}080.64 \text{ kWh/year}) \times (1/2.5) \times (\$.034/\text{kWh})$$
$$= \quad \$164.30/\text{year}$$

Total savings in refrigeration cost
$$= \quad \$634.78/\text{year} - \$164.30/\text{year}$$
$$= \quad \$470.48/\text{year}$$

Total savings in \$ = savings in kWh cost + savings in kW cost + savings in replacement cost + savings in heat removal cost

$$
\begin{aligned}
= \quad & (34{,}594.56 \text{ kWh/year}) \times (\$.034/\text{kWh}) + \\
& (7.56 \text{ kW/month}) \times (12 \text{ months/year}) \times \\
& (\$3.99/\text{kW}) - \$135.52/\text{year} \\
& + \$470.48/\text{year}
\end{aligned}
$$

$$= \quad \$1873.15/\text{year}$$

Implementation Cost:

Fixtures: (30 fixt.) $\times$ (\$91.55/fixt.):	\$2746.50
Lamps: (30 lamps) $\times$ (\$39.69/lamp):	\$1190.70
Labor (for adding HPS lamps):	
(30 fixt.) $\times$ (1 h/fixt.) $\times$ (\$15/h):	\$ 450.00
Labor (incandescent lamp removal):	
(68 fixt.) $\times$ (.5 h/fixt.) $\times$ (\$15/h):	\$ 510.00
Total:	\$4897.20

Simple Payback Period:

$$\text{SPP} = \frac{\text{implementation cost}}{\text{annual savings}} = \frac{\$4897.20}{\$1873.15/\text{year}} = 2.61 \text{ years}$$

This EMO has a SPP that is over two years, but many companies would still find it an attractive investment.

12.4.5 Sawdust Collection Control System
Summary:

A wood shop at a furniture factory has a sawdust collection system which collects sawdust from five machines. The system uses overhead hoods with vacuum motors, and exhausts the dust and air to the outside through ductwork. The dust collection system operates at all times the plant is in operation. By installing a damper and a control system to shut off the air collection system from any one of the machines not in use, significant savings will be realized.

Each of the five vacuum ducts must have a damper, and a speed control is required for the vacuum motor. Used in conjunction with a programmable controller, the damper corresponding to a specific machine will open that duct when the machine is turned on while the speed of the vacuum motor is increased proportionately. When no machines are in use, the motor is completely shut off. This EMO provides savings both in electricity to run the vacuum motor and in the gas used to heat the air that is being evacuated from the plant.

The cost-effectiveness analysis of this EMO follows.

Data:

Vacuum electric motor: 20-hp
   Operating time of motor: 8.5 hours a day
   Load on motor: 100%
   Efficiency of motor: 88%
Duct system: Exhausts 6500 ft^3 of air per minute
Efficiency of gas heating plant: 70%
Each machine is used approximately 2 hours/day.

Electric cost:
   Demand charge: $4.50/kW/month
   Energy charge: $0.02915/kWh
Natural gas cost: $3.50/10^6 Btu

Calculations:
- Energy Savings:

$$
\begin{aligned}
\text{Electricity cost} \ =\ & \text{demand charge + consumption charge} \\
=\ & (20\ hp) \times (0.746\ kW/hp) \times (1/.88) \\
& \times (12\ \text{months}) \times (\$4.50/kW/\text{month}) + \\
& (20\ hp) \times (0.746\ kW/hp) \times (1/.88) \\
& \times (8.5\ h/day) \times (250\ days/year) \\
& \times (\$0.02915/kWh) \\
=\ & \$915.55 + \$1050.23 \\
=\ & \$1965.78
\end{aligned}
$$

Electricity savings   =   demand savings + energy savings

demand savings $=$ (motor capacity) $\times$ (0.746 kW/hp)
$\times$ (1/efficiency) $\times$ (12 months/year)
$\times$ (cost/kW/month) $\times$ (% demand
reduction)

$=$ (20 HP) $\times$ (0.746 kW/hp) $\times$ (1/.88)
$\times$ (12 months/year) $\times$ ($4.50/kW/month)
$\times$ (3/5)

$=$ $549.33/year

(The percent demand reduction is taken to be 3/5 = 60% based on the assumption that no more than two machines will run at the same time for the 30 minute demand averaging interval.)

energy savings $=$ (motor capacity) $\times$ (0.746 kW/hp)
$\times$ (1/efficiency) $\times$ (6.5 h) $\times$
(250 days/year) $\times$ (cost/kWh)

$=$ (20 hp) $\times$ (0.746 kW/hp) $\times$ (1/.88)
$\times$ (6.5 h) $\times$ (250 days/year) $\times$
($0.02915/kWh)

$=$ $803.12/year

Total electricity savings $=$ ($549.33 + $803.12)/year
$=$ $1352.45/year

Gas cost $=$ (6500 ft^3/min) $\times$ (8.5 h/day) $\times$ (60 min/h) $\times$
(100 heating days/year) $\times$ (0.075 lb/ft^3)
$\times$ (.24 Btu/lb) $\times$ (65-45°F) $\times$
(1/.7) $\times$ ($3.50/10^6 Btu)

$=$ $596.70/yr

Gas savings from reduced heat loss
$=$ (air volume flow rate) $\times$ (operating hour
reduction) $\times$ (60 min/h) $\times$ (heating days/year)
$\times$ (air density) $\times$ (specific heat of air)
$\times$ (temperature difference) $\times$ (1/eff) $\times$
(gas cost)
$=$ (6500 ft^3/min) $\times$ (8.5 – 2) h/day $\times$ (60 min/h)
$\times$ (100 days/year) $\times$ (0.075 lb/ft^3)
$\times$ (.24 Btu/lb) $\times$ (65-45°) $\times$ (1/.7) $\times$ ($3.50/10^6 Btu)

$=$ $456.30/year

• **Total Cost Savings:**

Total annual cost savings = electricity cost savings + gas
cost savings

= $456 + $1352

= $1808

Implementation Cost Data:

Speed control:	$1500
Programmable controller:	550
Electric dampers (5 at $65)	325
Wire, switches	50
Installation (estimate)	300

Implementation Cost: $2725

Simple Payback Period:

$$SPP = \frac{implementation\,cost}{annual\,savings} = \frac{\$2725}{\$1808/year} = 1.5\ years$$

This EMO is also quite cost-effective.

12.5 TWENTY-FIVE
COMMON ENERGY MANAGEMENT OPPORTUNITIES

Through our combined energy management experiences with over 200 manufacturing plants and a review of the literature, we have found that a number of energy management opportunities (EMOs) have been used time after time. The astute energy manager must become familiar with these opportunities and be ready to apply them (*as well as others*) in energy management work.

Twenty-five of these common changes are summarized below. Most are process modifications, but a few are lighting and space conditioning oriented. The order of listing is not significant; the order would change depending on whether one were listing by frequency of occurrence, amount of savings, financial return on investment, etc.

1. *Switch to energy-efficient lamps.* Switch existing lamps to the energy-efficient ones, such as 34-watt energy-efficient fluorescent lamps for conventional 40-W ones, or replace T-12 lamps with T-8 or T-10 lamps.

2. *Switch to energy-efficient light sources.* Change to more efficient sources usually requiring fixture changes. Change from incandescent lights to fluorescent or from mercury vapor or fluorescent to high-pressure sodium for in-plant lighting, a very frequent conversion.

3. *Use night setback-setup.* Turn temperatures up or down at night when needs are reduced. Examples include large ovens that cannot be turned off, large refrigeration units where night operations involve less infiltration (fewer people going in and out), and space conditioning.

4. *Turn off equipment.* Turn off exhaust fans, ovens, motors, or any other equipment when not needed.

5. *Move air compressor intake to cooler locations.* Move air intakes from hot equipment rooms to cooler (often outside) locations. Efficiency improvements are large and paybacks attractive.

6. *Eliminate leaks in steam and compressed air systems.* Steam and compressed air leaks are very expensive and should be fixed. Technology exists for repairing leaks without shutting the equipment down. Night audits (when noise is minimized) often turn up large numbers of these leaks.

7. *Control excess air.* As shown earlier, careful control of combustion air can lead to significant energy savings.

8. *Optimize plant power factor.* Depending on the utility billing schedule and the company's power factor, large savings may be available through power factor improvement.

9. *Insulate bare tanks, vessels, lines, and process equipment.* Good savings are often available through insulation of process lines and tanks. Condensate return lines and tanks are often not insulated.

10. *Install storm windows, doors, and weather stripping.* Although these are often difficult to justify, sizable savings are sometimes available. This is especially true for large glass exposures in cold climates.

11. *Use energy-efficient electric motors.* When replacement is necessary or for new applications, energy-efficient motors can usually be justified. Electric utilities often provide rebates to customers who replace standard motors with energy-efficient models.

12. *Preheat combustion air.* Recuperators can save large amounts of energy and money. Sometimes they are highly cost effective.

13. *Reduce the pressure of compressed air and steam.* If the pressures have been overdesigned, a reduction will not harm the process. In such cases, large savings are possible.

14. *Insulate walls, ceilings, roofs, and doors.* Industrial plants are frequently poorly insulated. Insulation in dropped ceilings, on roofs or walls, and doors may be cost-justified.

15. *Recover heat from air compressor.* Larger air compressors reject large amounts of heat through air or water cooling. Proper design can allow this waste heat to be used for space conditioning in the winter and to be exhausted in warm weather. Sometimes the payback is very attractive.

16. *Insulate dock doors.* Plastic strips, dock bumpers, vestibules, or air screens all help block infiltration through large dock doors. If the space is heated and/or air-conditioned, the savings can be very large.

17. *Install economizers on air conditioners.* In some areas of the country, economizers can be very attractive. They allow the optimum use of outside air in air conditioning. Sometimes outside air can be used and the air conditioner turned off.

18. *Use radiant heat.* Sometimes infrared heaters can be used to spot-heat rather than heat entire areas. Infrared heat (like the sun) warms objects and people but not space. The payback can be very attractive.

19. *Return steam condensate to the boiler.* Returning hot condensate can yield dramatic savings in energy, water, and water conditioning costs. Return lines should probably be insulated.

20. *Change product design to reduce energy requirements.* Product redesign can often reduce the energy necessary in heat treating, cleaning, coating, painting, etc.

21. *Explore waste heat recovery for space exhaust systems.* Large amounts of exhaust in buildings that are heated and/or air-conditioned offer the potential for waste heat recovery.

22. *Install devices to improve heat transfer in boilers.* Turbulators and other devices designed to reap more energy out of the combustion process are often very cost effective.

23. *Reschedule operations to reduce peak demand.* Sometimes simple changes in equipment scheduling can dramatically reduce demand charges.

24. *Cover open heated tanks.* Covering open heated tanks can often lead to big energy savings. Floating balls, cantilevered tops, and rubber flaps have all been used as covers.

25. *Spot-ventilate or use air filters.* In welding areas or other areas where large amounts of ventilation are required, spot ventilation can often reduce the amount needed. Also, electrostatic or other types of air filters can sometimes allow reuse of the air. Savings are especially large if the space is heated and/or air-conditioned.

12.6 SUMMARY

This chapter has provided a suggested procedure for process improvement that is based on the industrial engineering concept of work simplification. It has also provide a detailed presentation of electric motor and drive system efficiency improvements. These are particularly important because of the large quantity of energy used in business and industry by electric motors. Some case studies of process energy management were also presented. The reader should not consider these examples as typical results but should use them as a starting point for potential analysis. Because process energy management can be intricate and complex, the energy manager must understand the entire process system and consider all of the impacts of any proposed changes.

Table 12-2. Mean frequency of occurrence of dry bulb temperature (°F) with mean coincident wet bulb temperature (°F) for each dry bulb temperature range.

Cooling Season

Temperature range (°F)	May					June					July				
	Observ./h, Cp			Total observ.	Mean coincident wet bulb (°F)	Observ./h, Gp			Total observ.	Mean coincident wet bulb (°F)	Observ./h, Gp			Total observ.	Mean coincident wet bulb (°F)
	02 to 09	10 to 17	18 to 01			02 to 09	10 to 17	18 to 01			02 to 09	10 to 17	18 to 01		
105-109							0	0	0	73		0	0	0	72
100-104							8	1	9	74		7		7	72
95-99		1		1	68		40	7	47	74	1	51	8	59	74
90-94	8	1		9	71	2	64	25	91	72	11	70	26	97	74
85-89	33	5		38	70	14	61	53	128	71	34	53	43	107	73
80-84	58	21	1	80	68	50	36	70	156	69	90	35	73	142	71
75-79	51	46	13	110	66	86	17	51	154	67	83	20	55	165	70
70-74	40	62	53	155	64	62	10	24	96	63	26	10	34	127	68
65-69	29	55	72	156	61	20	3	8	31	60	3	21	9	37	65
60-64	17	32	55	104	58	5	1	1	7	55	0		0	3	60
55-59	9	20	33	62	53	1	0	0	1	49				0	57
50-54	2	5	15	22	48										
45-49	0	1	5	6	43										
40-44		0	1	1	40										
35-39			0	0	35										
30-34															
25-29															

Table 12-2. (Continued)

Cooling Season

August

Temperature range (°F)	Observ./h, C_p 02 to 09	Observ./h, C_p 10 to 17	Observ./h, C_p 18 to 01	Total observ.	Mean coincident wet bulb (°F)
	2	0		2	73
	15	2		17	73
	37	7	0	44	73
	63	20	1	84	73
	64	39	7	110	72
	38	70	29	137	71
	20	64	82	166	69
	7	37	84	128	67
	2	8	37	47	63
	0	1	7	8	58
			1	1	55

September

Temperature range (°F)	Observ./h, G_p 02 to 09	Observ./h, G_p 10 to 17	Observ./h, G_p 18 to 01	Total observ.	Mean coincident wet bulb (°F)
			0	0	72
		11	0	11	71
		30	4	34	71
	1	48	13	62	70
	7	43	35	85	69
	30	42	56	128	67
	67	33	48	148	65
	53	18	46	117	62
	45	11	26	82	57
	27	3	8	38	53
	7	1	4	12	49
	3	0		3	46

October

Temperature range (°F)	Observ./h, G_p 02 to 09	Observ./h, G_p 10 to 17	Observ./h, G_p 18 to 01	Total observ.	Mean coincident wet bulb (°F)
		0		0	61
		6		6	64
		14	1	15	66
		30	5	35	65
	2	39	14	55	63
	13	44	37	94	63
	35	46	50	131	62
	54	34	53	141	59
	49	21	43	113	55
	48	8	27	83	51
	29	4	13	46	47
	12	1	3	16	42
	4	1	1	6	38
	1	1		2	34
	1		1	2	29
	1		0	1	26

Table 12-2. (Continued)
Cooling Season

Temperature range (°F)	November Observ./h, C_p 02 to 09	10 to 17	18 to 01	November Total observ.	November Mean coincident wet bulb (°F)	December Observ./h, C_p 02 to 09	10 to 17	18 to 01	December Total observ.	December Mean coincident wet bulb (°F)	January Observ./h, C_p 02 to 09	10 to 17	18 to 01	January Total observ.	January Mean coincident wet bulb (°F)
105-109															
100-104															
95-99															
90-94									0	60					
85-89		7		7	60		0		0	59					
80-84	2	25	4	31	60		0		0	55					
75-79	6	26	14	46	57		2	0	2	53		1		1	56
70-74	11	30	22	63	53		7	1	8	51		7	1	8	53
65-69	19	35	29	83	49	1	17	3	21	51	2	13	4	19	52
60-64	29	38	42	109	45	7	27	14	48	49	4	17	8	29	49
55-59	43	34	40	117	42	11	37	24	72	45	7	29	18	54	45
50-54	39	20	32	91	37	23	40	39	102	41	11	34	24	69	41
45-49	43	14	31	88	33	38	38	45	121	38	29	38	43	110	37
40-44	26	7	14	47	29	46	30	41	117	33	36	31	42	109	33
35-39	10	3	9	22	24	49	20	34	103	29	52	28	35	115	29
30-34	10	1	3	14	20	32	14	22	68	25	39	19	26	84	25
25-29	1	0	0	1	15	23	10	14	47	20	24	14	23	61	20
20-24	1			1	10	8	4	7	19	16	19	7	10	36	16
15-19						7	2	4	13	11	13	7	7	27	11
10-14						3	0	0	3	7	5	3	5	13	6
5-9						0		0	0	3	6	0		8	2
0-4											1	0		1	-2
-5,-1															

Table 12-2. (Continued)
Cooling Season

February					March					April					Annual (total all months)				
02 to 09	10 to 17	18 to 01	Total observ.	Mean coincident wet bulb (°F)	02 to 09	10 to 17	18 to 01	Total observ.	Mean coincident wet bulb (°F)	02 to 09	10 to 17	18 to 01	Total observ.	Mean coincident wet bulb (°F)	02 to 09	10 to 17	18 to 01	Total observ.	Mean coincident wet bulb (°F)
0	0		0	62	0	1	0	1	58		2		2	65		2	0	2	73
1	0	0	1	60	0	4	1	5	61	0	10	1	11	66		23	3	26	73
0	3	0	3	57	0	10	2	12	59	0	20	5	25	64	0	108	16	124	73
0	7	1	8	57	0	15	5	20	56	1	32	15	48	63	3	219	57	279	70
0	12	4	16	54	4	24	13	41	54	10	41	33	84	61	20	286	128	434	70
2	15	7	24	52	10	30	21	61	51	31	38	39	108	58	85	292	262	639	66
7	22	15	44	49	15	30	28	73	48	39	33	41	113	54	269	259	322	850	64
9	29	23	61	45	24	33	36	93	45	41	26	37	104	49	398	245	314	957	61
26	26	33	85	42	36	29	38	103	41	41	19	30	90	45	326	219	265	810	58
32	30	39	101	38	45	28	42	115	39	35	13	22	70	42	250	204	219	673	55
37	32	32	101	33	42	21	28	91	34	26	4	12	42	38	207	188	203	598	51
41	18	29	88	29	36	13	20	69	29	12	1	4	17	35	192	195	208	595	46
29	13	22	64	25	21	7	9	37	25	1	1	1	3	30	210	180	209	599	42
23	11	13	47	20	10	2	3	15	20	3			3	26	223	159	213	595	38
12	4	5	21	16	2	1	1	4	16						221	132	178	531	34
5	1	1	7	12	2	0	1	3	11						207	85	133	425	29
					1			1	8						132	57	89	278	25
					0			0	4						90	38	56	184	20
															42	17	25	84	16
															27	9	13	49	11
															10	3	5	18	6
															7	0	2	9	2
															1			1	-2

Table 12-3. Enthalpy data [Btu•h/lbm (air)]

March

| Temperature (°F) | | | | | Btu•hr |
Dry	Wet	h	$\Delta h_{(30-h)}$	Hours	1 lbm (air)
77	59	25.8	4.2	12	50.4
72	56	24	6	20	120
67	54	22.6	7.4	41	303.4
62	51	21	9	61	549
57	48	19.2	10.8	73	788.4
52	45	17.6	12.4	93	1153.2
47	41	15.8	14.2	103	1462.6
42	38	14.5	15.5	115	<u>1782.5</u>
					6209.5

April

| Temperature (°F) | | | | | Btu•hr |
Dry	Wet	h	$\Delta h_{(30-h)}$	Hours	1 lbm (air)
77	63	28.6	1.4	48	67.2
72	61	27.2	2.8	84	235.2
67	58	25	5	108	540
62	54	22.6	7.4	113	836.2
57	49	19.8	10.2	104	1060.8
52	45	17.6	12.4	90	1116
47	42	16.2	13.8	70	966
42	38	14.5	15.5	42	<u>651</u>
					5472.4

May

| Temperature (°F) | | | | | Btu•hr |
Dry	Wet	h	$\Delta h_{(30-h)}$	Hours	1 lbm (air)
77	66	no savings			
72	64	29.4	.6	155	93
67	61	27.2	2.8	156	436.8
62	58	25	5	104	520
57	53	22	8	62	496
52	48	19.2	10.8	22	237.6
47	43	16.8	13.2	6	79.2
42	40	15.2	14.8	1	<u>14.8</u>
					1877.4

June

Temperature (° F)					Btu•hr
Dry	Wet	h	$\Delta h_{(30-h)}$	Hours	1 lbm (air)
77	69	no savings			0
72	67	no savings			0
67	63	28.5	1.4	96	134.4
62	60	26.5	3.5	31	108.5
57	55	23.2	6.8	7	47.6
52	49	19.8	10.2	1	10.2
47				None	0
42				None	
					300.7

July

Temperature (° F)					Btu•hr
Dry	Wet	h	$\Delta h_{(30-h)}$	Hours	1 lbm (air)
77	70	no savings			0
72	68	no savings			0
67	65	no savings			0
62	60	26.5	3.5	3	10.5
57				None	0
52				None	0
47				None	0
42				None	0
					10.5

August

Temperature (° F)					Btu•hr
Dry	Wet	h	$\Delta h_{(30-h)}$	Hours	1 lbm (air)
77	69	no savings			0
72	67	no savings			0
67	63	28.6	1.4	47	65.8
62	58	25	5	8	40
57	55	23.2	6.8	1	6.8
52				None	0
47				None	0
42				None	0
					112.6

September

Temperature (° F)					Btu•hr
Dry	Wet	h	$\Delta h_{(30-h)}$	Hours	1 lbm (air)
77	67 no savings				0
72	65 no savings				0
67	62	28	2	117	234
62	57	24.6	5.4	82	442.8
57	53	22.2	7.8	38	296.4
52	49	19.8	10.2	12	122.4
47	46	18.4	11.6	3	34.8
42				None	0
					1130.4

October

Temperature (° F)					Btu•hr
Dry	Wet	h	$\Delta h_{(30-h)}$	Hours	1 lbm (air)
77	63	28.6	1.4	55	77
72	62	28	2	94	188
67	59	25.8	4.2	131	550.2
62	55	23.2	6.8	141	958.8
57	51	21	9	113	1017
52	47	18.8	11.2	83	929.6
47	42	16.2	13.8	46	634.8
42	38	14.5	15.5	16	248
					4603.4

November

Temperature (° F)					Btu•hr
Dry	Wet	h	$\Delta h_{(30-h)}$	Hours	1 lbm (air)
77	60	26.5	3.5	7	24.5
72	60	26.5	3.5	31	108.5
67	57	24.6	5.4	46	248.4
62	53	22.2	7.8	63	491.4
57	49	19.8	10.2	83	846.6
52	45	17.6	12.4	109	1351.6
47	42	16.2	13.8	117	1614.6
42	37	14	16	91	1456
					6141.6

$$\text{Total} \frac{Btu \bullet h}{1\ lbm\ (Air)}$$

March	6,209.5
April	5,472.4
May	1,877.4
June	300.7
July	10.5
August	112.6
September	1,130.4
October	4,603.4
November	<u>6,141.6</u>
	25,858.5

$$\frac{Btu \bullet h}{1\ lbm\ (air) \bullet year}$$

REFERENCES

1. Benjamin W. Niebel, *Motion and Time Study*, Ninth Edition, James D. Irwin, Homewood, IL, 1993.
2. Steve Nadel et. al., *Energy Efficient Motor Systems*, American Council for an Energy Efficient Economy (ACEEE), Washington, DC, 1991.
3. *Energy-Efficient Motor Selection Handbook*, Bonneville Power Administration, February, 1992. Prepared by the Washington State Energy Office.
4. *MotorMaster*, Washington State Energy Office, Olympia, WA, 1993.
5. Steve Nadel et. al., *Energy Efficient Motor Systems*, American Council for an Energy Efficient Economy (ACEEE), Washington, DC, 1991.
6. *Ibid.*
7. *Compressed Air Buyer's Guide*, Gardner-Denver Compressor Products, Quincy, IL, 1991.
8. *Your Guide to Reciprocating Compressors—A Brief Review of Theory and Practice*, Joy Manufacturing Company, Michigan City, IN, 1987.

BIBLIOGRAPHY

Brown, Harry L. et al, *Analysis of 108 Industrial Processes*, Fairmont Press, Atlanta, GA, 1985.
Dean, Norman L., *Energy Efficiency in Industry*, Ballinger Publishing Company, Cambridge, MA, 1980.

Gandhi, Nikhil, "Industrial Energy Management Success Using IE Techniques," *Proceedings of the Spring 1989 Industrial Engineering Institute Conference,* Toronto, Canada, May, 1989.

Hu, S. David, *Handbook of Industrial Energy Conservation,* Van Nostrand Reinhold Company, New York, NY, 1983.

Metals Fabrication Reference Guide, EPRI Center for Materials Fabrication, Battelle Laboratory, Columbus, OH, 1993.

Optimizing Energy Use in the Process Industries, Volumes 1-4, Electric Power Research Institute, Palo Alto, CA, 1990.

Payne, F. William, *Advanced Technologies: Improving Industrial Energy Efficiency,* Fairmont Press, Atlanta, GA, 1985.

Talbott, E.M., *Compressed Air Systems: A Guidebook on Energy and Cost Savings,* Second Edition, Fairmont Press, Atlanta, GA, 1993.

Turner, W.C., Senior ed., *Energy Management Handbook,* Second Edition, Fairmont Press, Atlanta, GA, 1993.

Turner, W.C., et al., Numerous Energy Analysis and Diagnostic Center Reports, Oklahoma Industrial Energy Management Program, Stillwater, Okla., 1980-1983.

Witte, L.C., Schmidt, P.S. and D.R. Brown, *Industrial Energy Management and Utilization,* Hemisphere Publishing Corporation, Washington, DC, 1988.

Chapter 13

Renewable Energy Sources and Water Management

13.0 INTRODUCTION

Renewable energy sources are those sources that replenish themselves and so are essentially inexhaustible, such as solar, wind and biomass energy. While renewable energy sources are not a major percentage of energy sources currently being utilized, their usage is expected to grow substantially, since they are typically less environmentally damaging than traditional energy sources. The future is likely to see more and more utilization of these renewable sources.

In this chapter we examine a selected subset of these potential sources in the following order: solar-active, solar-passive, solar-photovoltaic, wind, and refuse. Emphasis is on applications in the industrial and commercial environment. In the last part of the chapter we discuss management of another vital and renewable resource: water. Water use will likely cause a major crisis someday soon. The energy manager skilled in water management will be prepared to meet this challenge.

13.1 RENEWABLE ENERGY TECHNOLOGY

The largest portion of the contribution by renewable energy today comes from mature technologies that make use of biomass and hydropower resources. The newer technologies developed over the past two decades are beginning to enter the market and will provide an increasing share of renewable energy supplies in coming decades [1]. The percentage utilization of renewable resources has held relatively constant over the last ten years, and is around 8% of the U.S. national energy supply. Figure 13-1 shows the contributions of the various types of renewable energy sources.

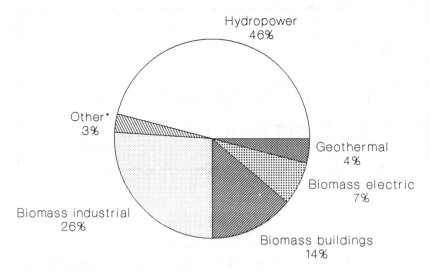

*Wind, alcohol fuels, solar thermal, PV

Figure 13-1. U.S. Renewable Energy Supply (1988): 7 quads; other includes wind, alcohol fuels, solar thermal, and photovoltaics. [1]

The contribution of renewable energy to the total energy requirements of the U.S. is expected to grow substantially over the next forty years. A broad-based study of the potential for renewable energy was conducted by five National Laboratories as part of the background for the National Energy Strategy [2]. This study projected that renewable energy sources would provide from 15% to 28% of our total energy supply in the year 2030. As concerns for the environment grow, the benefits of renewable sources in terms of reduced pollution emissions and reduced impacts from energy production will serve as positive factors in the growing use of these renewable energy technologies.

Since the focus of this chapter is on the use of renewable energy sources in commercial buildings and in industry, most of the discussion will be directed toward the use of active and passive solar systems for space heating, water heating, process heating and electricity generation. Wind energy will also be discussed as a source of electricity generation. Biomass and refuse will also be discussed, since they are sources of inexpensive fuel for many industries and some commercial buildings. Few businesses or industries directly operate hydroelectric or geothermal powered electric generation, so those sources will be covered only briefly.

13.2 SOLAR ENERGY

13.2.1 Solar Insolation

Approximately 430 Btu/h/ft^2 of solar energy hits the earth's atmosphere. Because of diffusion in the atmosphere and clouds, this is greatly reduced to somewhere around a maximum of 300 Btu/h/ft^2 on the earth's surface at 40°N latitude. This maximum, of course, only occurs at certain times of the day and year, so the average is significantly less. However, at this rate a set of collectors designed to develop 1×10^6 Btu/h of energy would have to be 3333 ft^2 in size without allowing for cloudiness, variances throughout the day, or collector efficiency. If the collector were tracking the sun throughout the day, it might be able to gather 8×10^6 Btu/day. Assuming it operates 365 days/year, the collector would be able to harvest an absolute maximum of

$$(8 \times 10^6 \text{ Btu/day})(365 \text{ days/year}) = 2920 \times 10^6 \text{ Btu/year}$$

At $4.00/10^6$ Btu, this energy would be worth \$11,680/year. The necessary collector space is 3333 ft^2, and the installed cost including controls might be around \$20/ft^2. Therefore, the cost of the proposed collectors would be around \$66,660, making the payback *under ideal conditions* somewhere around 5.7 years. Actual conditions would likely require a significantly larger collector, as will be shown below.

Detailed calculations of the amount of solar energy striking a surface located at a given latitude and tilted at a certain angle require knowledge of several angles, including the solar altitude angle, the solar azimuth angle, and the tilt angle. The values all vary with time of day, month, location, tilt of collector, etc. Tables have been developed to help a person determine the amount of solar energy available [3,4]. Table 13-1 is an example of such a table.

Example 13-1: A solar collector is to be located at Lincoln, Nebraska (approximately 40°N latitude). Find the actual energy available, the value of that energy, and the payback time for the cost of the solar collector.

Solution: Using Table 13-1, look up the data for Lincoln, NE. We find the solar radiation on a surface tilted at 40° to average about 0.6×10^6 Btu/ft^2/year. Assuming the collector is 70% efficient, the energy available is about 0.42×10^6 Btu/ft^2/year. Thus, the 3333 ft^2 collector discussed above would supply about:

Table 13-1. Average solar radiation for selected cities.

City	Slope		Average daily radiation (Btu/day•ft²)											
		Jan.	Feb.	March	Apr.	May	June	July	Aug.	Sept.	Oct.	Nov.	Dec.
Albuquerque, NM	Hor.	1134	1436	1885	2319	2533	2721	2540	2342	2084	1646	1244	1034
	30	1872	2041	2295	2411	2346	2390	2289	2318	2387	2251	1994	1780
	40	2027	2144	2319	2325	2181	2182	2109	2194	2369	2341	2146	1942
	50	2127	2190	2283	2183	1972	1932	1889	2028	2291	2369	2240	2052
	Vert.	1950	1815	1599	1182	868	754	795	1011	1455	1878	2011	1927
Atlanta, GA	Hor.	839	1045	1388	1782	1970	2040	1981	1848	1517	1288	975	740
	30	1232	1359	1594	1805	1814	1801	1782	1795	1656	1638	1415	1113
	40	1308	1403	1591	1732	1689	1653	1647	1701	1627	1679	1496	1188
	50	1351	1413	1551	1622	1532	1478	1482	1571	1562	1679	1540	1233
	Vert.	1189	1130	1068	899	725	659	680	811	990	1292	1332	1107
Boston, MA	Hor.	511	729	1078	1340	1738	1837	1826	1565	1255	876	533	438
	30	830	1021	1313	1414	1677	1701	1722	1593	1449	1184	818	736
	40	900	1074	1333	1379	1592	1595	1623	1536	1450	1234	878	803
	50	947	1101	1322	1316	1477	1461	1494	1448	1417	1254	916	850
	Vert.	895	950	996	831	810	759	791	857	993	1044	842	820
Chicago, IL	Hor.	353	541	836	1220	1563	1688	1743	1485	1153	763	442	280
	30	492	693	970	1273	1502	1561	1639	1503	1311	990	626	384
	40	519	716	975	1239	1425	1563	1544	1447	1307	1024	662	403
	50	535	723	959	1180	1322	1341	1421	1363	1274	1034	682	415
	Vert.	479	602	712	746	734	707	754	806	887	846	610	373

Location														
Ft/Worth, TX	Hor.	927	1182	1565	1078	2065	2364	2253	2165	1841	1450	1097	898
	30	1368	1550	1807	1065	1891	2060	2007	2097	2029	1859	1604	1388
	40	1452	1601	1803	1020	1755	1878	1845	1979	1995	1907	1698	1488
	50	1500	1614	1758	957	1586	1663	1648	1820	1914	1908	1749	1549
	Vert.	1315	1286	1196	569	728	679	705	890	1185	1459	1509	1396
Lincoln, NB	Hor.	629	950	1340	1752	2121	2286	2268	2054	1808	1329	865	629
	30	958	1304	1605	1829	2004	2063	2088	2060	2092	1818	1351	1027
	40	1026	1363	1620	1774	1882	1909	1944	1971	2087	1894	1450	1113
	50	1068	1389	1597	1679	1724	1720	1763	1838	2030	1922	1512	1170
	Vert.	972	1162	1156	989	856	788	828	992	1350	1561	1371	1100
Los Angeles, CA	Hor.	946	1266	1690	1907	2121	2272	2389	2168	1855	1355	1078	905
	30	1434	1709	1990	1940	1952	1997	2138	2115	2066	1741	1605	1439
	40	1530	1776	1996	1862	1816	1828	1966	2002	2037	1788	1706	1550
	50	1587	1799	1953	1744	1644	1628	1758	1845	1959	1791	1762	1620
	Vert.	1411	1455	1344	958	760	692	744	918	1230	1383	1537	1479
New Orleans, LA	Hor.	788	954	1235	1518	1655	1633	1537	1533	1411	1316	1024	729
	30	1061	1162	1356	1495	1499	1428	1369	1456	1490	1604	1402	1009
	40	1106	1182	1339	1424	1389	1309	1263	1371	1451	1626	1464	1058
	50	1125	1174	1292	1324	1256	1170	1137	1259	1381	1610	1490	1082
	Vert.	944	899	847	719	599	546	548	647	843	1189	1240	929
Portland, OR	Hor.	578	872	1321	1495	1889	1992	2065	1774	1410	1005	578	508
	30	1015	1308	1684	1602	1836	1853	1959	1830	1670	1427	941	941
	40	1114	1393	1727	1569	1746	1739	1848	1771	1680	1502	1020	1042
	50	1184	1442	1727	1502	1622	1594	1702	1673	1651	1539	1073	1116
	Vert.	1149	1279	1326	953	889	824	890	989	1172	1309	1010	1109

Source: Reproduced from Reference 2.

Solar energy = $(3333 \text{ ft}^2) \times (0.42 \text{ Btu}/\text{ft}^2/\text{yr})$

$= \underline{1400 \times 10^6 \text{ Btu}/\text{year}}$

At $\$4.00/10^6$ Btu, the energy value would be:

Energy value = $(\$4.00/10^6 \text{ Btu}) \times (1400 \text{ Btu} \times 10^6/\text{yr})$

$= \underline{\$5600/\text{year}}$

The time to pay back the cost of the collector—$66,660—would be:

Simple payback period = $(\$66,660)/(\$5600/\text{yr})$

$= \underline{11.9 \text{ years}}$

Considering the practical factors in this application lengthens the payback time substantially from the original 5.7 years determined earlier.

13.2.2 Solar Collectors

A solar collector is a device used to thermally collect, store, and move solar thermal energy. Essentially, solar collectors are heat exchangers that transfer the energy of incident solar radiation to sensible heat in a working fluid—liquid or air [5]. There are many different types of solar collectors, as shown in Figure 13-2.

13.2.2.1

A flat-plate solar collector generally consists of a shallow metal or wooden box which has a glass or plastic transparent cover, and which contains a black absorption plate that transfers heat to some fluid. The sun's shortwave radiation passes through the transparent cover, enters the collector and heats a fluid (usually water with or without antifreeze, or air). The hot fluid is then moved from the collector to the point of use or to storage for later use. A flat-plate collector almost always faces to the south (in the northern hemisphere) and is tilted at some angle. A typical flat-plate solar collector is illustrated in Figure 13-3.

A typical flat-plate solar collector application is given in Figure 13-4. Here, solar energy heats an ethylene glycol mixture that is pumped to a storage tank. The tank then heats water through a heat exchanger for alternative use as shown. In some applications, such as preheating boiler makeup water, the water itself can be pumped through the collector to a

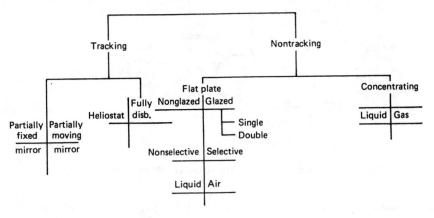

Figure 13-2. Types of solar collectors.

Figure 13-3. Flat-plate solar collector.

storage tank or directly to the boiler room. In such applications, care must be taken to prevent freezing; drain down provisions for the solar collector are usually employed.

Glazing:

The flat plate solar collector may be glazed or nonglazed. The most common *glazing* is tempered glass which allows the shortwave radiation

Figure 13-4. Typical flat-plate solar collector application.

of the sun to enter the collector but prevents the longer-waved reradiation from leaving. This produces a greenhouse effect and increases the efficiency of the collector but also increases the cost. Dual glazing would further cut down the heat loss while not appreciably restricting incoming solar energy, and is used for most high temperature flat plate collectors. Unglazed collectors are quite efficient at lower temperature applications such as swimming pool heaters, while glazed collectors are more efficient at higher temperatures. Typical flat-plate collector efficiencies with different glazings are given in Figure 13-5. T_{in} is the temperature of the water coming into the collector, and $T_{ambient}$ is the temperature outside the collector. The graph shows that the heating ability of the collector is greater with glazings that have higher insulating capability.

Selectivity:
       The *selectivity* of the collector absorbing surface is an important property that affects a collector's efficiency. The collector must absorb shortwave radiation readily and emit long-wave radiation stingily. Surfaces with high shortwave absorption and low long-wave emittance are *selective surfaces*. Selective surfaces perform better at higher temperatures than nonselective surfaces. A single glazed selective surface collector has efficiencies very similar to a double-glazed nonselective surface collector.

Transport medium:
       The *medium* chosen to move the thermal energy from the collector to the point of use or to storage can be either liquid or air. Each has advantages and disadvantages. It is much more difficult and expensive to move thermal energy with air than with liquid. In fact, the horsepower required to move the same amount of thermal energy may be 10 times higher for an air system than for a liquid system. Air systems also have lower heat transfer rates, so the system must be carefully designed to provide a

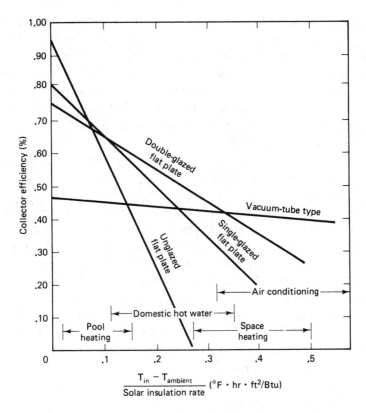

Figure 13-5. Typical flat-plate collector efficiencies

sufficiently large heat transfer surface. However, air does not freeze. In liquid systems, ethylene glycol or some other antifreeze must be used, or the system must have well-designed drain down controls. In addition, air systems do not have corrosion problems, and leaks do not present as much of a problem as with liquid systems.

The examples presented thus far have assumed the fluid is a liquid, but many applications are suitable for air. For example, an air solar collector could be used as an air preheater for an industrial furnace or boiler. Although this application is not widespread, it can be useful if the time of solar energy availability and the time of industrial heat use coincide (i.e., the energy is needed when the sun is shining).

13.2.2.2 Concentrating collectors

A need for temperatures of 250°F or higher usually requires a concentrating collector. The surface of a concentrating collector must be highly reflective, enabling concentration of the sun's rays on the heat absorption device. The heat transfer fluid can be a liquid or gas. A concentrating collector is usually also a tracking collector in order to keep the sun's rays focussed on a small surface. A typical design for a parabolic trough-type, tracking collector is shown in Figure 13-6. The collector can track in an east to west direction to follow the daily sun, in a north to south direction to follow the seasons, or both. Concentrating collectors that accurately track the sun's position are more efficient than those that do not track the sun's position as well.

Other types of concentrating, tracking collectors use movable mirrors that can concentrate the solar energy on a small surface that remains fixed. The *power tower* is such an application where the absorption surface or central receiver surface is located in a tower. Tracking mirrors are located on the ground around the base of the tower. These fully tracking mirrors are usually computer-controlled to concentrate the maximum amount of solar energy on the tower. Applications are mainly for steam generation used to produce electric power. This type of application takes a large amount of land area and requires careful maintenance. The most notable power tower is located in California, and is called Solar One [6]. It was operated by Southern California Edison Company up until 1989. This facility had a capacity of 10 MW, and successfully generated electric energy for almost ten years.

13.2.3 Solar Thermal Storage

One of the biggest obstacles to widespread solar utilization is that often the solar energy is not needed when it is available and it is not available when it is needed. For example, maximum heat is usually needed when the sun is not shining, especially at night. Also, solar energy flows cannot readily be regulated. When the sun shines, the collector usually delivers energy at its full capacity. Tracking collectors can be

Figure 13-6. Parabolic trough solar collector

programmed to turn away from the sun, and can be regulated, but this is not generally a cost-effective mode of operation.

For these reasons, many solar applications require some type of storage system. The system must both store energy for later use and regulate energy flows. Figure 13-4 depicts one possible liquid storage system. There are three basic types of storage systems:

- *Liquid.* Liquid storage systems normally utilize water or a water-antifreeze mixture. The storage capability is determined by the sensible heat capacity of the liquid. For water it is 1 Btu/lb/°F.

- *Rocks.* Used for air systems, rock storage uses the sensible heat content of rocks for storage. Typically, airflow is top to bottom for storage and bottom to top for use as needed.

- *Phase change materials.* The preceding systems utilize sensible heat. This system utilizes the larger latent heat in phase changes such as the melting of ice. The required storage volume is smaller, but the cost is higher. Eutectic salts are often used.

13.2.4 Applications of Solar Thermal Systems

Commercial facilities and industry have not yet incorporated the use of solar thermal energy systems on a large-scale basis. There are many solar applications throughout the country, but most replace only a small quantity of the traditional energy supply. The following are some applications of solar thermal energy in business and industry.

- *Solar-heated hot water.* One of the bright spots in the application of solar thermal systems today is solar-heated or augmented hot water. The hot water tank itself is the storage system (or at least a part of it), and the hot water is usually needed the entire year. In some parts of the country, solar water heaters are very cost effective. Hotels, motels and small businesses such as laundries are using solar-heated hot water. In the industrial sector, solar-heated makeup water for boilers and cleaning tanks and solar-augmented process feeds are additional process uses for solar heating.

- *Solar space heat.* Although solar space heating is feasible, active solar collectors for space heating are very seldom cost-effective. Because they are not used all year, they do not often save enough energy to justify their cost. Passive solar applications are often cost effective for minimizing the need for traditional fuel sources in providing space heating. This is discussed in the next section.

- *Solar recuperators.* Industrial furnaces require heat year-round, so combustion air preheating is a likely candidate for solar thermal energy. In installations where outside air might already be used, the application is very simple and probably requires no storage. When the sun shines, the air is preheated; otherwise it is not. However, such systems must be designed carefully or they may cause problems with burners and excess air control.

- *Solar detoxification.* One recent application of solar thermal energy that is rapidly growing in use in commercial and industrial facilities is the detoxification of hazardous wastes. These applications make use of the thermal energy and the high energy photons from solar energy that can more thoroughly decompose and destroy toxic chemicals [7].

- *Solar-heated asphalt storage tanks.* A company which had used portable propane burners to keep asphalt in their storage tanks hot switched to solar collectors and added insulation to the tanks. The energy savings and the convenience made this application attractive.

- *Solar air conditioning.* Solar air conditioning systems use the heat from a solar collector to drive an absorption chiller, and produce cool water or air. The cost effectiveness of these active systems is generally poor at this time, and the use of passive solar features in buildings and structures is a far more successful, and cost effective technology.

13.2.5 Passive solar systems

Passive solar systems result from design strategies and related technologies that use elements of the building structure—primarily glass and thermal mass—and building orientation to heat, cool, shade, and light buildings. Passive solar technologies include direct gain heating, radiative cooling, natural ventilation and economizer cycles, natural lighting, light shelves and shading systems.

13.2.5.1 Use of passive heating

A south-facing glass window serves as a passive solar collector of heat for a building, and the interior of the building serves as the heat storage device. Windows also provide significant amounts of natural lighting. Careful attention to passive solar energy utilization in building design can reduce energy costs significantly.

Example 13.2:

Consider a building located in Fort Worth, Texas that faces south. The building has a total wall area (minus glass) of 2000 ft² and a roof area of 3000 ft². The R values of the roof and walls are 22 and 12, respectively. There is 300 ft² (about 40% of the south wall) of south-facing, double-pane, overhung glass (R-1.85) which permits full sunlight for the 5-month heating season. There are 2363°F heating days in Fort Worth. The glass transmits 80% of the solar energy hitting it. How much of the building's heating needs does this passive solar feature provide?

Solution: From Table 13-1, for Fort Worth, TX, we find the total solar load on a vertical surface for the months of November, December, January, February, and March to be

$$(80\%)\,(300\ \text{ft}^2)\left[1509(30)+1396(31)+1315(31)+1286(28)+1196(31)\frac{\text{Btu}}{\text{ft}^2\bullet\text{year}}\right]$$

$$=48.58\times10^6\text{Btu/year}$$

The heat loss (HL) (assuming no setback) is

$$\text{HL walls}=\left(\frac{1\ \text{Btu}}{12\ \text{h}\bullet{}^\circ\text{F}\bullet\text{ft}^2}\right)\left(2000\ \text{ft}^2\right)\left(\frac{2363\ {}^\circ\text{F days}}{\text{year}}\right)\left(\frac{24\ \text{h}}{\text{day}}\right)$$

$$=9.45\times10^6\text{Btu/year}$$

$$\text{HL roof}=\left(\frac{1\ \text{Btu}}{18\ \text{h}\bullet{}^\circ\text{F}\bullet\text{ft}^2}\right)\left(3000\ \text{ft}^2\right)\left(\frac{2363{}^\circ\text{F days}}{\text{year}}\right)\left(\frac{24\ \text{h}}{\text{day}}\right)$$

$$=9.45\times10^6\ \text{Btu/year}$$

$$\text{HL glass}=\left(\frac{1\ \text{Btu}}{1/85\ \text{h}\bullet{}^\circ\text{F}\bullet\text{ft}^2}\right)\left(300\ \text{ft}^2\right)\left(\frac{2363{}^\circ\text{F days}}{\text{year}}\right)\left(\frac{24\ \text{h}}{\text{day}}\right)$$

$$=9.20\times10^6\text{Btu/year}$$

$$\text{total HL}=28.1\times10^6\text{Btu/year}$$

Percent heat supplied = 48.6/28.1 = <u>173%</u>

According to these calculations in Example 13-2, placing glass on 40% of the south-facing wall will allow solar energy to supply 170% of the total heat needed for the year. However, the following practical considerations are important:

- The heat loss by infiltration may be as large as the total heat loss through the roof and walls—especially if there is much exhaust air. Thus, the total heating load may be far greater than that initially stated.

- The sun does not shine on some days, so the heating plant will have to be designed as large as it would be without solar aid.

- On bright sunny days, the building might get too hot, so the glass area might have to be reduced or adjustable shading used.

- This building is fairly well insulated. Many manufacturing buildings have little insulation, and thus would need much larger amounts of heat to adequately warm the building.

- All glass is on the south wall. Some glass may be needed on the other walls.

- Fort Worth may not be typical of the rest of the country.

Nevertheless, passive solar energy can contribute a large percentage of the heating required in a facility. Passive solar should especially be considered in the design of manufacturing buildings whose hours of operation normally coincide well with sun hours. At night, the thermostat can be substantially reduced, whereas in homes, night setback cannot be used as readily.

13.2.6 Energy Efficiency in the Design of New Facilities

Building a new facility provides the industrial energy manager with numerous opportunities to incorporate energy efficiency into the facility design. The following sections include ways to use passive solar energy, to avoid unwanted solar loads, and to take advantage of renewable energy opportunities. For completeness, we have also included some other energy efficiency measures that should be considered in the design of a new facility.

13.2.6.1. The land.

Purchase property with good energy-efficiency characteristics. Do not choose property located in energy-intensive spots. Avoid areas that are too windy, areas on the tops of hills, and areas on north slopes. However, these spots may become attractive in the future as locations for utilizing other renewable energy sources.

Choose a location that is near energy supplies to minimize transmission losses as well as costs. Property that is near good transportation facilities will also save energy costs.

13.2.6.2. The building site.

Choose an energy-efficient building site for the facility. You should take advantage of unique spots that can use existing deciduous trees or other natural properties to provide shading from the summer sun and/or windbreaks in the winter. Hills can be utilized as berms to improve insulation instead of spending money to level them.

The building site should be close to major transportation facilities.

13.2.6.3. Facility orientation.

Orient the facility for energy conservation. The building should face south. The shorter dimension should run north to south and the longer, east to west. The manufacturing plant in Figure 13-7 below demonstrates this. This allows minimum sun exposure in the summer on the east and west walls. However, during the winter, because of the lower sun angle, the sun helps to heat the facility.

13.2.6.4. Underground construction.

Consider the use of underground structures. Use a large amount of backfill on northern and western walls. This will protect the facility from cold on the north side and heat on the west.

Partially submerge the entire structure. This will use the thermal mass of the ground to maintain a constant temperature in the facility with less energy input. Some facilities are completely submerged, such as the warehouses in the caverns under Kansas City.

13.2.6.5. Energy-conserving landscaping.

Much of the undesirable heat loss/gain can be prevented through proper landscaping [8,9]. Avoid asphalt or concrete areas around the building as much as possible; grasses, shrubs, and vines are much cooler in the summer.

Place deciduous trees strategically so they offer shade during the

Figure 13-7. Landscaping for energy conservation. (a) Deciduous trees: In summer, allow the early morning sun to penetrate and then protect; in winter, allow the sun to penetrate. (b) Hedge: Catch the late afternoon rays. (c) Deciduous trees: Provide shade in the afternoon. (d) Shrubs or tall hedge: In winter, block northwest winds; in summer, help divert southwest winds around the building.

summer yet allow sunlight to penetrate in the winter. Use vines and shrubs to offer additional shading. In fact, thick shrubs placed close to a building effectively increase the R value of the walls.

Use trees and shrubs as windbreaks and wind diverters. For example, evergreen trees or shrubs at the northwest corner of the building can break the cold winter winds and divert the summer breezes for better utilization. (See Figure 13-7 above)

13.2.6.6. Energy-efficient building envelope.

The energy-efficiency of the building envelope will significantly affect the energy use in a facility. Energy efficiency is easy to incorporate in the initial construction of a building, but very costly to retrofit. The following suggestions should be considered in the initial design:

- **Minimize the wall perimeter area.** Use regular-shaped buildings—square or rectangular. This minimizes the wall area and thus minimizes the heat loss.

- **Insulate the building well.** Install insulation in the ceiling and the walls as well as on the slab (if appropriate). Check for local recommended levels. Figure 13-8 illustrates recommended insulation placement for an underground manufacturing facility.

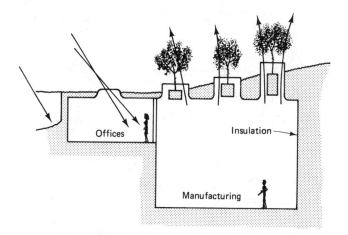

Figure 13-8. Recommended insulation placement

- **Use energy-efficiency considerations in window selection.** Use a minimum amount of glass—well placed. Avoid glass on northern and western sides. As discussed earlier, planning the southern exposure with proper solar influx is best.

  Consider installing insulating glass or storm windows. Double or triple glazed windows are cost effective in most parts of the country. Office areas are especially good candidates since they are typically heated and cooled to a greater extent than the production areas.

  With proper placement of windows, natural ventilation is also feasible. Therefore, consider the use of windows that can be opened to utilize natural ventilation.

  Consider the use of tinted window glass on all walls except the south one. The east and west walls are strong candidates for tinted glass.

Utilize drapes or outside partitions to further insulate windows and to reduce solar load when desired. Several examples of outside partitions are shown in Figure 13-9.

Figure 13-9. Outside partitions.

•   **Utilize overhangs or awnings.** On southern exposures, overhangs block the summer sun but allow the winter sun to enter as shown in Figure 13-10. An architect can tell you the proper amount of overhang to allow for sunlight in winter months. It varies with location.

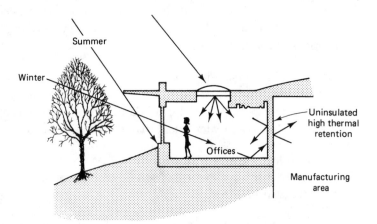

Figure 13-10. Energy conservation through the use of overhangs.

Passive solar systems (also shown in Figure 13-10) help to further reduce energy costs, and they provide attractive warm areas for personnel.

• **Design the roof carefully.** Use light colors in warm climates and dark in cold. Design the roof so it can be sprayed in the summer. Be careful not to flood the roof—it will leak eventually.

• **Engineer all wall openings for energy efficiency.** Minimize the number of openings. Caulk and weather-strip the doors and windows well.

Design the overhead doors with two position openings to match the truck size and/or use adjustable dock seals. Utilize insulating pads on the dock doors. Consider using adjustable dock pads.

Consider interlocking the doors and the heating units so that when the doors are open, the heat is off. Utilize automatic doors or various types of "see through" materials (plastic strips, plexiglass, etc.) for doors that must be used frequently. Air curtains are another option for minimizing heat loss.

Use good pedestrian doors to avoid using large dock doors for pedestrian traffic. Utilize revolving doors or entrance vestibules to minimize air infiltration.

13.2.6.7. Energy-efficient facility layout.

Locate the facilities within the plant to minimize the energy required to maintain personnel comfort. Departments with high personnel density should probably be located in southern exposure areas of the plant and not in northern or western areas. Figure 13-11 demonstrates one possible layout considering energy requirements only.

Avoid or minimize northern or western exposure for dock areas (shipping and packaging). In Figure 13-11 the shipping and receiving area is oriented so that the amount of northern exposure is minimized. The entry area for shipping and receiving should be on the east side.

Consider departmental or cost center metering of utilities. By doing this, each cost center can be held accountable for its energy consumption, and energy can become a part of the budgeting process. This requires extensive preconstruction planning to lay lines and install meters as needed. Then, arrange the facilities so that energy control is easy (e.g., lights and motors can be switched off in one location).

Plan the layout so the exhaust air from one area can be used in another; e.g., the hot air at ceiling height in one area may be used as the combustion air for a large furnace.

Figure 13-11. Hypothetical plant layout to minimize energy requirements for space conditioning.

13.2.6.8. Location of process equipment.

Locate all boilers, hot water tanks, and other heated tanks to minimize distribution distances and consequently energy loss in distribution. Design the boiler location and steam distribution system to facilitate return of condensate and/or reflashing or low-pressure steam.

Engineer waste heat recovery systems into the facility design. Waste heat is the one renewable energy source today whose utilization is frequently cost effective. It is much easier to incorporate feedstock preheating apparatus in the initial design rather than retrofitting. Boilers, furnaces, large motors being cooled, lighting fixtures, any cooling fluids, and compressors are just a few of the potential sources for waste heat recovery. Locate the waste heat producing equipment where it can be utilized and where the heat recovery equipment can be installed.

Locate air compressors so they can be maintained easily, they can use fresh cold air, and transmission losses are minimized. Utilize step-up air compressors to be able to reduce plant-wide pressures.

13.3 SOLAR-PHOTOVOLTAICS

Photovoltaics is the direct conversion of sunlight to direct current (dc) electricity through a photocell. Historically, photovoltaics has not been cost effective in competition with fossil fuel or electricity from a grid system, but the needs for photovoltaics in space and the subsequent research coupled with rising costs of traditional energy have pushed photovoltaics ahead. Although still not cost effective for replacing central station electric power plants, photovoltaics is cost effective for many applications where electricity is needed in remote areas. Each year, progress in research and development continually reduces the price of electricity from this source.

Photovoltaic cells are semiconductor devices which can convert the energy in photons of light into dc electrical energy. Most cells are made from single-crystal high-purity silicon and small amounts of trace elements such as boron and phosphorus. These elements are combined into a material with excess electrons (n-type semiconductor) connected to a material with insufficient electrons (p-type region). As long as the cell area is illuminated by light, electrical energy will be produced (see Figure 13-12).

Figure 13-12. Typical photovoltaic cell

The area where the semiconductor materials are joined together can be connected to a battery, or to other cells. Cells are then put together in modules, which can be in the form of flat plates or concentrators. Most cells generate electrical energy of about 0.5 volt and a current that varies with the area of the cell and amount of light. For flat plate modules, typical wattage is about 12.5 w/ft^2 with a conversion efficiency of 5 to 15%. Higher efficiencies can be obtained from concentrator modules.

To achieve higher voltages and currents, solar cells are combined in series and parallel, like batteries. Sets of cells are usually placed in series to generate the necessary voltage, and then multiple strings are placed in parallel to develop the desired current (see Figure 13-13). Because the time of need and the solar intensity do not always match, some type of storage device and voltage regulation is necessary. Normally, a chemical battery storage system fills this need. Finally, a backup generating system is often needed to allow for consecutive cloudy days.

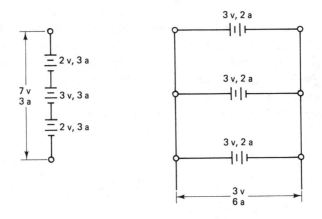

Figure 13-13. Photovoltaic hookups

Generally there are three types (sizes) of photovoltaic systems:

1. *Small* (1-10 kW). These photovoltaic systems are suitable for remote locations or other locations where conventional electricity may be costly. Examples are street lights, irrigation pumps, security lighting at construction sites, communications equipment, and field battery charging.

2. *Medium* (10-1000 kW). These photovoltaic systems could be used by larger industrial facilities and/or remote communities. They could also be used as supplemental or peaking power.

3. *Large* (more than 1000 kW). These photovoltaic systems are used in utility-owned large-scale power-generating stations often located in desert areas.

At their current costs, photovoltaic cells can provide electric energy at 25 to 30 cents per kWh. The largest use of photovoltaic cells today is in supplying remote electric power. Electric utilities are expanding their capabilities with photovoltaic generation. Pacific Gas and Electric Company recently installed over 1000 kW of photovoltaic generation. Southern California Edison Company has entered into a joint venture with Texas Instruments Company to produce photovoltaic cells called Spheral Solar Modules that produce power at 14 cents per kWh. Limited production is expected to begin in 1994.

Industry offers a number of applications for photovoltaics, but the actual usage has been limited in size and scope. Ideal applications would have the following characteristics:

1. The equipment operates on dc power.
2. Substantial sunlight is available.
3. The user can tolerate random losses of power.
4. Power is needed when the sun is shining.

Photovoltaics is one of the cleanest, most environmentally benign energy technologies. Many proposals have been made to combine the photovoltaic generation of electric energy with the production of hydrogen through electrolysis. The hydrogen would then be used as the energy supply to be marketed primarily for use in automobiles. Alternatively, the hydrogen could be stored and then burned at a later time to produce electric power when the sun was not shining.

13.4 INCORPORATING SOLAR FEATURES
INTO BUILDING DESIGN—AN EXAMPLE

The new building to be constructed for the Florida Solar Energy Center (FSEC) provides a good example of some of the areas for saving energy in a facility by utilizing solar energy features. The design objective for this building was to "incorporate the latest solar and conservation building techniques" [10]. Extensive simulations using the DOE-2 program (discussed in Chapter Nine) were conducted to compare alternative design features, and to estimate the energy savings from new design features.

One of the first design decisions was to locate the building on the site with the long axis facing north-south to minimize solar gains and reduce cooling loads, an important consideration in Florida. Both the roof and the walls of the building were coated with a low-absorbance material to further reduce cooling loads. Three-foot wide overhangs are used above the windows to reduce solar heat gains through the glass. Super-windows with spectrally selective coatings and double glazing were also chosen.

Daylighting was used extensively in order to minimize the energy devoted to artificial lighting. To enhance the use of daylighting, all offices are located on the perimeter wall of the main building. The north and south faces of the building utilize an extensive amount of glass. The windows are long, but only 4.5 feet high; the roof overhangs help limit the morning and afternoon solar heat gain, while still providing interior illumination. Above each window is a two-foot high "light shelf" which reflects additional illumination into the building. One of the buildings has a triangular-shaped section, and the windows on its southwest side have fixed shading devices to block direct sun while allowing reflected and diffuse light to enter. The top floors of the main building and the library contain "light wells" to introduce natural daylight to the interior.

Almost 400 photosensors were installed in the facility to measure light levels, and to provide inputs to the dimmable electronic ballasts to control the energy used by the T8 fluorescent lights. The electronic ballasts for the lights allow the power and illumination levels to be varied between 5% and 100% of their maximum output. The DOE-2 simulation showed that daylighting was able to reduce the energy needed for lighting by 46% over the entire year; mid-day reductions averaged 60% for all areas of the building. The payback for the 400 photosensors was less than six months.

The bottom-line energy savings resulting from this new building design using energy efficient and solar features is impressive: the overall energy savings compared to the base-case design was a 70% reduction in utility bills and a peak cooling demand reduction of 88%. Additional solar features such as solar water heating and photovoltaic cells for powering the lighting system are also under investigation. These features may even further reduce the energy use and energy costs for this facility.

13.5 WIND ENERGY

Much has been said about wind energy and its potential. Although the potential in the United States for wind energy is many times greater than the present consumption of electrical energy, this potential will never be obtained due to aesthetics, construction cost and radio-TV interference problems [11]. However, wind energy has proved to be the most cost-effective of the solar technologies, and 1500 MW of electric power generation is located in the state of California.

The power density of wind is given by

$$\frac{P}{A} = \frac{1}{2}\rho V^3$$

(13-1)

where   A = area normal to the wind
          ρ = density of air (about .075 lb/ft^3)
          V = velocity of air
          P = power contained in the wind

This can be rewritten as follows (K is a constant for correcting units):

$$\frac{P}{A} = KV^3$$

(13-2)

where   K = 5.08×10^{-3}

$$\frac{P}{A} = W/ft^2$$

$$V = mph$$

Unfortunately, only a small percentage of this power can be obtained. It can be shown that theoretically .5926 of the power can be extracted, but practically only 70% or so of that can be obtained. Consequently, only .70 × (.5926) or about 40% of the power is harvestable.

Example 13-3: Find the power in watts per square foot that can be produced from a 10-mph wind. How would this power change if the wind velocity were 20 mph?

Solution: Using Equation 13-2, the power per square foot is found as:

$$\frac{P}{A} = \left(5.08 \times 10^{-3}\right) \times \left(10^3\right) \times (.40) = 2 \ W/ft^2$$

If the wind speed increased to 20 mph, the power would become:

$$\frac{P}{A} = \left(5.08 \times 10^{-3}\right) \times \left(20^3\right) \times (.40) = 16.3\,\text{W/ft}^2$$

These figures demonstrate why it is necessary to find areas with consistent high wind velocities. The difference between a site with 10 mph winds and with 20 mph winds is a factor of 800%.

Just a difference in average wind speed of 3 mph is enough to change the wind characterization of a site. Good, excellent and outstanding wind sites have been described as having average wind velocities of 13, 16 and 19 mph, respectively [12].

Wind speed offers other problems. Too little wind will not initiate power output for most windmills. The wind speed must be above a "cut-in" speed unique to the windmill. Too much wind creates other problems that could conceivably destroy the windmill, so most windmills feather out at some high wind speed.

Industry has done almost nothing to utilize wind. Usually industry is located in congested areas where windmills would not be popular. Furthermore, industry is not likely to develop wind energy as an energy source until it is more cost effective.

3.6 BIOMASS AND REFUSE-DERIVED FUEL

13.6.1 Energy from Biomass

The term *biomass* includes all energy-producing materials that come from biological sources, such as wood or wood wastes, residues of wood-processing industries, food industry waste products, sewage or municipal solid waste, waste from food crops cultivated as energy sources, and other biological materials [13]. Both economic and environmental benefits can be achieved when biomass is used as a source of energy for space heating, process heat, or electricity production. If the biomass comes from a waste product, the cost is usually low; this source may have the added benefit of avoiding the cost of disposing of the waste. Utilizing waste biomass sources not only provides an inexpensive source of fuel, but also solves an environmental disposal problem as well.

Four main technologies are available for converting biomass to usable energy forms. The first is simply burning the biomass to provide direct heat for space and process heating, or for cooking. The second is burning the biomass to produce steam which is then used to generate electrical energy. The remaining two technologies involve converting the biomass to liquid or gas energy forms so that it can be transported to other locations for use.

13.6.2 Energy from Refuse-derived Fuel

Another area for renewable fuel utilization in industry is *refuse-derived fuel*. About 70% of the typical household refuse is combustible, but a higher percentage of industrial wastes is usually combustible. In fact, some industrial waste is a fairly high-quality fuel, as pulp and paper mills have proven for years. Although there are some successful applications of fuel derived from municipal waste, this book is concentrating on industrial—institutional energy management. In the rest of this section we will only discuss industrial-institutional applications.

Combusting waste reduces the volume of the waste by 80% or more which makes it easier to handle. Consequently, one of the big savings for refuse-derived fuel is in reduced disposal costs. In some industries, this disposal savings may be much larger than the economic value of the actual Btu content of the fuel.

Some typical heating values for various types of industrial wastes are shown in Figures 13-14 and 13-15. As shown in these tables, many sources of industrial waste have significant Btu content. Refuse-derived fuel is certainly a fuel source worth considering.

The following is a suggested procedure for analyzing waste fuel sources:

Step 1. Determine the heating value and quantity of the waste.

Step 2. Determine the technical feasibility of utilizing the waste (burning, pyrolysis, anaerobic digestion, etc.). Include necessary pollution control and ash disposal costs.

Step 3. Develop a system design including waste fuel handling, preparation, firing, and disposal. Estimate the cost involved.

Step 4. Perform an economic analysis including all incremental costs identified in step 3, and all savings over conventional fuels. Don't forget the savings in disposal costs and any tax incentives (tax credits or preferred depreciation schemes) from federal and state governments.

Figure 13-14. Wood waste characteristics.
(From *Instructions For Energy Auditors*. U.S. Dept. of Energy, 1978)

	Wood, avg, seasoned (%)	Wood waste, Douglas fir (%)	Hogged fuel, Douglas fir (%)	Sawdust (%)
Proximate analysis:				
Moisture	24.0	35.9	47.2	44.9
Volatile matter	65.5	52.5	42.9	44.9
Fixed carbon	9.5	11.1	8.9	9.5
Ash	1.0	.5	1.0	.7
Ultimate analysis:				
Hydrogen	7.2	8.0		
Carbon	37.9	33.5		
Nitrogen	.1	.1		
Oxygen	53.8	57.9		
Sulfur	–	0		
Ash	1.0	.5		
High heating value (Btu/lb)	6300	5800	4670	4910

Figure 13-15. Heating values of industrial waste fuels.
(From *Instructions For Energy Auditors*. U.S. Dept. of Energy, 1978)

	Heating value	
Fuel	Btu/lb	Btu/ft^3
Solid Fuels:		
Bagasse	3,600–6,500	
Bark	4,500–5,200	
Wood Waste	4,500–6,500	
Sawdust	4,500–7,500	
Coffee grounds	4,900–6,500	
Rice hulls	5,200–6,500	
Corn cobs	8,000–8,300	
Municipal refuse	4,500–6,500	
Industrial refuse	6,600–7,300	
Coal	8,000–24,000	
Liquid fuels:		
Black liquor (pulp mills)	4,000	
Dirty solvents	10,000–16,000	
Gasoline	20,700	
Industrial sludge	3,700–4,200	
Naphtha	20,250	
Naphthalene	18,500	
Oil waste	18,000	
Paints and resins	6,000–10,000	
Spent lubricant	10,000–14,000	
Sulfite liquor	4,200	
Oil	18,500	
Gas fuels:		
Coke, oven gas		500–1,000
Refinery gas		1,200–1,800
Natural gas		1,000

If you go through this procedure at least once for each alternative, you should make the most cost-effective decision.

At some stage, the analysis should include the incorporation of waste from another industry or a municipality so that economies of scale might be obtained. However, extreme care should be taken to ensure that these other sources are reliable in quantity, flow rate, and quality.

The technology for burning waste fuel is rapidly becoming more developed and the equipment is commercially available. Some of the technologies receiving attention include the following:

• Burning raw refuse in water well incinerators

• Shredding, pelletizing, or otherwise preparing fuel and burning in boilers

• Anaerobic digestion where waste is converted to gas in an oxygen-free atmosphere

• Pyrolysis or thermal decomposition of waste in the absence of oxygen

The following paragraphs summarize some applications of refuse-derived fuel:

• A large communications company converted a natural gas-fired boiler to burn methane from a nearby landfill. The plant's consumption of the landfill gas is expected to be between 200,000 and 300,000 million Btu per year. The plant expects to save around $60,000 per year from the use of the landfill gas. The cost of the conversion was $1 million, but it was paid by the supplier of the landfill gas [14].

• A county in North Carolina expanded its operation involving burning municipal waste to cogenerate electric power and steam. A 5 MW steam turbine generator was added to sell electric power to a major state utility, and to sell steam to a chemical and fertilizer firm located nearby. The firm will buy about 100,000 pounds per hour of steam from the county-owned and operated facility. The county calculated the payback time for the new addition at just under seven years [15].

• A dairy farm in Georgia installed a 190-hp engine running on biogas from an anaerobic digester to drive a 55 kW electric generator. Waste heat from the engine was used to increase the effectiveness of the digester by keeping it warm, and to dry the sludge from the digester prior to using it for bedding material. The overall efficiency of the cogeneration system was about 60%. The biogas from the digester had an energy content of about 600 Btu per cubic foot—or nearly 60% of the heat content of natural gas [16].

13.7 WATER MANAGEMENT

Water management is the efficient and effective use of water. Like energy management, the main goal is to improve profits or reduce costs. Water management is included in this text on energy management because water and energy utilization are intricately intertwined in most organizations. As water consumption goes up, so does energy consumption, and vice versa. Water management is included in this chapter on renewable energy sources because water is renewable and is often an efficient energy source. For example, water in cooling towers is a very inexpensive source of cooling.

Why worry about water? The sheer volume of water used in this country each day is staggering. Each person in the United States consumes directly and indirectly almost 90 gallons of water per day. At the same time, water costs are going up rapidly, and some sources are drying up (e.g., Midwestern aquifers).

Industry can save dramatic amounts of money and water through attention to its water-consuming equipment. In this section we will demonstrate how to save dollars and gallons of water in industrial water use.

There are three primary types of water users in industry. They are boilers, cooling towers, and process equipment.

13.7.1 Savings from Boilers
In boilers the primary ways to save are the following:

- *Blowdown.* Blowdown should be reduced to the minimum possible, which is determined by the feedwater quality and the amount of condensate return. Reducing blowdown also saves large amounts of energy.

- *Condensate return.* Condensate should be returned to the boiler whenever economically feasible. This saves water by reducing the amount of new makeup water needed and by reducing the amount of blowdown required. Returning the condensate also saves substantial amounts of energy and water treatment costs because condensate is essentially distilled water.

- *Steam leaks and steam traps.* As already demonstrated, steam leaks are extremely expensive energy losses, but significant amounts of water are also lost in steam leaks. An estimate of the amount of water lost can be obtained by dividing the annual loss in Btus by the enthalpy of evaporation. Stuck-open steam traps are steam leaks if the condensate is not returned and are still wasteful even if it is returned.

13.7.2 Water Savings from Cooling Towers

In cooling towers, the primary ways to save are the following:

- *Bleed.* Bleed in a cooling tower is almost identical to blowdown in a boiler. The purpose is to prevent impurity buildup. Bleed should be reduced to a minimum and reused if possible. Sometimes bleed can be used to water lawns or as rinse water makeup. However, careful attention must be paid to the chemicals in the water.

- *Sewerage charges.* Often sewerage charges are based on the amount of water consumed. Yet water consumed in a cooling tower does not go into the sewer. Negotiations with municipalities can often reduce the sewerage charge substantially if large cooling towers are present. Usually about 1% of the flow rate of water must evaporate for each 10°F drop in water temperature.

- *Preventive maintenance.* As baffles become broken or clogged with dirt and slime, the cooling capability drops dramatically. Since a 1°F drop in condensing water temperature can mean a 3 percent savings in electrical chiller input, preventive maintenance of cooling towers is important.

- *Tower water.* Often cooling tower water can be used directly for process cooling instead of using chilled water. When this is possible, large amounts of energy can be saved at the cost of higher water consumption. The trade-off is almost always cost effective.

13.7.3 Savings in Industrial Processes

It is much more difficult to generalize on ways to save in processes since water can be used in so many different ways in industrial processes. For example, water can be used to cool furnace walls, cool air compressors, wash, rinse, surface-treat, coat, test products, and cool molds and for a wide variety of other uses. The following paragraphs discuss some of the ways water and money can be saved in industrial processes:

- *Use water flow restrictors in shower heads and sinks.* As much as 60% savings in water and energy costs can be realized when flow restrictors are installed.

- *Recycle rinse water.* Often rinse water can be recycled by simple filtering or treatment. In one company, $30,000 and 30×10^6 gal of water were saved annually by simply running rinse water through a sand filter and reusing it.

- *Reuse cooling water.* Often air compressors, small chillers, and other equipment requiring cooling are cooled with once-through cooling water (i.e., water from the tap is run through the equipment one time and dumped into the sewer). In the same room, tap water may be used as boiler makeup water. Because the cooling water is hot, usually about 105°F, if the cooling water is used as boiler makeup water, significant amounts of water and energy can be saved. One company found it could save about $1300 and one million gallons of water per year for each air compressor by reusing the cooling water in other places or by recirculating it through a small cooling tower.

- *Reduce flow rates to the minimum necessary.* Usually, water flow rates are liberally set in washing, coating, or rinsing operations. By setting flow rates at minimum levels, significant water can be saved along with the energy required for pumping.

- *Cover open tanks. Often heated tanks are open at the top.* Floating balls, cantilevered tops, or flexible slit covers can be used to cover the tanks and reduce evaporation and heat loss. One plant saved about $12,000 per year in energy and water costs by covering their heated tanks.

There are many other ways to save dollars and water through water audits, but it is difficult to develop a general list. Each plant and each operation are unique and require individual engineering study. The preceding discussion should serve to stimulate some ideas.

In terms of total water studies, water audits at manufacturing locations uncovered the following potential:

- One plant saved $77,000 annually in energy and water cost and 32 × 10^6 gal of water.

- Another plant saved $20,000 annually in energy and water cost and 12 × 10^6 gal of water.

- A rubber hose manufacturer saved $31,000 in water cost alone and 50 × 10^6 gal of water per year.

- A metal cylinder manufacturer saved about $20,000/year and 36×10^6 gal of water.

The potential for real savings is large. In the future, water management will be critical as costs continue to climb and water sources dry up.

13.8 SUMMARY

In this chapter we analyzed alternative energy sources and water management opportunities. First, we examined solar energy options. Active and passive solar systems were studied, and photovoltaics was examined.

Active solar systems were found to be effective but expensive since they normally require backup and storage systems. Passive solar energy, which is cost effective today, offers substantial potential for reducing energy costs for heating; however, backup systems are usually needed here too. Finally, although the future is promising for the reduced cost of photovoltaics, present applications in industry are only for remote sites where other energy sources are too expensive.

Although wind energy is not very promising as an industrial energy source in the near future, refuse-derived fuel is cost-effective in many locations today. Savings result from reduced energy cost and reduced handling and disposal costs.

Water management was discussed in the last section. Water management can yield large dollar and water savings and be cost effective. Because water use is often intertwined with energy use, reductions in the amount of water used can often result in a concurrent energy savings.

REFERENCES
1. "The Potential of Renewable Energy," An Interlaboratory White Paper, SERI/TP-260-3674, U.S. Department of Energy, March, 1990.

2. *Ibid.*

3. *Climatic Atlas of the United States*, GPO, Washington, D.C., 1968.

4. *Introduction to Solar Heating and Cooling Designing and Sizing*, DOE/ CS-0011 UC 59A, B, C, Washington, D.C., Aug. 1978.

5. *Engineering Principles and Concepts for Active Solar Systems*, Solar Energy Research Institute, Hemisphere Publishing Corporation, Washington, DC, 1988.

6. R.W. Hallet and R.L. Gervais, *Design, Startup, and Operational Experiences of Solar One, in Progress in Solar Engineering*, Edited by D. Yogi Goswami, Hemisphere Publishing Corporation, Washington, DC, 1987.

7. D. Yogi Goswami, "Solar Energy and the Environment," *Proceedings of the Energy Systems and Ecology Conference*—ENSEE '93, Cracow, Poland, July 5-9, 1993.

8. John H. Parker, "Uses of Landscaping for Energy Conservation," *Star Project Report 78-012*, Florida Governor's Energy Office, Tallahassee, FL, January, 1981.

9. "Landscaping Can Reduce Energy Bills," *Grounds Keeping*, Intertec Publishing Company, February, 1993.

10. Danny S. Parker, "The Florida Solar Energy Center—Preliminary Analysis of Potential Efficiency Improvements," FSEC-RR-26-91, Florida Solar Energy Center, Cape Canaveral, FL, December, 1991.

11. W.C. Turner, Senior ed., *Energy Management Handbook*, Second Edition, Fairmont Press, Atlanta, GA, 1993.

12. "The Potential of Renewable Energy," An Interlaboratory White Paper, SERI/TP-260-3674, U.S. Department of Energy, March, 1990.

13. Ibid.

14. Amy Gahran, "Landfill Gas Cuts Boiler Fuel Bills at AT&T by $60K/yr.," *Energy User News*, January, 1993.

15. Mark Dunbar, "County Expands Steam Plant, Adds Waste-to-Energy Cogen.," *Energy user News*, October, 1990.

16. Doug M. Moore, "Biogas Powered Cogeneration System at Mathis Dairy," *Industrial Energy Conserver*, Industrial Energy Extension Service of Georgia Tech University, September, 1987.

BIBLIOGRAPHY

Goswami, D. Yogi, *Alternative Energy in Agriculture*, CRC Press, Boca Raton, FL, 1986.

Goswami, D. Yogi, Editor, *Progress in Solar Engineering*, Hemisphere Press, Washington, DC, 1987.

Instructions for Energy Auditors, U.S. Department of Energy, Washington, D.C., 1978.

Kreider, Jan F., *Solar Design: Components, Systems, and Economics*, Hemisphere Publishing Company, Washington, DC, 1989.

Kreith, Frank, *Economic Analysis of Solar Thermal Systems*, MIT Press, Cambridge, MA, 1988.

Meckler, Milton, *Innovative Energy Design for the 90's*, Fairmont Press, Atlanta, GA, 1993.

Myers, John, *Solar*, research report, Oklahoma Industrial Energy Management Program, Oklahoma State University, Stillwater, Oklahoma, 1981.

Rosenberg, Paul, *Alternative Energy Handbook*, Fairmont Press, Atlanta, GA, 1993.

Index